DSP Primer

DSP Primer

C. Britton Rorabaugh

McGraw-Hill

New York San Francisco Washington, D.C. Auckland Bogotá
Caracas Lisbon London Madrid Mexico City Milan
Montreal New Delhi San Juan Singapore
Sydney Tokyo Toronto

Library of Congress Cataloging-in-Publication Data

Rorabaugh, C. Britton.
    DSP primer / C. Britton Rorabaugh.
        p.   cm.
    Includes bibliographical references and index.
    ISBN 0-07-054004-7
    1. Signal processing—Digital techniques.   I. Title.
TK5102.9.R67   1998
621.382′2—dc21   98-36445
                             CIP

McGraw-Hill
A Division of The McGraw-Hill Companies

Copyright © 1999 by The McGraw-Hill Companies, Inc. All rights reserved. Printed in the United States of America. Except as permitted under the United States Copyright Act of 1976, no part of this publication may be reproduced or distributed in any form or by any means, or stored in a data base or retrieval system, without the prior written permission of the publisher.

1 2 3 4 5 6 7 8 9 0   DOC/DOC   9 0 3 2 1 0 9 8

P/N 134206-0
Part of ISBN 0-07-054004-7

The sponsoring editor for this book was Scott Grillo, the editing supervisor was ATLIS Graphics & Design, and the production supervisor was Pamela Pelton. It was set in Century Schoolbook.

Printed and bound by R. R. Donnelley & Sons Company.

McGraw-Hill books are available at special quantity discounts to use as premiums and sales promotions, or for use in corporate training programs. For more information, please write to the Director of Special Sales, McGraw-Hill, 11 West 19th Street, New York, NY 10011. Or contact your local bookstore.

> Information contained in this work has been obtained by The McGraw-Hill Companies, Inc. ("McGraw-Hill") from sources believed to be reliable. However, neither McGraw-Hill nor its authors guarantees the accuracy or completeness of any information published herein and neither McGraw-Hill nor its authors shall be responsible for any errors, omissions, or damages arising out of use of this information. This work is published with the understanding that McGraw-Hill and its authors are supplying information but are not attempting to render engineering or other professional services. If such services are required, the assistance of an appropriate professional should be sought.

 This book is printed on recycled, acid-free paper containing a minimum of 50% recycled, de-inked fiber.

To Joyce, Geoff, and Amber

Contents

Preface xvii

Chapter 1. Some Essential Preliminaries — 1

1.1 Time Series Analysis — 1
1.2 Digital Simulation — 2
1.3 Overview of Digital Signal Processing — 2
    Subdisciplines — 2

Chapter 2. Continuous-Time Signals and Their Spectra — 5

2.1 Mathematical Modeling of Signals — 5
    Steady-state signal models — 8
    Energy signals versus power signals — 10
2.2 Frequency Spectra of Periodic Signals: Fourier Series — 11
    Exponential form — 12
    Conditions of applicability — 14
    Properties of the Fourier series — 14
    Fourier series of a square pulse train — 16
    Parseval's theorem — 17
2.3 Transient Signals — 17
    Unit step — 17
    Unit impulse — 20
    Decaying exponential — 20
    Saturating exponential — 21
2.4 Fourier Transform — 21
2.5 Spectral Density — 24
    Energy spectral density — 24
    Power spectral density of a periodic signal — 25
    Reference — 25

Chapter 3. Noise — 27

3.1 Noise Processes — 27
    Autocorrelation and power spectral density — 28
    Linear filtering — 29
3.2 White Noise — 29
3.3 Noise Equivalent Bandwidth — 30

Contents

3.4	Uniform Distribution	31
	Software issues	32
	Minimal standard generator	35
3.5	Gaussian Distribution	35
	Error function	37
3.6	Simulation of White Gaussian Noise	39
	Method A	39
	Method B	39
	Variance for simulation of white noise	40
	Software	42
3.7	Thermal Noise	42
	Equivalent noise temperature	43
3.8	Bandpass Noise	43
	References	44

Chapter 4. Linear Systems — 47

4.1	Systems	47
	Linearity	48
	Time invariance	49
	Causality	50
4.2	Characterization of Linear Systems	51
	Impulse response	51
	Step response	53
4.3	Laplace Transform	53
4.4	Transfer Functions	55
4.5	Heaviside Expansion	57
	Simple pole case	57
4.6	Poles and Zeros	58
4.7	Magnitude, Phase, and Delay Responses	61
	Phase delay	61
	Group delay	62
4.8	Computer Representation of Polynomials and Transfer Functions	62
4.9	Computer Simulation of Analog Systems	64
	Sum-of-powers form	65
	Biquadratic form	67
	Polynomial expansions	68
	Software for simulation of analog filters	69
	References	69

Chapter 5. Classical Analog Filters — 71

5.1	Filter Fundamentals	71
	Magnitude response features of lowpass filters	71
	Scaling of lowpass filter responses	74
	Highpass filters	76
	Bandpass filters	76
5.2	Butterworth Filters	79
	Transfer function	79
	Frequency response	81
	Determination of minimum order	84
	Impulse response	84
	Step response	85
5.3	Chebyshev Filters	88
	Transfer function	89

Frequency response	95
Impulse response	96
Step response	100
5.4 Elliptical Filters	100
Parameter specification	100
Normalized transfer function	103
Denormalized transfer function	106
5.5 Bessel Filters	108
Transfer function	108
Frequency response	109
Group delay	112

Chapter 6. Foundations of Digital Signal Processing — 113

6.1 Digitization	113
Ideal sampling	115
Sampling rate selection	116
Instantaneous sampling	117
Natural sampling	120
Discrete-time signals	120
Notation	121
6.2 Discrete-Time Fourier Transform	123
Convergence conditions	125
Relationship to Fourier series	125
6.3 Discrete-Time Systems	126
Difference equations	126
Discrete convolution	127
6.4 Diagramming Discrete-Time Systems	127
Block diagrams	127
Signal flow graphs	130
6.5 Quantization	131
Fixed-point numeric formats	131
Floating-point numeric formats	133
Quantized coefficients	135
Quantization noise	136
References	142

Chapter 7. Transform Analysis of Discrete-Time Systems — 143

7.1 Region of Convergence	143
Finite-duration sequences	144
Infinite-duration sequences	145
Convergence of the unilateral z transform	146
7.2 Relationship Between the Laplace and z Transforms	147
7.3 System Functions	147
7.4 Common z-Transform Pairs and Properties	148
7.5 Inverse z Transform	149
Inverse z transform via partial fraction expansion	151
References	152

Chapter 8. Discrete Fourier Transform (DFT) — 153

8.1 Discrete Fourier Transform	153
Parameter selection	154
Periodicity	155

8.2 Properties of the DFT	155
Linearity	155
Time shifting	156
Frequency shifting	156
Even and odd symmetry	156
Real and imaginary properties	156
8.3 Applying the DFT	157
Short time-limited signals	157
Periodic signals	158
Long aperiodic signals	158
Software notes	159

Chapter 9. Fast Fourier Transforms — 163

9.1 Computational Complexity of the DFT	163
9.2 Decimation-in-Time Algorithms	163
Other variations of decimation-in-time FFT algorithms	171
9.3 Decimation-in-Frequency Algorithms	171
9.4 Prime Factor Algorithm	177

Chapter 10. Windows for Filtering and Spectral Analysis — 181

10.1 Rectangular Window	181
Software notes	183
Discrete-time window	183
Frequency windows and spectral windows	184
More software notes	184
10.2 Triangular Window	186
Discrete-time triangular window	187
Frequency and spectral windows	189
10.3 Window Software	190
10.4 von Hann Window	192
Discrete-time von Hann window	193
10.5 Hamming Window	194
Discrete-time Hamming windows	196
Computer generation of window coefficients	196
10.6 Dolph-Chebyshev Window	196
10.7 Kaiser Window	200
Computer generation of window coefficients	201
10.8 Impacts of Quantization	201
References	205

Chapter 11. FIR Filter Fundamentals — 207

11.1 Introduction to FIR Filters	207
Software design	207
11.2 Evaluating the Frequency Response of FIR Filters	208
Software design	209
11.3 Linear Phase FIR Filters	211
Software design	213
Frequency response of linear phase FIR filters	213
Software design	215
11.4 Structures for FIR Realizations	215
Direct form	215

Transposed direct form	216
Cascade form	217
Structures for linear phase FIR filters	218
11.5 Assessing the Impacts of Quantization and Finite Precision Arithmetic	221
Direct form realizations	221
Software design	222
References	223

Chapter 12. FIR Filter Design: Window Method — 225

12.1 Fourier Series Method	225
Properties of the Fourier series method	227
Highpass filters	229
Bandpass filters	230
Bandstop filters	232
12.2 Applying Windows to Fourier Series Filters	235
12.3 Impacts of Quantization	238
Coefficient quantization	239
Signal quantization	246

Chapter 13. FIR Filter Design: Frequency Sampling Method — 247

13.1 Introduction	247
13.2 Odd N versus Even N	248
Even N	252
13.3 Design Formulas	254
13.4 Frequency Sampling Design with Transition-Band Samples	255
Optimization	258
13.5 Optimization with Two Transition-Band Samples	258
Software notes	262
13.6 Optimization with Three Transition-Band Samples	264
13.7 Quantized Coefficients	266
References	268

Chapter 14. FIR Filter Design: Remez Exchange Method — 269

14.1 Chebyshev Approximation	269
Alternation theorem	271
14.2 Strategy of the Remez Exchange Method	271
Generating the desired response and weighting functions	274
14.3 Evaluating the Error	274
14.4 Selecting Candidate Extremal Frequencies	275
Testing $E(f)$ for $f=0$	276
Testing $E(f)$ within the passband and the stopband	276
Testing of $E(f)$ at the passband and stopband edges	276
Testing of $E(f)$ for $f=0.5$	277
Rejecting superfluous candidate frequencies	277
Deciding when to stop	277
14.5 Obtaining the Impulse Response	278
14.6 Using the Remez Exchange Method	278
Deciding on the filter length	278
14.7 Extension of the Basic Method	284
References	284

Chapter 15. IIR Filter Fundamentals — 287

- 15.1 Frequency Response of IIR Filters — 288
- 15.2 Structures for IIR Realizations — 288
  - Direct form — 288
  - Cascade form — 290
- 15.3 Assessing the Impacts of Quantization and Finite-Precision Arithmetic — 291
  - Direct form realizations — 292
- 15.4 Software Notes — 293

Chapter 16. IIR Filter Design: Invariance and Pole-Zero Placement Methods — 295

- 16.1 Impulse Invariance — 295
  - Programming considerations — 298
- 16.2 Step Invariance — 300
  - Programming considerations — 301
- 16.3 Matched z Transformation — 302

Chapter 17. IIR Filter Design: Bilinear Transformation — 305

- 17.1 Bilinear Transformation — 305
- 17.2 Factored Form of the Bilinear Transform — 306
- 17.3 Properties of the Bilinear Transformation — 309
  - Frequency warping — 310
- 17.4 Programming the Bilinear Transformation — 312
- 17.5 Computer Examples — 314
- 17.6 Quantization in IIR Filters Obtained via Bilinear Transformation — 318

Chapter 18. Multirate Signal Processing: Basic Concepts — 321

- 18.1 Decimation by Integer Factors — 322
- 18.2 Interpolation by Integer Factors — 322
- 18.3 Decimation and Interpolation by Non-Integer Factors — 325
- 18.4 Decimation and Interpolation of Bandpass Signals — 325
  - Quadrature modulation of bandpass signals — 326
  - Single sideband modulation — 328
  - Decimation via integer band sampling — 329
  - References — 332

Chapter 19. Structures for Decimators and Interpolators — 333

- 19.1 Decimator Structures — 333
- 19.2 Interpolators — 333
- 19.3 Polyphase FIR Interpolator Structures — 336
- 19.4 Polyphase FIR Decimator Structures — 337
- 19.5 Half-Band FIR Filters — 337
  - References — 340

Chapter 20. Advanced Multirate Techniques — 341

- 20.1 Multistage Decimators — 341
  - Software notes — 345

20.2	Multistage Interpolators	345
	Software notes	348
20.3	Multirate Implementation of Lowpass Filters	348
	Specifying the filters $h_1[n]$ and $h_2[n]$	349
	Software notes	350

Chapter 21. Random Signals and Sequences — 353

21.1	Random Sequences	353
21.2	Randomness and Probability	354
	Joint and conditional probabilities	354
	Independent events	355
21.3	Bernoulli Trials	356
21.4	Random Variables	356
	Cumulative distribution functions	357
	Properties of distribution functions	357
	Probability density function	358
21.5	Moments of a Random Variable	358
	Mean	358
	Mean of a function of an RV	358
	Moments	359
	Central moments	359
	Properties of variance	360
21.6	Relationships between RVs	360
	Statistical independence	361
21.7	Correlation and Covariance	361
21.8	Probability Densities for Functions of an RV	362
21.9	Random Processes	363
21.10	Autocorrelation and Autocovariance	365
	Stationarity	365
	Ergodicity	366
	Properties of autocorrelation functions	367
	Autocovariance	367
	Uncorrelated random processes	367
	Autocorrelation matrix	368
21.11	Power Spectral Density of Random Processes	368
21.12	Linear Filtering of Random Processes	369
21.13	Estimating the Moments for a Random Process	369
	Mean	370
	Correlation function	370
	Cross-correlation	371
21.14	Estimating the Correlation Matrix	372
21.15	Markov Processes	373
	Markov chains	374
	Classification of Markov chains	374
	State diagrams of Markov chains	375
	References	376

Chapter 22. Parametric Models of Random Processes — 379

22.1	Autoregressive-Moving Average Model	379
	Software notes	380
22.2	Autoregressive Model	380
	Yule-Walker equations	380

	Characterization of AR processes	381
	Software notes	382
22.3	Levinson Recursion	382
	Software notes	383
22.4	Moving Average Model	384
	Estimation of MA parameters	384
	Software notes	385
22.5	Estimation of ARMA Parameters	385
	Modified Yule-Walker equations	386
	Software notes	387
	References	387

Chapter 23. Linear Prediction — 389

23.1	Linear Estimation	389
23.2	Linear Predictive Filtering	390
23.3	Autocorrelation Method	391
	Software notes	393
	More software notes	394
23.4	Covariance Method	396
	Cholesky decomposition	397
	Software notes	399
23.5	Lattice Filters	399
	References	402

Chapter 24. Adaptive Filters — 403

24.1	Adaptive Linear Combiner	404
24.2	Properties of the Performance Surface	405
	Gradient	409
	Gradient estimation	411
24.3	Constructing Test Cases for Adaptive Filters	411
	Test signals from regular random processes	411
	Test signals from predictable random processes	417
	Test signals from mixed random processes	421
24.4	Method of Steepest Descent	425
	Stability	427
24.5	Performance Measures	429
	Learning curve	429
	Misadjustment and excess MSE	429
24.6	The LMS Algorithm	431
24.7	Recursive Least-Squares Algorithm	436
	References	439

Chapter 25. Classical Spectral Estimation — 441

25.1	Introduction to Periodograms	441
25.2	Daniell Periodogram	443
25.3	Bartlett Periodogram	446
	Software notes	447
25.4	Windowing and Other Issues	450
25.5	Welch Periodogram	454

25.6	Correlograms	455
	References	456

Chapter 26. Modern Spectral Estimation — 457

26.1	Yule-Walker Method	457
26.2	Burg Method	460
26.3	RLS Method	463
26.4	LMS Method	464
26.5	Spectral Estimation of Noisy AR Processes	466

Chapter 27. Speech Processing — 469

27.1	Speech Signals	469
27.2	Cepstral Analysis	471
	Computation of the real cepstrum	473
	Formant estimation	475
27.3	Nonlinear Quantization of Speech Signals	477
	A-law companding	478
	A-law expansion	481
	μ-law companding	483
	μ-law expansion	486
	References	488

Appendix A. Mathematical Tools — 489

A.1	Exponentials and Logarithms	489
	Exponentials	489
	Logarithms	490
	Decibels	491
A.2	Complex Numbers	491
	Operations on complex numbers in rectangular form	492
	Polar form of complex numbers	492
	Operations on complex numbers in polar form	493
	Logarithms of complex numbers	494
A.3	Trigonometry	494
	Phase shifting of sinusoids	495
	Trigonometric identities	495
	Euler's identities	497
	Series and product expansions	497
	Orthonormality of sine and cosine	499
A.4	Derivatives	501
	Derivatives of polynomial ratios	501
A.5	Integration	502
A.6	Dirac Delta Function	503
	Distributions	505
	Properties of the delta distribution	506
A.7	Sinc Function	506
A.8	Bessel Functions	507
	Modified Bessel functions of the first kind	507
	Identities for modified Bessel functions	507
	Evaluation of Bessel functions	508

A.9 Matrix Algebra	**508**
Basic definitions	508
Matrix arithmetic	510
Matrix transformations	511
Hermitian transpose	511
References	512

Index 513

Preface

If you're going to own only one book on digital signal processing, this is the one to have. If you already own several, you need this book anyway—it contains quite a lot of useful information not available in any other book. I wrote this book for individuals faced with the need to select and apply DSP techniques to real-world problems—it is not intended as an academic text. Dozens of mathematical methods, processing algorithms, and design procedures are presented in a concise step-by-step format that makes it easy to quickly select and implement an appropriate technique for the problem at hand. Numerous software modules on the included CD-ROM allow the techniques to be tested immediately on a computer, before more application-specific implementations are undertaken.

These programs were written using Microsoft Visual C++ 5.0, and they are provided in both executable and source forms. It is unlikely that the executables will run "as-is" on every possible Windows platform, and they of course will not run under Unix or VMS. Therefore, I have also provided source code so that you may tailor the programs to fit your favorite hardware environments. I have included the computer programs primarily to illustrate how the various methods presented in the book actually work—the programs are not being offered as a full-featured software "product" and they are not intended for incorporation directly into application-specific implementations. I expect that you will use these programs to become familiar with various processing techniques and gain insight into their operation before designing an implementation that is optimized for your specific application.

The field of digital signal processing is vast, and even a very efficient summary of the professional and academic literature would run to at least 10,000 pages. However, certain core areas account for perhaps 85 percent of the "deployed" DSP currently being used in actual applications. In this book, I have attempted to cover all of these core areas: digital filtering, discrete transform techniques, digital spectrum analysis, multirate signal processing, statistical signal processing, adaptive filtering, and speech processing. I hope I have been successful in providing good practical introductions to these areas and shown how their techniques are typically applied.

Britt Rorabaugh

DSP Primer

Chapter 1

Some Essential Preliminaries

Digital signal processing (DSP) involves the digital representation of signals and the use of digital processors to analyze, synthsize, or modify such signals. The digital processors employed can be general purpose computers, specialized programmable processors optimized for performing DSP operations, or dedicated digital hardware designed to perform specific processing tasks. The processing that can be performed upon a digital signal seems to come in an infinite variety of forms, and it is the study of these forms that fills the bulk of the DSP literature. DSP is also concerned with the conversion of analog signals into digital form and the conversion of digital signals into analog form. There is considerable overlap between DSP and the related areas of *time-series analysis* and *digital simulation*. One view of the world has it that DSP was born when computers became powerful enough to allow techniques from time-series analysis to be applied to the real-time processing of signals.

1.1 Time Series Analysis

Time-series analysis is a mathematical discipline that has been around since the days when engineers depended more on intuition than on mathematics. In 1738, Daniel Bernoulli discovered a general solution for the wave equation for the vibrating musical string, and in 1822 Jean Baptiste Joseph Fourier extended these results by asserting that almost any arbitrary function $f(x)$ could be represented as an infinite summation of sine and cosine terms

$$f(x) = \sum_{n=1}^{\infty}(A_n \cos n\alpha x + B_n \sin n\alpha x)$$

Determination of the A_n and B_n for a given function $f(x)$ is called *harmonic analysis*. During the mid-nineteenth century harmonic analysis began to be applied to the study of sound, sunspot activity, weather, and tidal variations.

As increasing processor speeds allow more and more of the techniques traditionally considered as belonging to time-series analysis to be used in real-time or near real-time applications, it is becoming harder and harder to distinguish between what is time-series analysis and what is DSP. There are still specific things (such as implementation of digital filters with quantized coefficients) that clearly fall within the DSP category, and there are some things (such as the analysis of cycles in stock market performance) that are generally regarded as falling within the time-series analysis category. DSP engineers will embrace whatever they can from the field of time-series analysis, but mathematicians involved in time-series analysis are not likely to get into the details of designing special-purpose digital processing hardware. There will always be a part of DSP that remains distinct from time-series analysis, but it is likely, turf battles between university mathematics and engineering faculties notwithstanding, that in the future time-series analysis may be considered to be a subset of DSP. The point to all of this is that DSP practitioners should be aware that time-series analysis still exists as a separate discipline, with its own rich body of literature that can be mined for many useful techniques and insights that may not yet appear in the DSP literature.

1.2 Digital Simulation

In certain application areas it is common practice to use digital computers to simulate the behavior of analog signal processing systems. If the processors employed are fast enough, it is possible to use a simulation of an analog signal processing system to directly perform DSP. However, within DSP are processing techniques significantly less computationally burdensome than digital simulation of an analog system. The key difference lies in the sampling rates required. Practical DSP systems can use sampling rates as low as 2.5 to 4 times the highest signal frequency, but digital simulation of analog systems typically require sampling rates many times higher. In DSP systems the sampling rate needs to be only high enough to avoid unacceptable levels of aliasing (see Sec. 6.1). In digital simulations, the sampling rate typically needs to be high enough so that the first backwards difference is a reasonably good approximation of the derivative.

1.3 Overview of Digital Signal Processing

Since its beginnings some 35 years ago, DSP has grown explosively in terms of both its techniques and the applications in which these techniques are employed.

Subdisciplines

Like many other fields of endeavor, DSP has grown so large that there are a number of recognized subdisciplines.

1. **Digital filtering.** This is one of the oldest areas in DSP. The digital implementation of traditionally analog filtering functions drove the development of early DSP techniques. The *bilinear transform* (Chap. 17) is still widely used to construct digital approximations of classical analog filters (Chap. 5).

2. **Fast Fourier transform analysis.** The *fast Fourier transform* (FFT) algorithms (Chap. 9) are simply computationally efficient ways to evaluate the *discrete Fourier transform* (DFT) (Chap. 8). The selection and fitting of FFT algorithms to the specific requirements of a particular application sometimes requires significant mathematical agility. However, the FFT itself is usually a means to an end rather than an end in its own right. The major applications of the FFT include *finite impulse response* (FIR) filter design (Chap. 13), *spectral analysis* (Chap. 25), and *fast convolution*.

3. **Multirate signal processing.** There are many situation in which it is advantageous to be able to change the sampling rate within a DSP system. Multirate signal processing (Chaps. 18–20) is the area of DSP concerned with the techniques that can be used to efficiently increase or decrease the sampling rate without introducing unacceptable distortion into a digitized signal.

4. **Statistical signal processing.** The statistical approach to signal processing focusses on the characterization of signals as random processes (Chaps. 21–22) and the use of techniques such as *linear prediction* (Chap. 23) that are based on such characterizations.

5. **Adaptive filtering.** Adaptive filtering can be viewed as a subdiscipline of statistical signal processing that makes use of digital filters with coefficients that can change over time (Chap. 24). However, adaptive filtering is so different from the rest of the statistical processing techniques that it is usually treated separately.

6. **Digital spectral analysis.** Spectral analysis involves much more than simply using the FFT to compute the DFT for a signal sequence. Classical spectral estimation (Chap. 25) involves several different smoothing, averaging, and windowing techniques; modern spectral estimation techniques (Chap. 26) are intimately intertwined with statistical signal processing and adaptive filtering techniques.

7. **Speech processing.** Digital processing of speech signals might appear to be just an applications area rather than a subdiscipline of DSP *per se*, but the peculiar characteristics of speech signals have given rise to a number of processing techniques that collectively constitute a distinct subdiscipline within DSP.

Chapter 2

Continuous-Time Signals and Their Spectra

DSP is based on the use of mathematical functions to represent or model real-world electronic signals as illustrated in Fig. 2.1. Actual signals can be complicted phenomena, and their exact behavior is impossible to describe completely. However, simple mathematical models can describe signals well enough to yield useful results that can be applied in a variety of practical situations. The distinction between a signal and its mathematical representation is not always rigidly observed in signal-processing literature: Properties of these models are often presented as properties of the signals themselves. Nevertheless, mathematical models of signals are crucial to DSP. This chapter is devoted to a review of the mathematical techniques that are used to model and analyze signals throughout this book. The appendices present some of the purely mathematical material upon which this chapter is based.

2.1 Mathematical Modeling of Signals

Mathematical models of signals are generally categorized as either *steady-state* or *transient* models. The typical voltage output from an oscillator is sketched in Fig. 2.2. This signal exhibits three different parts: a *turn-on transient* at the beginning; an interval of *steady-state operation* in the middle; and a *turn-off transient* at the end. It is possible to formulate a single mathematical expression that describes all three parts, but for most uses, such an expression would be unnecessarily complicated. In cases where the primary concern is steady-state behavior, simplified mathematical expressions that ignore the transients will often be adequate. The steady-state portion of the oscillator output can be modeled as a sinusoid that theoretically exists for all time. This seems to contradict the fact that the oscillator output exists for a limited time interval between turn-on and turn-off. However, the apparent contradiction is not really a problem; over the interval of steady-state operation that we are interested in,

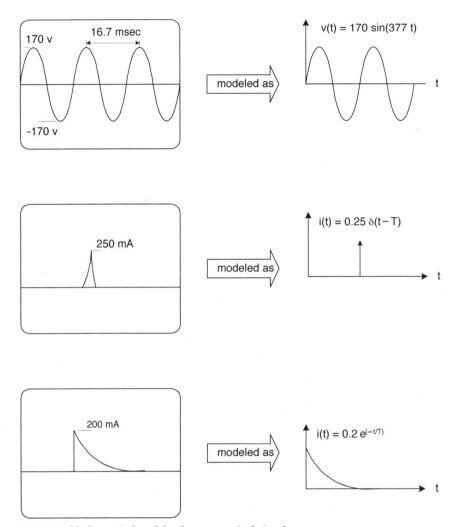

Figure 2.1 Mathematical models of some practical signals.

the mathematical sine function accurately describes the behavior of the oscillator's output voltage. Allowing the mathematical model to assume that the steady-state signal exists over all time greatly simplifies matters since the transients' behavior can be excluded from the model. In situations where the transients are important, they can be modeled as exponentially saturating and decaying sinusoids, as shown in Figs. 2.3 and 2.4. In Fig. 2.3, the saturating exponential envelope continues to increase, but it never quite reaches the steady-state value. Likewise, the decaying exponential envelope of Fig. 2.4 continues to decrease, but it never quite reaches zero. In this context, the steady-state value is sometimes called an *asymptote*, or the envelope can be

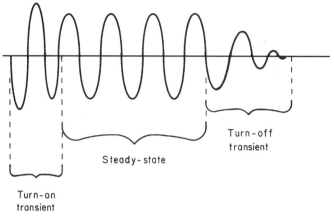

Figure 2.2 Typical output of an audio oscillator.

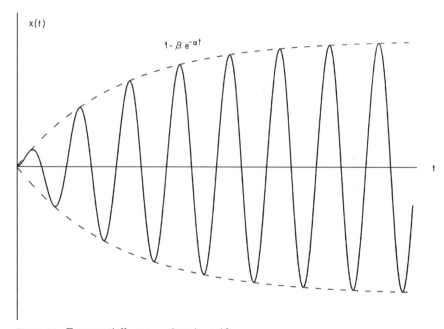

Figure 2.3 Exponentially saturating sinusoid.

said to *asymptotically* approach the steady-state value. Steady-state and transient models of signal behavior inherently contradict each other, and neither constitutes a "true" description of a particular signal. The formulation of the appropriate model requires an understanding of the signal to be modeled and of the implications that a particular choice of model will have for the intended application.

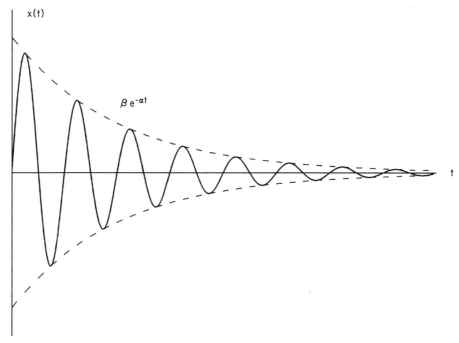

Figure 2.4 Exponentially decaying sinusoid.

Steady-state signal models

Generally, steady-state signals are limited to just sinusoids or sums of sinusoids. This includes virtually all periodic signals of practical interest, since periodic signals can be resolved into sums of weighted and shifted sinusoids using the Fourier analysis techniques presented in Sec. 2.2.

Periodicity. Sines, cosines, and square waves are all periodic functions. The characteristic that makes them periodic is the way in which each of the complete waveforms can be formed by repeating a particular cycle of the waveform over and over at regular intervals, as shown in Fig. 2.5.

Def. 2.1 A function $x(t)$ is periodic with a period of T if and only if $x(t + nT) = x(t)$ for all integer values of n.

Functions that are not periodic are called *aperiodic*, and functions that are almost periodic are called *quasi-periodic*.

Symmetry. A function can exhibit a certain symmetry regarding its position relative to the origin.

Def. 2.2 A function $x(t)$ is said to be *even*, or to exhibit *even symmetry*, if for all t, $x(t) = x(-t)$.

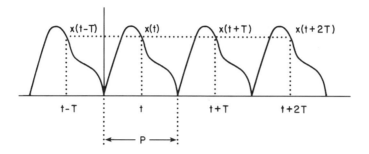

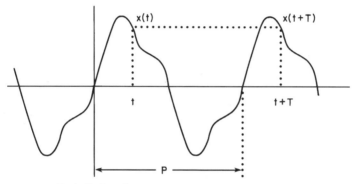

Figure 2.5 Periodic functions.

Def. 2.3 A function $x(t)$ is said to be *odd*, or to exhibit *odd symmetry*, if for all t, $x(t) = -x(-t)$.

An even function is shown in Fig. 2.6, and an odd function is shown in Fig. 2.7.

Symmetry may appear at first to be something that is not particularly useful in practical applications where the definition of time zero is often somewhat arbitrary. Symmetry considerations play an important role in Fourier analysis, however especially the discrete forms of Fourier analysis used in DSP. Some functions are neither odd nor even, but any periodic function can be resolved into a sum of an even function and an odd function as given by

$$x(t) = x_{\text{even}}(t) + x_{\text{odd}}(t)$$

where $x_{\text{even}}(t) = \frac{1}{2}[x(t) + x(-t)]$
$x_{\text{odd}}(t) = \frac{1}{2}[x(t) - x(-t)]$

Addition and multiplication of symmetric functions will obey the following rules:

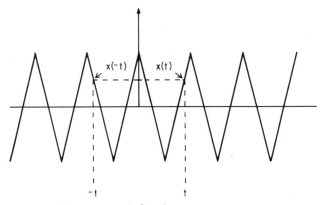

Figure 2.6 Even-symmetric function.

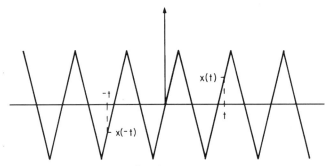

Figure 2.7 Odd-symmetric function.

Even + even = even
Odd + odd = odd
Odd × odd = even
Even × even = even
Odd × even = odd

Energy signals versus power signals

It is a common practice to deal with mathematical functions representing abstract signals as though they are either voltages across a 1-Ω resistor, or currents through a 1-Ω resistor. Since, in either case, the resistance has an assumed value of unity, the voltage and current for any particular signal will be numerically equal—thus obviating the need to select one viewpoint over the other. Thus, for a signal $x(t)$, the instantaneous power $p(t)$ dissipated in the 1-Ω resistor is simply the squared amplitude of the signal

$$p(t) = |x(t)|^2 \qquad (2.1)$$

regardless of whether $x(t)$ represents a voltage or a current. To emphasize the fact that the power given by Eq. (2.1) is based on unity resistance, it is often referred to as the *normalized power*. The total energy of the signal $x(t)$ is then obtained by integrating the right-hand side of Eq. (2.1) over all time:

$$E = \int_{-\infty}^{\infty} |x(t)|^2 \, dt \qquad (2.2)$$

and the average power is given by

$$P = \lim_{T \to \infty} \frac{1}{T} \int_{-T/2}^{T/2} |x(t)|^2 \, dt \qquad (2.3)$$

A few texts (for example, [1]) equivalently define the average power as

$$P = \lim_{T \to \infty} \frac{1}{2T} \int_{-T}^{T} |x(t)|^2 \, dt \qquad (2.4)$$

If the total energy is finite and nonzero, $x(t)$ is referred to as an *energy signal*. If the average power is finite and nonzero, $x(t)$ is referred to as a *power signal*. Note that a power signal has infinite energy, and an energy signal has zero average power; thus the two categories are mutually exclusive. Periodic signals and most random signals are power signals, while most deterministic aperiodic signals are energy signals.

2.2 Frequency Spectra of Periodic Signals: Fourier Series

Periodic signals can be resolved into linear combinations of phase-shifted sinusoids using the *Fourier series* (FS), which is given by

$$x(t) = \frac{a_0}{2} + \sum_{n=1}^{\infty} [a_n \cos(n\omega_0 t) + b_n \sin(n\omega_0 t)] \qquad (2.5)$$

where

$$a_0 = \frac{2}{T} \int_{-T/2}^{T/2} x(t) \, dt \qquad (2.6)$$

$$a_n = \frac{2}{T} \int_{-T/2}^{T/2} x(t) \cos(n\omega_0 t) \, dt \qquad (2.7)$$

$$b_n = \frac{2}{T} \int_{-T/2}^{T/2} x(t) \sin(n\omega_0 t)\, dt \tag{2.8}$$

$T =$ period of $x(t)$

$\omega_0 = \dfrac{2\pi}{T} = 2\pi f_0 =$ fundamental radian frequency of $x(t)$

Upon application of the appropriate trigonometric identities, Eq. (2.5) can be put into the following alternative form:

$$x(t) = c_0 + \sum_{n=1}^{\infty} c_n \cos(n\omega_0 t - \theta_n) \tag{2.9}$$

where the c_n and θ_n are obtained from a_n and b_n using

$$c_0 = \frac{a_0}{2} \tag{2.10}$$

$$c_n = \sqrt{a_n^2 + b_n^2} \tag{2.11}$$

$$\theta_n = \tan^{-1}\left(\frac{b_n}{a_n}\right) \tag{2.12}$$

Examination of Eq. (2.5) reveals that a periodic signal contains only a dc component plus sinusoids whose frequencies are integer multiples of the original signal's *fundamental frequency*. (For a fundamental frequency of f_0, $2f_0$ is the *second harmonic*, $3f_0$ is the *third harmonic*, and so on.) Theoretically, periodic signals will contain an infinite number of harmonically related sinusoidal components. In the real world, however, periodic signals will contain a finite number of measurable harmonics. Consequently, pure mathematical functions are only approximately equal to the practical signals which they model.

Exponential form

The trigonometric form of the FS given by Eq. (2.5) makes it easy to visualize periodic signals as summations of sine and cosine waves, but mathematical manipulations are often more convenient when the series is in the exponential form given by

$$x(t) = \sum_{n=-\infty}^{\infty} c_n \exp(j2\pi n f_0 t) \tag{2.13}$$

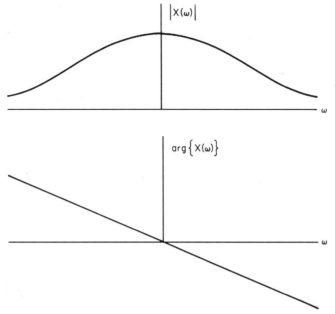

Figure 2.8 Magnitude and phase spectra.

where

$$c_n = \frac{1}{T} \int_T x(t) \exp(-j2\pi n f_0 t)\, dt \tag{2.14}$$

The integral notation used in Eq. (2.14) indicates that the integral is to be evaluated over one period of $x(t)$. In general, the values of c_n are complex, and are often presented in the form of a magnitude spectrum and phase spectrum, as shown in Fig. 2.8. The magnitude and phase values plotted in such spectra are obtained from c_n using

$$|c_n| = \sqrt{[\text{Re}(c_n)]^2 + [\text{Im}(c_n)]^2} \tag{2.15}$$

$$\theta_n = \tan^{-1}\left[\frac{\text{Im}(c_n)}{\text{Re}(c_n)}\right] \tag{2.16}$$

The complex c_n of Eq. (2.14) can be obtained from the a_n and b_n of Eqs. (2.7) and (2.8) using

$$c_n = \begin{cases} \frac{a_n + jb_n}{2} & n < 0 \\ a_0 & n = 0 \\ \frac{a_n - jb_n}{2} & n > 0 \end{cases} \tag{2.17}$$

Conditions of applicability

The FS can be applied to almost all periodic signals of practical interest. However, there are some functions for which the series will not converge. The FS coefficients are guaranteed to exist and the series will converge uniformly if $x(t)$ satisfies the following conditions:

1. $x(t)$ is a single-valued function.
2. $x(t)$ has, at most, a finite number of discontinuities within each period.
3. $x(t)$ has, at most, a finite number of extrema within each period.
4. $x(t)$ is absolutely integrable over a period:

$$\int_T |x(t)|\, dt < \infty \qquad (2.18)$$

These conditions are often called the *Dirichlet conditions* in honor of Peter Gustav Lejeune Dirichlet (1805–1859) who first published them in the 1828 issue of *Journal für die reine und angewandte Mathematik* (commonly known as *Crelle's Journal*). In applications where it is sufficient for the FS coefficients to be convergent in the mean, rather than uniformly convergent, it suffices for $x(t)$ to be integrable square over a period:

$$\int_T |x(t)|^2\, dt < \infty \qquad (2.19)$$

For most engineering purposes, the FS is usually assumed to be identical to $x(t)$ if conditions 1 through 3 plus either Eq. (2.18) or Eq. (2.19) are satisfied.

Properties of the Fourier series

A number of useful FS properties are listed in Table 2.1. For ease of notation, the coefficients c_n corresponding to $x(t)$ are denoted as $X(n)$, and the c_n corresponding to $y(t)$ are denoted as $Y(n)$. In other words, the FS representations of $x(t)$ and $y(t)$ are given by

$$x(t) = \sum_{n=-\infty}^{\infty} X(n) \exp\left(\frac{j2\pi nt}{T}\right) \qquad (2.20)$$

$$y(t) = \sum_{n=-\infty}^{\infty} Y(n) \exp\left(\frac{j2\pi nt}{T}\right) \qquad (2.21)$$

where T is the period of both $x(t)$ and $y(t)$. In addition to the properties listed in Table 2.1, the FS coefficients exhibit certain symmetries. If and only if $x(t)$ is real, the corresponding FS coefficients will exhibit even symmetry in their

TABLE 2.1 Properties of the Fourier Series

Property	Time function	Transform
1. Homogeneity	$ax(t)$	$aX(n)$
2. Additivity	$x(t) + y(t)$	$X(n) + Y(n)$
3. Linearity	$ax(t) + by(t)$	$aX(n) + bY(n)$
4. Multiplication	$x(t)y(t)$	$\sum_{m=-\infty}^{\infty} X(n-m)Y(m)$
5. Convolution	$\frac{1}{T}\int_0^T x(t-\tau)y(\tau)\,d\tau$	$X(n)Y(n)$
6. Time shifting	$x(t-\tau)$	$\exp\left(\frac{-j2\pi n\tau}{T}\right)X(n)$
7. Frequency shifting	$\exp\left(\frac{-j2\pi mt}{T}\right)x(t)$	$X(n-m)$

NOTE: $x(t)$, $y(t)$, $X(n)$, and $Y(n)$ are as given in Eqs. (2.20) and (2.21).

real part and odd symmetry in their imaginary part:

$$\text{Im}[x(t)] = 0 \Leftrightarrow \begin{aligned} \text{Re}[X(-n)] &= \text{Re}[X(n)] \\ \text{Im}[X(-n)] &= \text{Im}[X(n)] \end{aligned} \quad (2.22)$$

Equation (2.22) can be rewritten in more compact form as

$$\text{Im}[x(t)] = 0 \Leftrightarrow X(-n) = X^*(n) \quad (2.23)$$

where the superscript asterisk indicates complex conjugation. Likewise, for purely imaginary $x(t)$, the corresponding FS coefficients will exhibit odd symmetry in their real part and even symmetry in their imaginary part:

$$\text{Re}[x(t)] = 0 \Leftrightarrow X(-n) = -[X^*(n)] \quad (2.24)$$

If and only if $x(t)$ is, in general, complex with even symmetry in the real part and odd symmetry in the imaginary part, then the corresponding FS coefficients will be purely real:

$$x(-t) = x^*(t) \Leftrightarrow \text{Im}[X(n)] = 0 \quad (2.25)$$

If and only if $x(t)$ is, in general, complex with odd symmetry in the real part and even symmetry in the imaginary part, then the corresponding FS coefficients will be purely imaginary:

$$x(-t) = -[x^*(t)] \Leftrightarrow \text{Re}[X(n)] = 0 \quad (2.26)$$

In terms of the amplitude and phase spectra, Eq. (2.23) means that, for real signals, the amplitude spectrum will have even symmetry and the phase spectrum

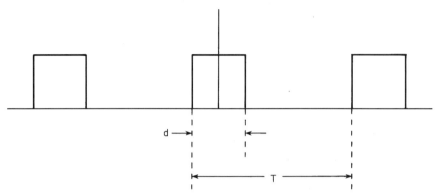

Figure 2.9 Square pulse train.

will have odd symmetry. If $x(t)$ is both real and even, then both Eqs. (2.23) and (2.25) apply. In this special case, the FS coefficients will be both real and even-symmetric. At first glance, it may appear that real even-symmetric coefficients are in contradiction with the expected odd-symmetric phase spectrum; but, in fact, there is no contradiction. For all the positive real coefficients, the corresponding phase is, of course, zero. For each of the negative real coefficients, we can choose a phase value of either plus or minus 180°. By appropriate selection of positive and negative values, odd symmetry in the phase spectrum can be maintained.

Fourier series of a square pulse train

Consider the square pulse train shown in Fig. 2.9. The FS representation of this signal is given by

$$x(t) = \sum_{n=-\infty}^{\infty} c_n \exp\left(\frac{j2\pi nt}{T}\right) \qquad (2.27)$$

where

$$c_n = \frac{TA}{T} \operatorname{sinc}\left(\frac{nT}{T}\right)$$

Since the signal is both real and even-symmetric, the FS coefficients are real and even-symmetric, as shown in Fig. 2.10. The corresponding magnitude spectrum will be even, as shown in Fig. 2.11a. Appropriate selection of ±180° values for the phase of negative coefficients will allow an odd-symmetric phase spectrum to be plotted as in Fig. 2.11b.

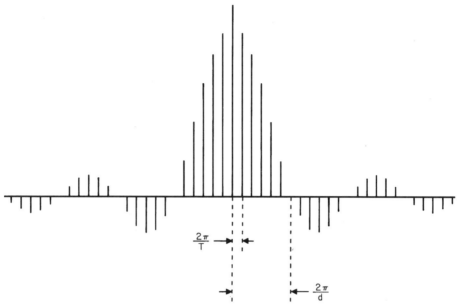

Figure 2.10 Fourier series for a square pulse train.

Parseval's theorem

The average power (normalized for 1 Ω) of a real-valued periodic function of time can be obtained directly from the FS coefficients by using Parseval's theorem:

$$P = \frac{1}{T} \int_T |x(t)|^2 \, dt$$

$$= \sum_{n=-\infty}^{\infty} |c_n|^2 = c_0^2 + \sum_{n=1}^{\infty} \frac{1}{2} |2c_n|^2 \qquad (2.28)$$

2.3 Transient Signals

There are several transient signals, modeled by aperiodic functions, that are frequently encountered in signal-processing work. These include the *unit step*, *unit impulse*, and *exponentials*, which are discussed in the following sections.

Unit step

Consider the circuit shown in Fig. 2.12. When the switch is open, the voltage between terminals A and B is 0. When the switch is closed, the voltage between A and B will jump abruptly to 9 V. An abrupt change in level such as this is represented mathematically as a *step function*. Figure 2.13 shows a *unit step* which shifts its level from zero to one at time zero. This function can be

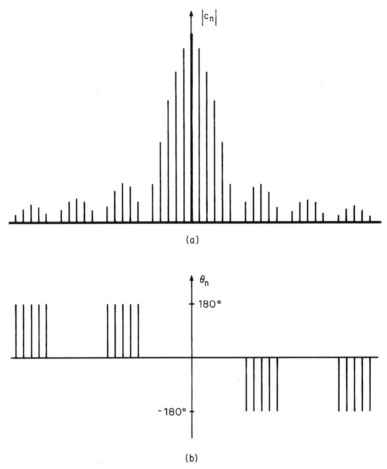

Figure 2.11 Fourier series spectra for a square pulse train: (*a*) amplitude spectrum and (*b*) phase spectrum.

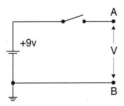

Figure 2.12 Circuit used in discussion of the unit step function.

multiplied by a constant gain factor and time-shifted in order to represent virtually any abrupt dc level shift of practical interest. In signal-processing and linear systems literature, the unit step is often denoted as $u_1(t)$. As depicted in Fig. 2.14, the integral of the unit step is the *unit ramp*. The derivative of the unit step is the *unit impulse*.

Continuous-Time Signals and Their Spectra

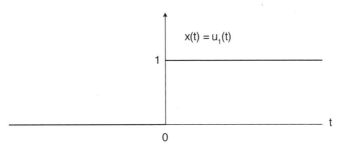

Figure 2.13 Unit step function.

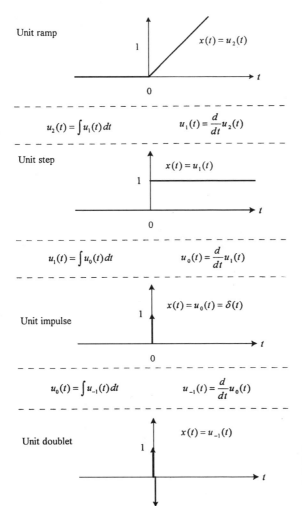

Figure 2.14 Elementary time functions: (a) unit ramp, (b) unit step, (c) unit impulse, and (d) unit doublet.

Unit impulse

When a switch is opened or closed in a circuit containing reactive components (i.e., capacitors or inductors), a spike of voltage or current as shown in Fig. 2.14c may be produced. Although this spike has a finite amplitude and nonzero rise time and fall time, it is often convenient to represent it mathematically as an *impulse function*. As described in Sec. A.6 of App. A, an impulse is often loosely described as having a zero width and an infinite amplitude at the origin such that the total area under the impulse is finite. In the case of a *unit impulse*, the area under the impulse is equal to unity.

As with the unit step, the unit impulse can be time-shifted and multiplied by a constant gain in order to represent almost any spiking phenomenon of practical interest. The unit impulse is denoted as either $u_0(t)$ or $\delta(t)$. As depicted in Fig. 2.14, the integral of the unit impulse is the unit step, and the derivative of the unit impulse is the *unit doublet*.

Decaying exponential

Consider the circuit shown in Fig. 2.15. The battery is initially connected across the capacitor, and the voltage across the capacitor will be 9 V. If the switch is moved from position A to position B, the voltage across the capacitor terminals will decay as shown. Mathematically, the shape of this decaying waveform is described as a *decaying exponential* of the form

$$y = \beta \exp(-\alpha t)$$

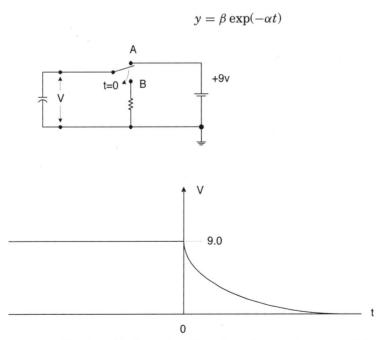

Figure 2.15 Circuit used in discussion of decaying and saturating exponentials: (*a*) diagram of the circuit, and (*b*) voltage decaying across the capacitor when the switch is moved to position B.

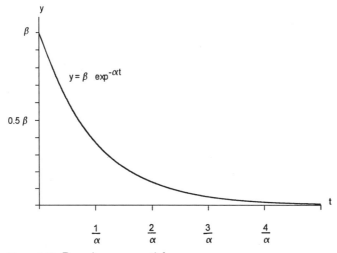

Figure 2.16 Decaying exponential.

A decaying exponential is shown in Fig. 2.16. As time increases, the amplitude of the function approaches closer and closer to—but never quite reaches—zero. The function is said to *asymptotically* approach zero, and the horizontal line at $y = 0$ is called an *asymptote*.

Saturating exponential

Referring again to the circuit in Fig. 2.15, let us assume that the switch is in position B and the voltage across the capacitor is zero. If the switch is moved to position A, the voltage across the capacitor will begin to increase as a saturating exponential of the form

$$y = 1 - \beta \exp(-\alpha t)$$

A saturating exponential is shown in Fig. 2.17. The function approaches an asymptote at $y = \beta$.

2.4 Fourier Transform

The *Fourier transform* is defined as

$$X(f) = \int_{-\infty}^{\infty} x(t) e^{-j 2\pi f t} \, dt \qquad (2.29)$$

or, in terms of the radian frequency $\omega = 2\pi f$

$$X(\omega) = \int_{-\infty}^{\infty} x(t) e^{-j\omega t} \, dt \qquad (2.30)$$

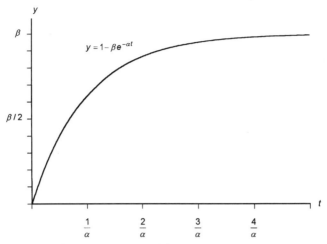

Figure 2.17 Saturating exponential.

The *inverse transform* is defined as

$$x(t) = \int_{-\infty}^{\infty} X(t)e^{j2\pi ft}\, df \qquad (2.31)$$

$$= \frac{1}{2\pi}\int_{-\infty}^{\infty} X(t)e^{j\omega t}\, d\omega \qquad (2.32)$$

There are a number of different shorthand notations for indicating that $x(t)$ and $X(f)$ are related via the Fourier transform. Some of the more common notations include the following:

$$X(f) = \mathcal{F}[x(t)] \qquad (2.33)$$

$$x(t) = \mathcal{F}^{-1}[X(f)] \qquad (2.34)$$

$$x(t) \xleftrightarrow{FT} X(f) \qquad (2.35)$$

$$x(t) \underset{IFT}{\overset{FT}{\rightleftarrows}} X(f) \qquad (2.36)$$

The notations of Eqs. (2.35) and (2.36), while more difficult to typeset, offer the flexibility of changing the letters FT to FS, DFT, or DTFT to indicate respectively, Fourier series, discrete Fourier transform, and discrete-time Fourier transform. Strictly speaking, the equality shown in Eq. (2.34) is incorrect, because the inverse transform of $X(f)$ is only guaranteed to approach $x(t)$ in the sense of convergence in the mean. Nevertheless, the notation of Eq. (2.34) appears often throughout the engineering literature. To facilitate comparison with

TABLE 2.2 Properties of the Fourier Transform

Property	Time function, $x(t)$	Transform, $X(f)$
1. Homogeneity	$ax(t)$	$aX(f)$
2. Additivity	$x(t) + y(t)$	$X(f) + Y(f)$
3. Linearity	$ax(t) + by(t)$	$aX(f) + bY(f)$
4. Differentiation	$\dfrac{d^n}{dt^n} x(t)$	$(j2\pi f)^n X(f)$
5. Integration	$\int_{-\infty}^{t} x(\tau)\,d\tau$	$\dfrac{X(f)}{j2\pi f} + \dfrac{1}{2} X(0)\delta(f)$
6. Frequency shifting	$e^{-j2\pi f_0 t} x(t)$	$X(f + f_0)$
7. Sine modulation	$x(t)\sin(2\pi f_0 t)$	$\tfrac{1}{2}[X(f - f_0) + X(f + f_0)]$
8. Cosine modulation	$x(t)\cos(2\pi f_0 t)$	$\tfrac{1}{2}[x(f - f_0) - X(f + f_0)]$
9. Time shifting	$x(t - \tau)$	$e^{-j\omega\tau} X(f)$
10. Time convolution	$\int_{-\infty}^{\infty} h(t - \tau) x(\tau)\,d\tau$	$H(f)X(f)$
11. Multiplication	$x(t)y(t)$	$\int_{-\infty}^{\infty} X(\lambda) Y(f - \lambda)\,d\lambda$
12. Time and frequency scaling	$x\left(\dfrac{t}{a}\right) \quad a > 0$	$aX(af)$
13. Duality	$x(t)$	$x(-f)$
14. Conjugation	$x^*(t)$	$X^*(-f)$
15. Real part	$\text{Re}[x(t)]$	$\tfrac{1}{2}[X(f) + X^*(-f)]$
16. Imaginary part	$\text{Im}[x(t)]$	$\dfrac{1}{2j}[X(f) - X^*(-f)]$

the Laplace transform, the frequency domain function often is written as $X(j\omega)$ rather than $X(\omega)$. We can write

$$X(j\omega) = \int_{-\infty}^{\infty} x(t) e^{-j\omega t}\,dt$$

and realize that this is identical to the two-sided Laplace transform defined by Eq. (4.16), with $j\omega$ substituted for s. A number of useful Fourier transform properties are listed in Table 2.2. A number of frequently encountered Fourier transform pairs are listed in Table 2.3.

TABLE 2.3 Some Common Fourier Transform Pairs

Pair no.	$x(t)$	$X(\omega)$	$X(f)$
1	1	$2\pi\delta(\omega)$	$\delta(f)$
2	$u_1(t)$	$\dfrac{1}{j\omega} + \pi\delta(w)$	$\dfrac{1}{2\pi f} + \dfrac{1}{2}\delta(f)$
3	$\delta(t)$	1	1
4	t^n	$2\pi j^n \delta^{(n)}(\omega)$	$\left(\dfrac{j}{2\pi}\right)^n \delta^{(n)}(f)$
5	$\sin\omega_0 t$	$j\pi[\delta(\omega+\omega_0) - \delta(\omega-\omega_0)]$	$\dfrac{j}{2}[\delta(f+f_0) - \delta(f-f_0)]$
6	$\cos\omega_0 t$	$\pi[\delta(\omega+\omega_0) - \delta(\omega-\omega_0)]$	$\dfrac{1}{2}[\delta(f+f_0) + \delta(f-f_0)]$
7	$e^{-at}u_1(t)$	$\dfrac{1}{j\omega + a}$	$\dfrac{1}{j2\pi f + a}$
8	$u_1(t)e^{-at}\sin\omega_0 t$	$\dfrac{\omega_0}{(a+j\omega)^2 + \omega_0^2}$	$\dfrac{2\pi f_0}{(a+j2\pi f)^2 + (2\pi f_0)^2}$
9	$u_1(t)e^{-at}\cos\omega_0 t$	$\dfrac{a+j\omega}{(a+j\omega)^2 + \omega_0^2}$	$\dfrac{a+j2\pi f}{(a+j2\pi f)^2 + (2\pi f_0)^2}$
10	$\begin{cases} 1 & \|t\| \le \frac{1}{2} \\ 0 & \text{elsewhere} \end{cases}$	$\text{sinc}\left(\dfrac{\omega}{2\pi}\right)$	$\text{sinc}\, f$
11	$\text{sinc}\, t \triangleq \dfrac{\sin\pi t}{\pi t}$	$\begin{cases} 1 & \|\omega\| \le \pi \\ 0 & \text{elsewhere} \end{cases}$	$\begin{cases} 1 & \|f\| \le \frac{1}{2} \\ 0 & \text{elsewhere} \end{cases}$
12	$\begin{cases} at\exp(-at) & t > 0 \\ 0 & \text{elsewhere} \end{cases}$	$\dfrac{a}{(a+j\omega)^2}$	$\dfrac{a}{(a+j2\pi f)^2}$
13	$\exp(-a\|t\|)$	$\dfrac{2a}{a^2 + \omega^2}$	$\dfrac{2a}{a^2 + 4\pi^2 f^2}$
14	$\text{signum}\, t \triangleq \begin{cases} 1 & t > 0 \\ 0 & t = 0 \\ -1 & t < 0 \end{cases}$	$\dfrac{2}{j\omega}$	$\dfrac{1}{j\pi f}$

2.5 Spectral Density

Energy spectral density

The *energy spectral density* of an energy signal is defined as the squared magnitude of the signal's Fourier transform:

$$S_e(f) = |X(f)|^2$$

Analogous to the way in which Parseval's theorem relates the FS coefficients to the average power of a power signal, *Rayleigh's energy theorem* relates the

Fourier transform to the total energy of an energy signal as follows:

$$E = \int_{-\infty}^{\infty} |x(t)|^2 \, dt = \int_{-\infty}^{\infty} S_e(f) \, df = \int_{-\infty}^{\infty} |X(f)|^2 \, df \qquad (2.37)$$

In many texts where $x(t)$ is assumed to be real-valued, the absolute-value signs are omitted from the first integrand in Eq. (2.37).

Power spectral density of a periodic signal

The *power spectral density* (PSD) of a periodic signal is defined as the squared magnitude of the signal's line spectrum obtained via either a Fourier series or a Fourier transform with impulses. Using the Dirac delta notational conventions of the latter, the PSD is defined as

$$S_p(f) = \frac{1}{T^2} \sum_{n=-\infty}^{\infty} \delta\left(f - \frac{n}{T}\right) \left|X\left(\frac{n}{T}\right)\right|^2 \qquad (2.38)$$

where T is the period of the signal $x(t)$. Parseval's theorem as given by Eq. (2.28) can be restated in the notation of Fourier transform spectra as

$$P = \frac{1}{T^2} \sum_{n=-\infty}^{\infty} \left|X\left(\frac{n}{T}\right)\right|^2$$

Reference

1. Haykin, Simon, *Communication Systems*, 2d ed., Wiley, New York, 1983.

Chapter

3

Noise

Noise is present in virtually all signals. In some situations it is negligible; in other situations it all but obliterates the signal of interest. Removing unwanted noise from signals has historically been a driving force behind the development of signal processing technology, and it continues to be a major application for both analog and digital signal processing systems. The goal of this chapter is to provide a practical treatment of noise that provides sufficient information to support the design and use of DSP systems in applications that involve additive noise. Chapter 21 contains a review of probability and random-variable theory that is used for a more in-depth mathematical characterization of noise and random signals, which supports the study of statistical signal processing and spectral estimation (Chaps. 22 through 26).

3.1 Noise Processes

Let us assume that we want to make a single measurement of the voltage across the output terminals of a benchtop dc power supply. The nominal output voltage of the supply is 5.000 volts. If we measure to 3 digits of precision, we may read a value of 4.997 volts. Or, we may read a value of 5.001 volts. Or maybe 5.013 volts. Mathematically we can represent this measurement M as the sum of two parts:

$$M = V + E$$

where the *deterministic* part V is the "true" output voltage which we might assume to be exactly 5.000 volts, and the *random* or *stochastic* part E is the measurement error which we assume to be small relative to V. Typically, E will be treated as a *random variable* whose characterization includes a *distribution*, a *mean*, and a *variance*. For the example under discussion, E might be gaussian-distributed with zero mean and some non-zero variance, say 0.02. From this information alone it would be possible to calculate the probability of reading a measurement value that falls within any particular range, for example, 4.995 to

5.005. Each time the measurement is repeated, it is possible to read a different value.

If we were to repeat the measurement at regular intervals we would obtain a sequence of values that vary over time. (This sequence constitutes a discrete-time noise signal which could be subjected to processing in a digital filter or in a spectral estimation algorithm.) Each individual sample in this sequence can be viewed as a random variable, but the sequence of samples has additional properties beyond the properties of the individual samples. These additional properties are captured when the repeated measurement process is treated mathematically as a *random process*. These additional properties deal with the relationships between individual measurement samples. In a *wideband* noise process, the measurement values from one sample to the next exhibit little or no correlation, while in a narrowband noise process, the measurement values from one sample to the next can exhibit significant correlation.

Autocorrelation and power spectral density

All of the noise sequences considered in Chapters 3 through 20 will be assumed to be realizations of ergodic, stationary, discrete-time random processes. The sample-to-sample correlation of such a process can be characterized by the time-averaged autocorrelation over any one of the realizations of the process

$$R_x[k] = \lim_{M \to \infty} \frac{1}{2M+1} \sum_{n=-M}^{M} x[n]x^*[n-k]$$

The *power spectral density* (PSD), $S_x(f)$, and the *autocorrelation function* (ACF), $R_x(\tau)$, of an ergodic stationary random process $x(t)$ comprise a Fourier transform pair.

$$S_x(f) = \mathcal{F}\{R_x(\tau)\} = \int_{-\infty}^{\infty} R_x(\tau) \exp(-j2\pi ft)\, dt \qquad (3.1)$$

$$R_x(\tau) = \mathcal{F}^{-1}\{S_x(f)\} = \int_{-\infty}^{\infty} S_x(f) \exp(j2\pi ft)\, df \qquad (3.2)$$

In the case of a discrete-time random process having a discrete-lag autocorrelation function $R_x[k]$, the relationship between the ACF and PSD involves the discrete-time Fourier transform

$$S_x(f) = \sum_{k=-\infty}^{\infty} R_x[k] \exp(-j2\pi kfT)$$

$$R_x[k] = \int_{-\infty}^{\infty} S_x(f) \exp(j2\pi kfT)\, df$$

Linear filtering

Application of a noise process $x(t)$ to the input of a linear time-invariant filter will produce a different noise process $y(t)$ at the filter output. Some statistical properties of the output can be determined directly from the filter's impulse response and the statistical properties of the input. For the assumed case of $x(t)$ being ergodic and stationary, the mean of the output can be expressed as

$$\mu_y = \mu_x H(0)$$

where μ_x is the mean of the input process and $H(0)$ is the filter's transfer function evaluated at zero frequency. For an input with power spectral density of $S_x(f)$, the output will have a PSD given by

$$S_y(f) = |H(f)|^2 S_x(f)$$

When the input noise process to a linear filter is a gaussian random process, the resulting output noise process is also gaussian.

3.2 White Noise

White noise is an idealized noise process having a power spectral density that is constant over all frequencies as shown in Fig. 3.1. The term *white* is used to describe such noise based on the fact that the spectrum of white light is constant over all frequencies in the visible range. In a similar vein, the term *pink noise* refers to noise in which the PSD has higher values at lower frequencies and slopes downward with increasing frequency, similar to the way the spectrum of pink light is tilted across the visible spectrum. Most noise processes tend to be lowpass in nature, so there is no corresponding "violet noise" PSD that slopes upward with increasing frequency.

By convention, the constant value of the two-sided PSD for white noise is usually denoted as $N_0/2$. The factor of 2 in the denominator is a convenience so that the noise power passed by an ideal *lowpass filter* (LPF) having a bandwidth of B will be equal to $N_0 B$. In a similar vein, the constant value of the one-sided PSD for white noise is denoted as N_0. When the noise generated within a system is modeled as white noise applied to the system's input, the value of N_0 is simply kT where k is Boltzmann's constant and T is the equivalent noise temperature of the system.

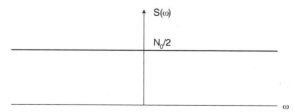

Figure 3.1 Power spectral density of white noise.

As discussed in Sec. 3.1, the PSD and ACF for a random process form a Fourier transform pair. Using pair 3 from Table 2.3, we conclude that the ACF of white noise is an impulse or delta function located at the origin.

$$\frac{N_0}{2}\delta(\tau) \overset{\text{FT}}{\leftrightarrow} \frac{N_0}{2}$$

Physically, an impulse at the origin signifies that the white noise has infinite average power. Obviously, no noise process in the real world can have infinite average power, which is why the word *idealized* appears in the opening sentence of this section. Nevertheless, the concept of white noise is a mathematical convenience that finds widespread use in theoretical work. Once white noise is (conceptually) passed through a filter of finite bandwidth, the objectionable attribute of infinite average power disappears. Although it theoretically need not be so, whenever a white noise process is assumed, it is almost always assumed to be a gaussian process and is called *white gaussian noise* (WGN) or *additive white gaussian noise* (AWGN).

Example 3.1 Ideal lowpass filtering of white noise is a classic example which is presented in numerous texts [1–5]. The transfer function of the ideal LPF is given by

$$H(f) = \begin{cases} 1 & |f| \leq B \\ 0 & \text{elsewhere} \end{cases}$$

If white noise with a PSD of $N_0/2$ is applied to the input of such a filter, the PSD of the output noise process will be given by

$$S(f) = \begin{cases} \frac{N_0}{2} & |f| \leq B \\ 0 & \text{elsewhere} \end{cases}$$

Using pair 11 from Table 2.3, we find the autocorrelation of the output noise is

$$R(\tau) = N_0 B \operatorname{sinc}(2B\tau)$$

The average power is equal to $R(\tau)$ evaluated at $\tau = 0$ or simply $N_0 B$—an intuitively pleasing result.

When white noise is subjected to filtering, the output noise process is referred to as *bandlimited white noise*. To avoid the verbal tap dancing needed to deal with the infinite average power of ideal white noise, it may be helpful to consider band-limited white noise first and then simply think of ideal white noise as a limiting process that is approached as the bandwidth approaches infinity.

3.3 Noise Equivalent Bandwidth

The magnitude response $|H(f)|$ of an arbitrary LPF is sketched (solid line) in Fig. 3.2. If zero-mean white noise having a (two-sided) PSD of $N_0/2$ is applied to the input of such a filter, the resulting output will have a finite average power

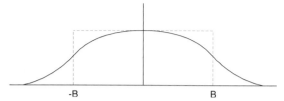

Figure 3.2 Illustration of noise equivalent bandwidth.

that is given by

$$N = \frac{N_0}{2} \int_{-\infty}^{\infty} |H(f)|^2 \, df \qquad (3.3)$$

Figure 3.2 also shows (dashed line) the magnitude response of an ideal LPF. The *noise equivalent bandwidth* of the arbitrary filter is defined as the value of B for which the ideal filter and arbitrary filter produce the same output power from identical white noise inputs. The average output noise power from the ideal filter is

$$N = N_0 B \, H^2(0) \qquad (3.4)$$

The value of B for which the two filters produce the same output noise power is found by equating Eqs. (3.3) and (3.4) and solving for B to yield

$$B = \frac{\int_{-\infty}^{\infty} |H(f)|^2 \, df}{2H^2(0)} = \frac{\int_0^{\infty} |H(f)|^2 \, df}{H^2(0)}$$

3.4 Uniform Distribution

A uniformly distributed random variable x has a probability density function given by

$$p(x) = \begin{cases} \frac{1}{b-a} & a \le x \le b \\ 0 & \text{elsewhere} \end{cases} \qquad (3.5)$$

The mean is given by

$$\mu = \frac{a+b}{2} \qquad (3.6)$$

and the variance is given by

$$\sigma^2 = \frac{(b-a)^2}{12} \qquad (3.7)$$

The characteristic function $\phi_x(\omega)$ of a uniform random variable x with PDF as in Eq. (3.5) is given by

$$\phi_x(\omega) = \frac{\exp(j\omega b) - \exp(j\omega a)}{j\omega(b-a)} \qquad (3.8)$$

Software issues

Many different random number generators can be used to produce sequences of uniform deviates. Some of these generators are better than most—and some are notoriously bad. This section is not meant to provide exhaustive coverage of all the possibilities. Chapter 3 of Knuth's classic work [6] remains the most comprehensive readily available treatment of random number generators. A much more concise and narrowly focussed treatment can be found in Press et al. [7]. In this section we explore only two of the many possible generators.

The first generator is one that is sure to find widespread use because it appears in the standard library that accompanies many C/C++ compilers. There is no true standard for a C/C++ random number generator but, by including source code for an easily implemented example within the standard, the standards committee has virtually guaranteed that this generator will appear in C/C++ libraries for years to come. The example code is provided in Listing 3.1. This generator is of the linear congruential type and uses the following recursion for generating the seed sequence

$$x_{n+1} = ax_n + c \,(\mathrm{mod}\, m) \qquad (3.9)$$

where $a = 1103515245$, $c = 12345$, and $m = 2^{32}$. (Note: The code provided in the standard uses the name next for the seed value, reserving the name seed for the initial value of the seed that is passed into srand.) The generator's output value is an integer value from 0 to RAND_MAX inclusive, where RAND_MAX is defined to be 32767. This output value is obtained by shifting the new seed value 16 bits to the right and then taking the 15 rightmost bits after the shift. In the example published within the standard, the 16-place shift is accomplished via dividing the seed by $2^{16} = 65536$. Keeping only the 15 rightmost bits of the shifted value is accomplished by computing the remainder modulo $2^{15} = 32768$. An alternative approach would be to use the shift operator followed by a bit-wise ANDing with a 15-bit mask as in

```
return ((unsigned int) (next >> 16) & 32767);
```

There are several different schools of though regarding how a random number generator should be packaged. The "module" approach taken in Listing 3.1 is midway between a stand-alone function and a full class implementation. This approach is object-like in that it has state memory and separate functions for initializing the state and for obtaining the next number in the pseudorandom sequence. However, because of the way that state memory must be implemented using a global variable, there can be only one instance of this generator in any given program. Even if random numbers are used at several different places in a program, they must all be drawn from the generator. This can sometimes be an inconvenience. For example, let us assume that to evaluate a filter design we need to simulate a noisy digital signal that randomly switches between 0 and 1. We run the simulation several times, using the same bit sequence each time. However, the sequence of added noise values is different each time the simulation is run. In this situation, it would be convenient to use one generator for the bit values and a second generator for the noise values. The bit generator would be given the same initial seed value each time the program is run, but the noise generator would be given a different initial seed value each time the program is run.

Listing 3.1

```
static unsigned long int next=1;

int rand(void)
{
 next = next * 1103515245 + 12345;
 return (unsigned int) (next/65536)  }

void srand( unsigned int seed)
{
 next = seed;
}
```

Listings 3.2 and 3.3 provide a full class implementation of a generator that uses the same algorithm as the generator in Listing 3.1. Using this class, several different generators can be instantiated in the same program, with each instance being given a different initial seed value. A third alternative would be to move the state memory out of the generator routine and transfer to the calling program the responsibility for maintaining this memory. This function approach has been used most often for random number generators implemented in FORTRAN. However, there is no reason why this approach cannot also be used in a C/C++ implementation, as shown in Listing 3.4.

Listing 3.2

```
#ifndef _SRAND_H_
#define _SRAND_H_

class srand
{
 public:

  srand();  // default constructor sets seed=1

  srand(int seed);  // constructor for user-supplied initial seed

  unsigned int rand(void);

 private:

  long int Seed;
};

#endif
```

Listing 3.3

```
#include "srand.h"
#include <iostream.h>

srand::srand()
      :Seed(1)
{
};

srand::srand(int seed)
{
 Seed = seed;
}

unsigned int srand::rand(void)
{
 Seed = Seed * 1103515245 + 12345;
 return((unsigned int)((Seed >> 16) & 32767));
}
```

Listing 3.4

```
// function implementation of rand

unsigned int rand( long *seed )
{
 *seed = (*seed) * 1103515245 + 12345;
 return((unsigned int) ((*seed >> 16) & 32767));
}
```

It is not possible to select one of these three different approaches as being the best—each approach has different strengths and weaknesses. The *class* approach allows for multiple instances within the same program, but it runs slower than the other two approaches. Furthermore, this approach is reasonable only for C++. A C implementation using the class approach is possible, but it is somewhat cumbersome. The *function* approach also allows for multiple instances within the same program, is reasonably fast, but violates principles of good objected-oriented design by making the calling program responsible for the care and feeding of the seed which is actually an attribute of the generator. The *module* approach is reasonably fast, keeps all generator attributes inside the module, but does not conveniently support multiple instances. The module approach is, however, the approach chosen by the C standards committee for its example of how to implement a portable random number generator. Although this particular generator is widely available, it is not a particularly good choice for many applications. As do all linear congruential generators, this generator has a tendency to exhibit sequential correlation on successive calls.

Minimal standard generator

Park and Miller [8] have proposed a "Minimal Standard" uniform random number generator which uses the multiplicative congruential algorithm

$$x_{n+1} = ax_n (\mathrm{mod}\, m) \tag{3.10}$$

where $a = 7^5 = 16807$ and $m = 2^{31} - 1 = 2147483647$. Equation (3.10) cannot be implemented directly in a high-level language using 32-bit integers. An implementation in C or C++ can be devised if an arithmetic trick called *Schrage's algorithm* is employed, as described in Press et al. [7]. Briefly, Schrage's algorithm provides a means for computing $ax \bmod m$ using only 32-bit integers even when the product ax is larger than 32 bits. Specifically, $ax \bmod m$ can be calculated as

$$ax \bmod m = a(x \bmod q) - r \lfloor z/q \rfloor \tag{3.11}$$

when the right-hand side evaluates to a non-negative value and where

$$q = \lfloor m/a \rfloor, \quad r = m \bmod a, \text{ and } m = aq + r$$

If the right-hand side of Eq. (3.11) evaluates to a negative value, simply add m to obtain the correct value for $ax \bmod m$. To apply Schrage's algorithm to the calculation of Eq. (3.10), use the values $q = 127773$ and $r = 2836$. A class implementation of the Minimal Standard generator is provided in file uni_rand.cpp.

3.5 Gaussian Distribution

A zero-mean, unity variance gaussian random variable y has a probability density function given by

$$p(y) = \frac{1}{\sqrt{2\pi}} \exp \frac{-y^2}{2} \tag{3.12}$$

A sketch of Eq. (3.12) is shown in Fig. 3.3. The corresponding cumulative distribution function is obtained by integrating Eq. (3.12).

$$P(y \leq Y) = \frac{1}{\sqrt{2\pi}} \int_{-\infty}^{Y} e^{-y^2/2}\, dy \tag{3.13}$$

A sketch of Eq. (3.13) is shown in Fig. 3.4. The integral in Eq. (3.13) cannot be evaluated in closed form; but it occurs so often in engineering and science

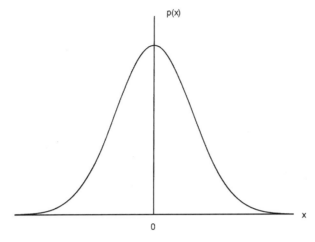

Figure 3.3 Gaussian probability density function.

that a special function, called the *error function* and denoted as erf, has been defined as:

$$\operatorname{erf} x = \frac{1}{\sqrt{2\pi}} \int_0^x e^{-y^2/2}\, dy \qquad (3.14)$$

Closed-form approximations for the error function have been established (see below), and values for erf x along with other closely related functions have been extensively tabulated. Thus $P(y \leq Y)$ can be obtained as:

$$P(y \leq Y) = \frac{1}{2} + \operatorname{erf} y \qquad (3.15)$$

For a gaussian random variable of mean μ variance σ^2, the PDF given in

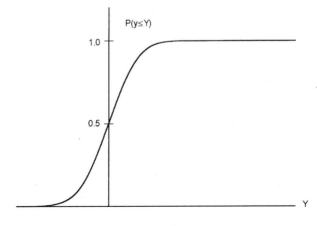

Figure 3.4 Cumulative distribution function for a gaussian random variable.

Eq. (3.12) can be scaled by σ and shifted by μ to yield

$$p(y) = \frac{1}{\sigma\sqrt{2\pi}}\exp\left(\frac{-(y-\mu)^2}{2\sigma^2}\right)$$

The characteristic function $\phi_y(\omega)$ of the gaussian random variable y is given by

$$\phi_y(\omega) = \exp\frac{j\mu\omega - \sigma^2\omega^2}{2}$$

The gaussian distribution is named in honor of Johann Karl Friedrich Gauss (1777–1855), a German mathematician who is often regarded as the greatest mathematician of all time. A gaussian random variable is also called a *normal variate*. Engineering literature tends to favor the use of "gaussian," while mathematical literature favors "normal," perhaps due to the many other things to which mathematicians attach the name of Gauss. A normal variate of mean μ and variance σ^2 is sometimes denoted as $N(\mu;\sigma^2)$.

Error function

The error function of x, written as erf x, is defined by

$$\text{erf } x \triangleq \frac{2}{\sqrt{\pi}}\int_0^x \exp(-u^2)\,du \tag{3.16}$$

The complementary error function of x, written as erfc x, is defined by

$$\text{erfc } x \triangleq \frac{2}{\sqrt{\pi}}\int_x^\infty \exp(-u^2)\,du$$

$$= 1 - \text{erf } x \tag{3.17}$$

The integral in Eq. (3.16) cannot be solved in closed form, but numerically computed values have been extensively tabulated. A short table of values for erf x is presented in Table 3.1. (A more extensive table can be found in [9].) Although erfc x cannot be evaluated in closed form, analytical expressions for upper and lower bounds have been established.

$$\text{erfc } x > \left(1 - \frac{1}{2x^2}\right)\frac{\exp(-x^2)}{x\sqrt{\pi}} \tag{3.18}$$

$$\text{erfc } x < \frac{\exp(-x^2)}{x\sqrt{\pi}} \tag{3.19}$$

For values of $x \geq 2$, both Eqs. (3.18) and (3.19) closely approximate erfc x.

TABLE 3.1 Values of the Error Function

x	erf x	x	erf x
0.00	0.0000000	1.00	0.8427008
0.05	0.0563720	1.05	0.8624361
0.10	0.1124629	1.10	0.8802051
0.15	0.1679960	1.15	0.8961238
0.20	0.2227026	1.20	0.9103140
0.25	0.2763264	1.25	0.9229001
0.30	0.3286268	1.30	0.9340079
0.35	0.3793821	1.35	0.9437622
0.40	0.4283924	1.40	0.9522851
0.45	0.4754817	1.45	0.9596950
0.50	0.5204999	1.50	0.9661051
0.55	0.5633234	1.55	0.9716227
0.60	0.0638561	1.60	0.9763484
0.65	0.6420293	1.65	0.9803756
0.70	0.6778012	1.70	0.9837904
0.75	0.7111556	1.75	0.9866717
0.80	0.7421010	1.80	0.9890905
0.85	0.7706681	1.85	0.9911110
0.90	0.7969082	1.90	0.9927904
0.95	0.8208908	1.95	0.9941793
1.00	0.8427008	2.00	0.9953223

The following expansions may be useful in work involving the error function.

$$\int_0^x \text{erf}(y)\,dy = x\,\text{erf}(x) - \frac{1}{\sqrt{\pi}}[1 - \exp(-x^2)]$$

$$\text{erf}(x) = \frac{2}{\sqrt{\pi}} \exp(-x^2) \sum_{n=0}^{\infty} \frac{2^{2n+1} x^{2n+1}(n+1)!}{(2n+2)}$$

$$\text{erf}(x) = \frac{2}{\sqrt{\pi}} \sum_{n=0}^{\infty} \frac{(-1)^n x^{2n+1}}{n!(2n+1)}$$

Values of erf for negative x are obtained using the identity $\text{erf}(-x) = -\text{erf}(x)$.

3.6 Simulation of White Gaussian Noise

The pseudorandom number generators commonly available in most high-level programming languages provide sequences of numbers that are uniformly distributed on the interval [0,1). The simulation of noise for evaluating digital filter performance usually requires gaussian-distributed numbers rather than uniformly distributed numbers. Using one of the two methods presented in this section, it is possible to take values from a sequence of independent, uniformly distributed numbers and from them generate a sequence of independent pseudorandom numbers that are approximately gaussian-distributed.

Method A

1. Generate a pseudorandom value U_1 uniformly distributed on [0,1).
2. Generate a second pseudorandom value U_2 which is uniform on [0,1) and independent of U_1.
3. Compute

$$G_1 = \cos(2\pi U_2)\sqrt{-2\sigma^2 \ln(U_1)}$$

$$G_2 = \sin(2\pi U_2)\sqrt{-2\sigma^2 \ln(U_1)}$$

The resulting G_1 and G_2 will be independent, zero-mean gaussian-distributed random variates each with variance σ^2. A histogram based on 10,000 samples of G_1 is shown in Fig. 3.5 along with a sketch of the corresponding theoretical probability density function.

Method B

1. Generate two pseudorandom values U_A and U_B which are independent and uniformly distributed on [0,1).

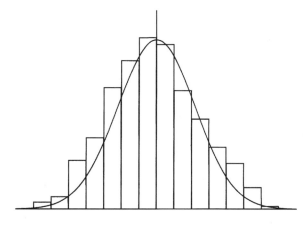

Figure 3.5 Histogram of outputs from a gaussian-distributed pseudorandom number generator.

2. Compute

$$U_1 = 1 - 2U_A$$

$$U_2 = 1 - 2U_B$$

3. Compute

$$S = U_1^2 + U_2^2$$

If $S \geq 1$, then go back to step 1; otherwise continue on to step 4.

4. Compute

$$G_1 = U_1 \sqrt{\frac{-2\sigma^2 \ln(S)}{S}}$$

$$G_2 = U_2 \sqrt{\frac{-2\sigma^2 \ln(S)}{S}}$$

The resulting G_1 and G_2 will be independent, zero-mean gaussian-distributed random variables each with variance σ^2.

Variance for simulation of white noise

In system performance specifications and theoretical analyses of signal processing systems, levels of white gaussian background noise are often specified in terms of the constant spectral density N_0. However, generation of gaussian-distributed pseudorandom numbers for simulating noise requires that a noise variance σ^2 be specified. To compare computer simulations against specifications and theoretical results, we need to establish a relationship between N_0 and σ^2.

The variance of a random noise process equals the process autocorrelation at lag zero:

$$\sigma^2 = R(0) \tag{3.20}$$

In the case of white noise (which is, recall, just an unrealizable but convenient idealization), $R(0)$ is infinite. However, in the analysis of communication system performance, we will usually be concerned with the noise power contained within some particular bandwidth of interest. Given a white noise density of N_0, the noise power in a bandwidth B is

$$P = \sigma^2 = N_0 B \tag{3.21}$$

To satisfy the uniform sampling theorem and to prevent aliasing (see Sec. 8.1), the sampling rate must be chosen such that

$$R_s \geq \begin{cases} 2B & \text{for real sampling} \\ B & \text{for complex sampling} \end{cases} \tag{3.22}$$

Thus

$$\sigma^2 = N_0 B \leq \begin{cases} \frac{N_0 R_S}{2} & \text{for real sampling} \\ N_0 R_S & \text{for complex sampling} \end{cases} \quad (3.23)$$

If the noise is critically sampled to preserve whiteness, then the equalities in Eqs. (3.22) and (3.23) will hold.

Example 3.2 Consider a simulation of data transmission at 1200 bits/sec via a 4-ary FSK signal in additive white gaussian noise of spectral density N_0. The simulation program will generate 32 real samples during each symbol interval. Assume that the noise is critically sampled and that the signal tone has unity amplitude. Find the noise generator variance σ^2 which corresponds to $E_b/N_0 = 5$ dB, where E_b denotes energy per transmitted bit.

solution During each symbol interval, the energy in the signal tone is given by

$$E = \frac{A^2 T}{2} \quad (3.24)$$

where A is the signal tone amplitude and T is the symbol duration. Since each 4-ary symbol conveys 2 bits, the stated data rate of 1200 bits/sec corresponds to a symbol duration of $T = 1/600$. Thus, $E = 1/1200$. The energy per bit is one-half of the energy per symbol:

$$E_b = \frac{E}{2} = \frac{A^2 T}{4} = \frac{1}{2400} \quad (3.25)$$

The sampling rate is simply the symbol rate multiplied by the number of samples per symbol:

$$R_s = (32)(600) = 19200 \quad (3.26)$$

From Eq. (3.23)

$$N_0 = \frac{2\sigma^2}{R_s} = \frac{\sigma^2}{9600} \quad (3.27)$$

Thus

$$\frac{E_b}{N_0} = \frac{1/2400}{\sigma^2/9600} = \frac{4}{\sigma^2} \quad (3.28)$$

The specified E_b/N_0 of 5 dB is easily converted into a numeric value using

$$\frac{E_b}{N_0} = 10^{[(E_b/N_0)\,\text{dB}/10]} = 10^{0.5}$$

Substituting this into Eq. (3.28) we obtain

$$\sigma^2 = 4 \times 10^{10-0.5}$$
$$= 1.2649$$

Software

We can take several different approaches to implement the methods described above for generating normally distributed random numbers. Since both of the methods operate by transforming uniform random numbers, one of the first design choices that must be made is whether to provide the gaussian generator with a built-in means for generating uniform numbers, or to have the gaussian generator call a separate uniform generator of the sort presented in Section 3.4.

Calling a separate uniform generator results in code which is easier to test and maintain, but which incurs a significant speed penalty due to overhead associated with additional calls to the separate generator. In many simulations, a random number generator may be called millions of times, and even a slight increase in per-call overhead can add up quickly. One possible way to provide both easy maintenance and higher execution speeds would be to implement a uniform generator as an inline function for use by non-uniform generators that need a source of uniform random numbers. The gaussian random number generator provided in file gausrand.cpp takes the approach of calling a separate uniform generator.

3.7 Thermal Noise

The random motion of electrons in a conductor causes an electrical noise called *thermal noise* which is gaussian-distributed with zero mean. The variance is given by

$$\sigma^2 = E[v^2] = \frac{2R(\pi kT)^2}{3h} \tag{3.29}$$

where k = Boltzmann's constant $\approx 1.380622 \times 10^{-23}$ J/K
h = Planck's constant $\approx 6.626196 \times 10^{-34}$ J$\cdot$s
T = temperature, kelvin
R = resistance, ohm

The spectral density of the mean-square noise voltage is given by

$$S(f) = \frac{2Rh|f|}{\exp[h|f|/(kT)] - 1} \tag{3.30}$$

For $|f| \ll kT/h$, Eq. (3.30) can be approximated as:

$$S(f) \approx 2RkT\left(1 - \frac{h|f|}{2kT}\right) \tag{3.31}$$

For frequencies up to approximately 10^{12} Hz, this expression is nearly constant and can be further simplified to

$$S(f) = 2RkT \tag{3.32}$$

In a bandwidth of B Hz, the mean-square value of the noise voltage is given by

$$E[v^2] = 2BS(f) = 4kTRB \tag{3.33}$$

A noisy resistor can be modeled as an ideal resistor in series with a noise voltage source having a mean-square voltage as given by Eq. (3.33). The *available power* of a source is defined as the power delivered to a load which is matched to the impedance of the source. Thus the available power will be the power delivered by the resistor and voltage source combination to a load resistance R. The load current will be $v/(2R)$, and the voltage drop across the load will be $v/2$. Thus the available power is equal to $v^2/(4R)$. Using the expected value of v^2 given by Eq. (3.33), we find that the available power is simply kTB watts.

Equivalent noise temperature

All actual systems generate noise internally. *Equivalent noise temperature* is one way of characterizing the amount of noise generated in a linear two-port system. Imagine that it is possible to obtain a noiseless version of the system under consideration. Conceptually, we could connect a resistor to the input and then adjust the temperature of the resistor until the thermal noise generated by the resistor causes the available noise power at the output of the idealized system to exactly equal the noise power at the output of the actual system. The temperature of the resistor then defines the equivalent noise temperature of the system. Note that this definition assumes that the resistor is matched to the input impedance of the system.

3.8 Bandpass Noise

Consider a bandpass noise process $n(t)$ which has a power spectral density as depicted in Fig. 3.6. Such a process can be expressed in quadrature form relative to a carrier of frequency f_c and phase ψ:

$$n(t) = n_c(t)\cos(2\pi f_c t) - n_s(t)\sin(2\pi f_c t) \tag{3.34}$$

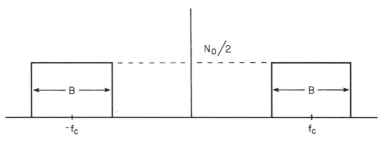

Figure 3.6 Spectrum of bandpass noise produced by ideal filtering of white noise.

The *envelope* of $n(t)$ is given by

$$A(t) = \sqrt{n_c^2(t) + n_s^2(t)} \tag{3.35}$$

The *instantaneous phase* of $n(t)$ is given by

$$\theta(t) = \tan^{-1}\left(\frac{n_s(t)}{n_c(t)}\right)$$

It can be shown that the quadrature form representation of Eq. (3.34) exhibits the following properties:

1. If $n(t)$ is a bandpass process of bandwidth $2B$ centered at f_c, the *inphase* component $n_c(t)$ and *quadrature* component $n_s(t)$ are each lowpass processes of (one-sided) bandwidth B. Specifically, if $n(t)$ has a power spectral density $S_n(f)$ which occupies frequencies in the interval $(f_c - B) \le |f| \le (f_c + B)$, then the power spectral densities $S_c(f)$ and $S_s(f)$ of the inphase and quadrature component, respectively, are given by

$$S_c(f) = S_s(f) = \begin{cases} S_n(f + f_c) + S_n(f - f_c) & |f| \le B \\ 0 & \text{elsewhere} \end{cases}$$

2. The two components $n_c(t)$ and $n_s(t)$ are uncorrelated.
3. If $n(t)$ has zero mean, then $n_c(t)$ and $n_s(t)$ each have zero mean.
4. If $n(t)$ has zero mean, then the variances of $n_c(t)$ and $n_s(t)$ will each equal the variance of $n(t)$.

$$\sigma_n^2 = \sigma_c^2 = \sigma_s^2$$

5. If $n(t)$ is wide-sense stationary, then $n_c(t)$ and $n_s(t)$ are jointly wide-sense stationary.

If $n(t)$ is a bandpass gaussian process, then the following properties will hold in addition to the properties given above.

1. The inphase component $n_c(t)$ and the quadrature component $n_s(t)$ will be jointly gaussian, as well as statistically independent.
2. The envelope $A(t)$ will be a random process in which the ensemble at any given instant is Rayleigh distributed.
3. The phase $\theta(t)$ will be a random process in which the ensemble at any given instant is uniformly distributed from zero to 2π.

References

1. Haykin, S. *Communication Systems*, 2d ed., Wiley, New York, 1983.
2. Carlson, A. B. *Communication Systems: An Introduction to Signals and Noise in Electrical Communication*, McGraw-Hill, New York, 1968.
3. Taub, H. and D. L. Schilling. *Principles of Communication Systems*, 2d ed., McGraw-Hill, New York, 1986.

4. Cooper, G. R. and C. D. McGillem. *Modern Communication and Spread Spectrum*, McGraw-Hill, New York, 1986.
5. Proakis, J. G. *Digital Communications*, McGraw-Hill, New York, 1983.
6. Knuth, D. K. *The Art of Computer Programming, Vol. 2: Seminumerical Algorithms*, Addison-Wesley, Reading, Mass., 1981.
7. Press, W. H. et al. *Numerical Recipes*, Cambridge University Press, Cambridge, 1986.
8. Park, S. K. and K. W. Miller. *Communications of the ACM*, vol. 31, pp. 1192–1201, 1988.
9. Abramowitz, M. and I. A. Stegun. *Handbook of Mathematical Functions*, National Bureau of Standards, Appl. Math Series 55, 1966.

Chapter 4

Linear Systems

4.1 Systems

Within the context of signal processing, a *system* is something that accepts one or more input signals and operates upon them to produce one or more output signals. Filters, amplifiers, and digitizers are some of the systems used in various signal processing applications. When signals are represented as mathematical functions, it is convenient to represent systems as *operators* that operate upon input functions to produce output functions. Two alternative notations for representing a system H with input x and output y are given in Eqs. (4.1) and (4.2). Note that x and y can each be scalar valued or vector valued.

$$y = H[x] \qquad (4.1)$$
$$y = Hx \qquad (4.2)$$

This book uses the notation of Eq. (4.1) as this is less likely to be confused with multiplication of x by a value H.

A system H can be represented pictorially in a flow diagram as shown in Fig. 4.1. For vector-valued x and y, the individual components are sometimes explicitly shown, as in Fig. 4.2a or lumped together, as shown in Fig. 4.2b. Sometimes, to emphasize their vector nature, the input and output are drawn as in Fig. 4.2c.

In different presentations of system theory, the notational schemes used exhibit some variation. The more precise treatments (such as [1]) use x or $x(\cdot)$ to denote a function of time defined over the interval $(-\infty, \infty)$. A function defined over a more restricted interval such as $[t_0, t_1)$ would be denoted as $x_{[t_0, t_1)}$. The notation $x(t)$ is reserved for denoting the value of x at time t. Less precise treatments (such as [2]) use $x(t)$ to denote both functions of time defined over $(-\infty, \infty)$ and the value of function x at time t. When not evident from context, words of explanation must be included to indicate which particular meaning is intended. Using the less precise notational scheme, Eq. (4.1) could be

Figure 4.1 Pictorial representation of a system.

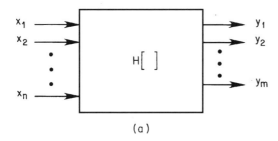

(a)

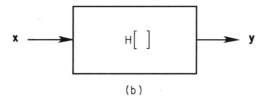

(b)

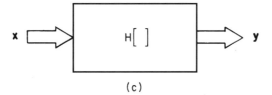

(c)

Figure 4.2 Pictorial representations of a system with multiple inputs and outputs.

rewritten as

$$y(t) = H[x(t)] \tag{4.3}$$

While it appears that the precise notation should be the more desireable, the relaxed conventions exemplified by Eq. (4.3) dominate the engineering literature.

Linearity

If the relaxed system H is *homogeneous*, multiplying the input by a constant gain is equivalent to multiplying the output by the same constant gain, and the

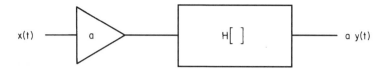

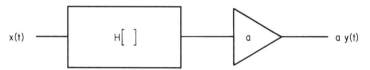

Figure 4.3 Homogeneous system.

two configurations shown in Fig. 4.3 are equivalent. Mathematically stated, the relaxed system H is homogeneous if, for constant a

$$H[ax] = aH[x] \qquad (4.4)$$

If the relaxed system H is *additive*, the output produced for the sum of two input signals is equal to the sum of the outputs produced for each input individually, and the two configurations shown in Fig. 4.4 are equivalent. Mathematically stated, the relaxed system H is additive if

$$H[x_1 + x_2] = H[x_1] + H[x_2] \qquad (4.5)$$

A system that is both homogeneous and additive is said to exhibit or satisfy the principle of *superposition*. A system that exhibits superposition is called a *linear system*. Under certain restrictions, additivity implies homogeneity. Specifically, the fact that a system H is additive implies that

$$H[\alpha x] = \alpha H[x] \qquad (4.6)$$

for any rational α. Any real number can be approximated with arbitrary precision by a rational number; therefore, additivity implies homogeneity for real a provided that

$$\lim_{\alpha \to a} H[\alpha x] = H[ax] \qquad (4.7)$$

Time invariance

The characteristics of a *time-invariant* system do not change over time. A system is said to be *relaxed* if it is not still responding to any previously applied input. Given a relaxed system H such that

$$y(t) = H[x(t)] \qquad (4.8)$$

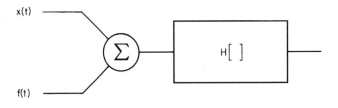

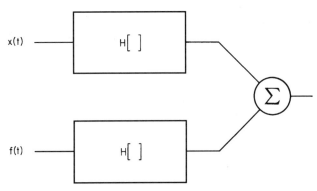

Figure 4.4 Additive system.

then H is time invariant if and only if

$$y(t - \tau) = H[x(t - \tau)] \tag{4.9}$$

for any τ and any $x(t)$. A time-invariant system is also called a *fixed* system or *stationary* system. A system that is not time invariant is called a *time-varying* system, *variable* system, or *nonstationary* system.

Causality

In a *causal* system, the output at time t can depend only upon the input at times t and prior. Mathematically stated, a system H is causal if and only if

$$H[x_1(t)] = H[x_2(t)] \quad \text{for } t \leq t_0 \tag{4.10}$$

given that

$$x_1(t) = x_2(t) \quad \text{for } t \leq t_0$$

A *noncausal* or *anticipatory* system is one in which the present output depends upon future values of the input. Noncausal systems occur in theory, but they cannot exist in the real world. This is unfortunate, since we often discover that some especially desireable frequency responses can be obtained only from noncausal systems. However, causal realizations can be created for those noncausal systems in which the present output depends at most upon past, present, and a finite extent of future inputs. In such cases, a causal realization is obtained by simply delaying the output of the system for a finite interval until all the required inputs have entered the system and are available for determination of the output.

4.2 Characterization of Linear Systems

A linear system can be characterized by a differential equation, step response, impulse response, complex-frequency-domain system function, or a transfer function. The relationships among these various characterizations are given in Table 4.1.

Impulse response

The *impulse response* of a system is the output response produced when a unit impulse $\delta(t)$ is applied to the input of a previously relaxed system. This is an especially convenient characterization of a linear system, since the response $y(t)$ to any continuous-time input signal $x(t)$ is given by

$$y(t) = \int_{-\infty}^{\infty} x(\tau) \, h(t, \tau) \, d\tau \tag{4.11}$$

where $h(t, \tau)$ denotes the system's response at time t to an impulse applied at time τ. The integral in Eq. (4.11) is sometimes referred to as the *superposition integral*. The particular notation used indicates that, in general, the system is time varying. For a time-invariant system, the impulse response at time t depends only upon the time delay from τ to t, and we can redefine the

TABLE 4.1 Relationships among Characterizations of Linear Systems

Starting with	Perform	To obtain
Time domain differential equation relating $x(t)$ and $y(t)$	Laplace transform	Complex frequency domain System function
	Compute $y(t)$ for $x(t) =$ unit impulse	Impulse response $h(t)$
	Compute $y(t)$ for $x(t) =$ unit step	Step response $a(t)$
Step response $a(t)$	Differentiate with respect to time	Impulse response $h(t)$
Impulse response $h(t)$	Integrate with respect to time	Step response $a(t)$
	Laplace transform	Transfer function $H(s)$
Complex frequency domain system function	Solve for $H(s) = Y(s)/X(s)$	Transfer function $H(s)$
Transfer function $H(s)$	Inverse Laplace transform	Impulse response $h(t)$

impulse response to be a function of a single variable and denote it as $h(t - \tau)$. Equation (4.11) then becomes

$$y(t) = \int_{-\infty}^{\infty} x(\tau) \, h(t - \tau) \, d\tau \tag{4.12}$$

Via the simple change of variables $\lambda = t - \tau$, Eq. (4.12) can be rewritten as

$$y(t) = \int_{-\infty}^{\infty} x(t - \lambda) \, h(\lambda) \, d\lambda \tag{4.13}$$

If we assume that the input is 0 for $t < 0$, the lower limit of integration can be changed to zero; if we further assume that the system is causal, the upper limit of integration can be changed to t, yielding

$$y(t) = \int_{0}^{t} x(\tau) \, h(t - \tau) \, d\tau = \int_{0}^{t} x(t - \lambda) \, h(\lambda) \, d\lambda \tag{4.14}$$

The integrals in Eq. (4.14) are known as *convolution integrals*, and the equation indicates that $y(t)$ equals the *convolution* of $x(t)$ and $h(t)$. It is often more compact and convenient to denote this relationship as

$$y(t) = x(t) \otimes h(t) = h(t) \otimes x(t) \tag{4.15}$$

Various texts use different symbols, such as stars or asterisks, in place of $\otimes$ to indicate convolution. The asterisk is probably favored by most printers, but in some contexts its usage to indicate convolution could be confused with the complex-conjugation operator. A typical system's impulse response is sketched in Fig. 4.5.

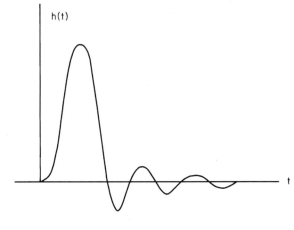

Figure 4.5 Impulse response of a typical system.

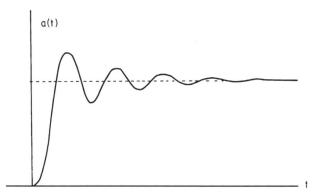

Figure 4.6 Step response of a typical system.

Step response

The *step response* of a system is the output signal produced when a unit step $u(t)$ is applied to the input of the previously relaxed system. Since the unit step is simply the time integration of a unit impulse, it can be shown easily that the step response of a system can be obtained by integrating the impulse response. A typical system's step response is shown in Fig. 4.6.

4.3 Laplace Transform

The *Laplace transform* is a useful technique for transforming differential equations into algebraic equations that can be more easily manipulated to obtain desired results.

In most signal processing applications, the functions of interest are usually (but not always) functions of time. The Laplace transform of a time function $x(t)$ is usually denoted as $X(s)$ or $\mathcal{L}[x(t)]$, and is defined by

$$X(s) = \mathcal{L}[x(t)] = \int_{-\infty}^{\infty} x(t)e^{-st}\,dt \qquad (4.16)$$

The complex variable s is usually referred to as *complex frequency* and is of the form $\sigma + j\omega$, where σ and ω are real variables sometimes referred to as *neper frequency* and *radian frequency*, respectively. The Laplace transform for a given function $x(t)$ is obtained by simply evaluating the given integral. Some mathematics texts (such as [3]) denote the time function with an uppercase letter and the frequency function with a lowercase letter. However, the use of lowercase for time functions is almost universal within the engineering literature.

If we transform both sides of a differential equation in t using the definition Eq. (4.16), we obtain an algebraic equation in s that can be solved for the desired quantity. The solved algebraic equation can then be transformed back into the time domain by using the inverse Laplace transform. The inverse Laplace

TABLE 4.2 Laplace Transform Pairs

Ref. no.	$x(t)$	$X(s)$
1	1	$\dfrac{1}{s}$
2	$u_1(t)$	$\dfrac{1}{s}$
3	$\delta(t)$	1
4	t	$\dfrac{1}{s^2}$
5	t^n	$\dfrac{n!}{s^{n+1}}$
6	$\sin \omega t$	$\dfrac{\omega}{s^2 + \omega^2}$
7	$\cos \omega t$	$\dfrac{s}{s^2 + \omega^2}$
8	e^{-at}	$\dfrac{1}{s+a}$
9	$e^{-at} \sin \omega t$	$\dfrac{\omega}{(s+a)^2 + \omega^2}$
10	$e^{-at} \cos \omega t$	$\dfrac{s+a}{(s+a)^2 + \omega^2}$

TABLE 4.3 Properties of the Laplace Transform

Property	Time function	Transform
1. Homogeneity	$af(t)$	$aF(s)$
2. Additivity	$f(t) + g(t)$	$F(s) + G(s)$
3. Linearity	$af(t) + bg(t)$	$aF(s) + bG(s)$
4. First derivative	$\dfrac{d}{dt} f(t)$	$sF(s) - f(0)$
5. Second derivative	$\dfrac{d^2}{dt^2} f(t)$	$s^2 F(s) - sf(0) - \dfrac{d}{dt} f(0)$
6. kth derivative	$\dfrac{d^{(k)}}{dt^k} f(t)$	$s^k F(s) - \sum_{n=0}^{k-1} s^{k-1-n} f^{(n)}(0)$
7. Integration	$\int_{-\infty}^{t} f(\tau)\, d\tau$	$\dfrac{F(s)}{s} + \dfrac{1}{s} \left(\int_{-\infty}^{t} f(\tau)\, d\tau \right)_{t=0}$
	$\int_{0}^{t} f(\tau)\, d\tau$	$\dfrac{F(s)}{s}$
8. Frequency shift	$e^{-at} f(t)$	$X(s+a)$
9. Time shift right	$u_1(t-\tau) f(t-\tau)$	$e^{-\tau s} F(s) \quad a > 0$
10. Time shift left	$f(t+\tau), \quad f(t) = 0 \text{ for } 0 < t < \tau$	$e^{\tau s} F(s)$
11. Convolution	$y(t) = \int_{0}^{t} h(t-\tau) x(\tau)\, d\tau$	$Y(s) = H(s) X(s)$

transform is defined by

$$x(t) = \mathcal{L}^{-1}[X(s)] = \frac{1}{2\pi j} \int_C X(s)e^{st}\,ds \qquad (4.17)$$

where C is a contour of integration chosen to include all singularities of $X(s)$. The inverse Laplace transform for a given function $X(s)$ can be obtained by evaluating the given integral. However, this integration is often a major chore: when tractable, it will usually involve application of the residue theorem from the theory of complex variables. Fortunately, in most cases of practical interest, direct evaluation of Eqs. (4.16) and (4.17) can be avoided by using some well-known transform pairs, as listed in Table 4.2, along with a number of transform properties listed in Table 4.3.

4.4 Transfer Functions

The *transfer function* $H(s)$ of a system is equal to the Laplace transform of the output signal divided by the Laplace transform of the input signal:

$$H(s) = \frac{Y(s)}{X(s)} = \frac{\mathcal{L}[y(t)]}{\mathcal{L}[x(t)]} \qquad (4.18)$$

It can be shown that the transfer function is also equal to the Laplace transform of the system's impulse response:

$$H(s) = \mathcal{L}[h(t)] \qquad (4.19)$$

Therefore,
$$y(t) = \mathcal{L}^{-1}\{H(s)\mathcal{L}[x(t)]\} \qquad (4.20)$$

Equation (4.20) presents an alternative to the convolution defined by Eq. (4.14) for obtaining a system's response $y(t)$ to any input $x(t)$, given the impulse response $h(t)$. Simply perform the following steps:

1. Compute $H(s)$ as the Laplace transform of $h(t)$.
2. Compute $X(s)$ as the Laplace transform of $x(t)$.
3. Compute $Y(s)$ as the product of $H(s)$ and $X(s)$.
4. Compute $y(t)$ as the inverse Laplace transform of $Y(s)$. (The Heaviside expansion presented in Sec. 4.5 is a convenient technique for performing the inverse transform operation.)

A transfer function defined as in Eq. (4.18) can be put into the form

$$H(s) = \frac{P(s)}{Q(s)} \qquad (4.21)$$

where $P(s)$ and $Q(s)$ are polynomials in s. For $H(s)$ to be stable and realizable in the form of a lumped-parameter network, the following conditions must be satisfied:

1. The coefficients in $P(s)$ must be real.
2. The coefficients in $Q(s)$ must be real and positive.
3. The polynomial $Q(s)$ must have a nonzero term for each degree of s from the highest to lowest, unless all even-degree terms or all odd-degree terms are missing.
4. If $H(s)$ is the voltage ratio or current ratio (i.e., the input and output are either both voltages or both currents), the maximum degree of s in $P(s)$ cannot exceed the maximum degree of s in $Q(s)$.
5. If $H(s)$ is a transfer impedance (i.e., the input is a current and the output is a voltage) or a transfer admittance (i.e., the input is a voltage and the output is a current), then the maximum degree of s in $P(s)$ can exceed the maximum degree of s in $Q(s)$ by at most 1.

Note that conditions 4 and 5 establish only upper limits on the degree of s in $P(s)$; in either case, the maximum degree of s in $P(s)$ may be as small as zero. Also note that these are necessary, but not sufficient, conditions for $H(s)$ to be a valid transfer function. A candidate $H(s)$ satisfying all these conditions may still not be realizable as a lumped parameter network.

Example 4.1 Consider the following alleged transfer functions:

$$H_1(s) = \frac{s^2 - 2s + 1}{s^3 - 3s^2 + 3s + 1} \tag{4.22}$$

$$H_2(s) = \frac{s^4 + 2s^3 + 2s^2 - 3s + 1}{s^3 + 3s^2 + 3s + 2} \tag{4.23}$$

$$H_3(s) = \frac{s^2 - 2s + 1}{s^3 + 3s^2 + 1} \tag{4.24}$$

Equation (4.22) is cannot be a transfer function because the coefficient of s^2 in the denominator is negative. If Eq. (4.23) is intended as a voltage- or current-transfer ratio, it is not acceptable because the degree of the numerator exceeds the degree of the denominator. However, if Eq. (4.23) represents a transfer impedance or transfer admittance, it may be valid since the degree of the numerator exceeds the degree of the denominator by just 1. Equation (4.24) is not acceptable because the term for s is missing from the denominator.

A system's transfer function can be manipulated to provide a number of useful characterizations of the system's behavior. These characterizations are listed in Table 4.4 and examined in more detail in subsequent sections.

Some authors, such as [4]. use the term *network function* in place of *transfer function*.

TABLE 4.4 System Characterizations Obtained from the Transfer Function

Starting with	Perform	To obtain
Transfer function $H(s)$	Compute roots of $H(s)$ denominator	Pole locations
	Compute roots of $H(s)$ numerator	Zero locations
	Compute $|H(j\omega)|$ over all ω	Magnitude response $A(\omega)$
	Compute $\arg[H(j\omega)]$ over all ω	Phase response $\theta(\omega)$
Phase response $\theta(\omega)$	Divide by ω	Phase delay $\tau_p(\omega)$
	Differentiate with respect to ω	Group delay $\tau_g(\omega)$

4.5 Heaviside Expansion

The Heaviside expansion provides a straightforward computational method for obtaining the inverse Laplace transform of certain types of complex-frequency functions. The function to be inverse transformed must be expressed as a ratio of polynomials in s, where the order of the denominator polynomial exceeds the order of the numerator polynomial. If

$$H(s) = K_0 \frac{P(s)}{Q(s)} \tag{4.25}$$

where

$$Q(s) = \prod_{k=1}^{n}(s - s_k)^{m_k} = (s - s_1)^{m_1}(s - s_2)^{m_2} \cdots (s - s_n)^{m_n} \tag{4.26}$$

then inverse transformation via the Heaviside expansion yields

$$\mathcal{L}^{-1}[H(s)] = K_0 \sum_{r=1}^{n} \sum_{k=1}^{m_r} [K_{rk} t^{m_r - k} \exp(s_r t)] \tag{4.27}$$

where

$$K_{rk} = \frac{1}{(k-1)!(m_r - k)!} \frac{d^{k-1}}{ds^{k-1}} \left[\frac{(s - s_r)^{m_r} P(s)}{Q(s)} \right]_{s=s_r} \tag{4.28}$$

A method for computing the derivative in Eq. (4.28) can be found in Sec. A.4 of Appendix A.

Simple pole case

The complexity of the expansion is significantly reduced for the case of $Q(s)$ having no repeated roots. The denominator of Eq. (4.25) is then given by

$$Q(s) = \prod_{k=1}^{n}(s - s_k) = (s - s_1)(s - s_2) \cdots (s - s_n) \qquad s_1 \neq s_2 \neq \cdots s_n \tag{4.29}$$

Inverse transformation via the Heaviside expansion then yields

$$\mathcal{L}^{-1}[H(s)] = K_0 \sum_{r=1}^{n} K_r e^{s_r t} \tag{4.30}$$

where

$$K_r = \left[\frac{(s - s_r)P(s)}{Q(s)}\right]_{s=s_r} \tag{4.31}$$

The Heaviside expansion is named for Oliver Heaviside (1850–1925), an English physicist and electrical engineer who was the nephew of Charles Wheatstone (as in Wheatstone bridge).

4.6 Poles and Zeros

As pointed out previously, the transfer function for a realizable linear time-invariant system can always be expressed as a ratio of polynomials in s:

$$H(s) = \frac{P(s)}{Q(s)} \tag{4.32}$$

The numerator and denominator can each be factored to yield

$$H(s) = H_0 \frac{(s - z_1)(s - z_2)(s - z_3) \cdots (s - z_m)}{(s - p_1)(s - p_2)(s - p_3) \cdots (s - p_n)} \tag{4.33}$$

where the roots $z_1, z_2, \ldots, z_m$ of the denominator are called *zeros* of the transfer function, and the roots $p_1, p_2, \ldots, p_n$ of the denominator are called *poles* of the transfer function. Together, poles and zeros can be collectively referred to as *critical frequencies*. Each factor $(s - z_i)$ is called a *zero factor*, and each factor $(s - p_j)$ is called a *pole factor*. A repeated zero appearing n times is called either an *nth-order zero* or a *zero of multiplicity n*. Likewise, a repeated pole appearing n times is called either an *nth-order pole* or a *pole of multiplicity n*. Nonrepeated poles or zeros are sometimes described as *simple* or *distinct* to emphasize their nonrepeated nature.

Example 4.2 Consider the transfer function given by

$$H(s) = \frac{s^3 + 5s^2 + 8s + 4}{s^3 + 13s^2 + 59s + 87} \tag{4.34}$$

The numerator and denominator can be factored to yield

$$H(s) = \frac{(s + 2)^2(s + 1)}{(s + 5 + 2j)(s + 5 - 2j)(s + 3)} \tag{4.35}$$

Examination of (4.35) reveals that

$s = -1$ is a simple zero.
$s = -2$ is a second-order zero.
$s = -5 + 2j$ is a simple pole.
$s = -5 - 2j$ is a simple pole.
$s = -3$ is a simple zero.

A system's poles and zeros can be depicted graphically as locations in a complex plane, as shown in Fig. 4.7. In mathematics, the complex plane itself is called the *gaussian plane*, while a plot depicting complex values as points in the plane is called an *Argand diagram* or a *Wessel-Argand gaussian diagram*. In the 1798 transactions of the Danish academy, Caspar Wessel (1745–1818) published a technique for a graphical representation of complex numbers, and in 1806 Jean Robert Argand published a similar technique. Geometric interpretation of complex numbers played a central role in the doctoral thesis of Gauss.

Pole locations can provide convenient indications of a system's behavior as indicated in Table 4.5. Furthermore, poles and zeros possess the following properties that can sometimes be used to expedite the analysis of a system:

1. For real $H(s)$, complex or imaginary poles and zeros will each occur in complex conjugate pairs that are symmetric about the σ axis.

2. For $H(s)$ having even symmetry, the poles and zeros will exhibit symmetry about the $j\omega$ axis.

3. For nonnegative $H(s)$, any zeros on the $j\omega$ axis will occur in pairs.

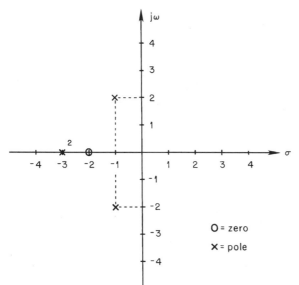

Figure 4.7 Plot of pole and zero locations.

TABLE 4.5 Impact of Pole Locations upon System Behavior

Pole type	Corresponding natural response component	Corresponding description of system behavior
Single real, negative	Decaying exponential	Stable
Single real, positive	Divergent exponential	Divergent instability
Real pair, negative, unequal	Decaying exponential	Overdamped (stable)
Real pair, negative equal	Decaying exponential	Critically damped (stable)
Complex conjugate pair with negative real parts	Exponentially decaying sinusoid	Underdamped (stable)
Complex conjugate pair with zero real parts	Sinusoid	Undamped (marginally stable)
Complex conjugate pair with positive real parts	Exponentially saturating sinusoid	Oscillatory instability

In many situations, it is necessary to determine the poles of a given transfer function. For some systems, such as Chebyshev filters or Butterworth filters, explicit expressions have been found for evaluation of pole locations. For other systems, such as Bessel filters, the poles must be found by numerically solving for the roots of the transfer function's denominator polynomial. Several root-finding algorithms appear in the literature, but I have found the *Laguerre method* to be the most useful for approximating pole locations. The approximate roots can be subjected to small-step iterative refinement or polishing as needed.

Algorithm 4.1 Laguerre method for approximating one root of a polynomial $P(z)$.

1. Set z equal to an initial guess for the value of a root. Typically, z is set to zero so that the smallest root will tend to be found first.
2. Evaluate the polynomial $P(z)$ and its first two derivatives $P'(z)$ and $P''(z)$ at the current value of z.
3. If $P(z)$ evaluates to zero or to within some predefined epsilon of zero, exit with the current value of z as the root. Otherwise, continue on to step 4.
4. Compute a correction term Δz using

$$\Delta z = \frac{N}{F \pm \sqrt{(N-1)(NG - G^2)}}$$

where $F \triangleq P'(z)/P(z)$, $G \triangleq F^2 - P''(z)/P(z)$, and the sign in the denominator is taken so as to minimize the magnitude of the correction (or, equivalently, so as to maximize the denominator).

5. If the correction term Δz has a magnitude smaller than some specified fraction of the magnitude of z, then take z as the value of the root and terminate the algorithm.

6. If the algorithm has been running for a while (for example, six iterations) and the correction value has increased since the previous iteration, then take z as the value of the root and terminate the algorithm.
7. If the algorithm was not terminated in Step 3, 5, or 6, then subtract Δz from z and go back to step 2.

A function LaguerreMethod that implements Algorithm 4.1 is provided in file laguerre.cpp.

4.7 Magnitude, Phase, and Delay Responses

A system's *steady-state response* $H(j\omega)$ can be determined by evaluating the transfer function $H(s)$ at $s = j\omega$:

$$H(j\omega) = |H(j\omega)|e^{j\theta(\omega)} = H(s)|_{s=j\omega} \qquad (4.36)$$

The *magnitude response* is simply the magnitude of $H(j\omega)$:

$$|H(j\omega)| = (\{\text{Re}[H(j\omega)]\}^2 + \{\text{Im}[H(j\omega)]\}^2)^{1/2} \qquad (4.37)$$

It can be shown that

$$|H(j\omega)| = H(s)H(-s)|_{s=j\omega} \qquad (4.38)$$

If $H(s)$ is available in factored form as given by

$$H(s) = H_0 \frac{(s-z_1)(s-z_2)\cdots(s-z_m)}{(s-p_1)(s-p_2)\cdots(s-p_n)} \qquad (4.39)$$

then the magnitude response can be obtained by replacing each factor with its absolute value evaluated at $s = j\omega$:

$$|H(j\omega)| = H_0 \frac{|j\omega - z_1| \cdot |j\omega - z_2| \cdot |j\omega - z_3| \cdots\cdots |j\omega - z_m|}{|j\omega - p_1| \cdot |j\omega - p_2| \cdot |j\omega - p_3| \cdots\cdots |j\omega - p_n|} \qquad (4.40)$$

The *phase response* $\theta(\omega)$ is given by

$$\theta(\omega) = \tan^{-1}\left\{\frac{\text{Im}[H(j\omega)]}{\text{Re}[H(j\omega)]}\right\} \qquad (4.41)$$

Phase delay

The *phase delay* $\tau_p(\omega)$ of a system is defined as

$$\tau_p(\omega) = \frac{-\theta(\omega)}{\omega} \qquad (4.42)$$

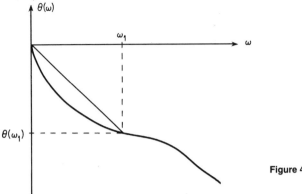

Figure 4.8 Phase delay.

where $\theta(\omega)$ is the phase response defined in Eq. (4.41). When evaluated at any specific frequency ω_1, Eq. (4.42) will yield the time delay experienced by a sinusoid of frequency ω_1 passing through the system. Some authors define $\tau_p(\omega)$ without the minus sign shown in the numerator of Eq. (4.42). As illustrated in Fig. 4.8, the phase delay at a frequency ω_1 is equal to the negative slope of a secant drawn from the origin to the phase response curve at ω_1. Phase delay is sometimes called *carrier delay*.

Group delay

The *group delay* $\tau_g(\omega)$ of a system is defined as

$$\tau_g(\omega) = \frac{-d}{dt}\theta(\omega) \qquad (4.43)$$

where $\theta(\omega)$ is the phase response defined in Eq. (4.41). In the case of a modulated carrier passing through the system, the modulation envelope will be delayed by an amount equal to τ_g. If the group delay is not constant over the entire bandwidth of the signal, the envelope will be distorted. As shown in Fig. 4.9, the group delay at a frequency ω_1 is equal to the negative slope of a tangent to the phase response at ω_1. Group delay is also called *envelope delay*.

4.8 Computer Representation of Polynomials and Transfer Functions

Consider a polynomial of degree N in sum-of-powers form

$$P(x) = \sum_{n=0}^{N} c_n x^n$$

In the computer, such a polynomial can be represented using an array, for example, `complex c[]`, to hold the coefficients c_n for $n = 0, 1, \ldots, N$. To multiply

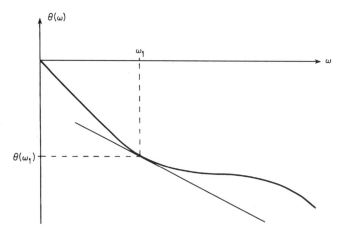

Figure 4.9 Group delay.

such a polynomial by a constant value, we need only to multiply each coefficient by this value as in

```
for( int n=0; n<=N; n++)
    { c[n] *= const_val; }
```

To multiply the polynomial by x, we only need to shift the contents of the array upwards by one location so that c[N] moves to c[N+1], c[N-1] moves to c[N] and so on until c[0] moves to c[1]. The original contents of c[0] are then replaced by zero:

```
for( int n=N; n>=0; n--)
    { c[n+1] = c[n]; }
c[0] = complex(0.0,0.0);
```

Using these rules for multiplying a polynomial by a constant and multiplying a polynomial in x by x, we can easily automate the process of performing the multiplications indicated in the numerator or denominator of Eq. (4.56) to obtain the sum-of-powers form of Eq. (4.45) or the biquadratic form of Eq. (4.55).

Let us define a class CmplxPolynomial for representing a polynomial with complex-valued coefficients. Clearly the data members of this class must include an integer (for example, Deg_Of_Poly) for holding the degree of the polynomial and a complex-valued array (for example, Cmplx_Coeff[]) of length Deg_Of_Poly+1 for holding the coefficients. But what about member functions? What should the default constructor for this class do? Does it make any sense to have a polynomial of degree 0? Yes it does. A constant can be viewed as a polynomial of degree 0, and for complete generality our class should support this idea. Therefore, the default constructor will create a "zero" polynomial of degree 0 that has a single coefficient with a value of 0.

There should also be a copy constructor that can be used to make a duplicate of an existing `CmplxPolynomial` object. Binomial factors of the form $(s + a_n)$ or $(a_n s + b_n)$ occur so frequently that the `CmplxPolynomial` class should have a constructor specifically designed for creating and initializing binomials. The class also will need member functions for assignment, multiplication, and publishing the polynomial coefficients to an output stream. An implementation of class `CmplxPolynomial` that incorporates all of these features is provided in file `cmpxpoly.cpp`.

In addition to `CmplxPolynomial`, it will be convenient to have a class, for example, `Polynomial`, that can be used to represent polynomials with real-valued coefficients. Such polynomials are used in the biquadratic form of Eq. (4.55) and in the sum-of-powers form of Eq. (4.45). `Polynomial` should include real-valued versions of each of the functions in `CmplxPolynomial`.

`CmplxPolynomial` objects often are used to multiply together terms with complex-valued coefficients to produce an ultimate result having all real-valued coefficients. This result is in the form of a `CmplxPolynomial` object that has zero-valued imaginary parts for all of its coefficients. In simulations of analog filters, it proves convenient to handle such a result as an object of type `Polynomial` rather than as an object of type `CmplxPolynomial`. Therefore, we need to provide a conversion constructor that will construct a `Polynomial` object and initialize it to be numerically equivalent to a given `CmplxPolynomial` object that happens to have zero-valued imaginary parts for all of its coefficients. However, there is nothing to prevent a user from attempting to create a `Polynomial` object from a `CmplxPolynomial` object whose coefficients do not have zero-valued imaginary parts. There are two strategies for dealing with such attempted misuse. We can blindly set each coefficient in the newly created `Polynomial` equal to the real part of the corresponding coefficient in `CmplxPolynomial`, without regard to the imaginary part of the coefficient. Or we can choose to check each coefficient in `CmplxPolynomial` and issue an error if the imaginary part of any coefficient is not very close to zero. (In theory, the imaginary parts should be exactly zero, but due to the effects of finite-precision arithmetic, they rarely are exactly zero in any `CmplxPolynomial` that has been built up by multiplying a number of binomial terms having complex coefficients.) The first approach is faster, and the second approach is safer. Since the polynomial conversions will only be done once or twice as part of the setup for a filter simulation, speed is not a driving concern. Therefore, the second approach was selected for the class implementation provided in file `poly.cpp`.

4.9 Computer Simulation of Analog Systems

Certain types of digital filters are obtained via transformations that are performed upon analog *prototype* filters. In assessing the performance of these digital filters it often is convenient to compare their responses to certain discrete-time input signals against the prototype filters' responses to the corresponding continuous time input signals. Computer simulation of the analog prototype filters provides a convenient way to make such comparisons.

Sum-of-powers form

For all of the "traditional" analog filter families, the filter's transfer function can be expressed as a ratio of two polynomials in the complex-frequency variable s:

$$H(s) = H_0 \frac{\sum_{m=0}^{M} \alpha_m s^m}{\sum_{n=0}^{N} \beta_n s^n} \quad (4.44)$$

where the coefficients α_m and β_n are real.

Consider a filter with a transfer function given by

$$H(s) = \frac{Y(s)}{X(s)} = \frac{a_2 s^2 + a_1 s + a_0}{s^3 + b_2 s^2 + b_1 s + b_0} \quad (4.45)$$

Recalling that multiplication by s in the Laplace domain corresponds to differentiation with respect to time in the time domain, we could use Eq. (4.45) directly to build an implementation of the filter. However, this implementation would contain differentiators, which are difficult to build in hardware and which tend to be extremely noisy when implemented in software. Integration is a much less difficult operation to implement. Therefore we can divide each term in the numerator and denominator of Eq. (4.45) by s^3 to obtain

$$H(s) = \frac{Y(s)}{X(s)} = \frac{a_2 s^{-1} + a_1 s^{-2} + a_0 s^{-3}}{1 + b_2 s^{-1} + b_1 s^{-2} + b_0 s^{-3}} \quad (4.46)$$

Now we make use of a well known trick and split $H(s)$ into two parts

$$H(s) = \frac{Y(s)}{U(s)} \cdot \frac{U(s)}{X(s)}$$

where

$$\frac{Y(s)}{U(s)} = a_2 s^{-1} + a_1 s^{-2} + a_0 s^{-3} \quad (4.47)$$

and

$$\frac{X(s)}{U(s)} = 1 + b_2 s^{-1} + b_1 s^{-2} + b_0 s^{-3} \quad (4.48)$$

Solving Eq. (4.47) for $Y(s)$ and Eq. (4.48) for $U(s)$, we obtain

$$Y(s) = a_2 s^{-1} U(s) + a_1 s^{-2} U(s) + a_0 s^{-3} U(s) \quad (4.49)$$

and

$$U(s) = X(s) - b_2 s^{-1} U(s) - b_1 s^{-2} U(s) - b_0 s^{-3} U(s) \quad (4.50)$$

If we were to take the inverse Laplace transform of Eqs. (4.49) and (4.50) [assuming that $u(t) = 0$ for $t < 0$] we would obtain

$$y(t) = a_2 \int_0^t u(\tau)\,d\tau + a_1 \int_0^t \int_0^{\tau_2} u(\tau_1)\,d\tau_1 d\tau_2 + a_0 \int_0^t \int_0^{\tau_3} \int_0^{\tau_2} u(\tau_1)\,d\tau_1 d\tau_2 d\tau_3 \quad (4.51)$$

and

$$u(t) = x(t) - b_2 \int_0^t u(\tau)\,d\tau - b_1 \int_0^t \int_0^{\tau_2} u(\tau_1)\,d\tau_1 d\tau_2 - b_0 \int_0^t \int_0^{\tau_3} \int_0^{\tau_2} u(\tau_1)\,d\tau_1 d\tau_2 d\tau_3 \quad (4.52)$$

The notation in these equations is becoming quite cumbersome—further manipulations will be simplified if we substitute $w'''(t) \equiv \frac{d^3}{dt^3} w(t)$ for $u(t)$ in Eqs. (4.51) and (4.52) to obtain

$$y(t) = a_2 w''(t) + a_1 w'(t) + a_0 w(t) \quad (4.53)$$

and

$$w'''(t) = x(t) - b_2 w''(t) - b_1 w'(t) - b_0 w(t) \quad (4.54)$$

Equation (4.54) now provides a recipe for generating $w'''(t)$ from $x(t)$ using only integrators, adders, and constant multipliers as shown in Fig. 4.10. Likewise, Eq. (4.53) provides a recipe for generating $y(t)$ from $w'''(t)$ using only integrators, adders, and constant multipliers as shown in Fig. 4.11. Notice that in both Fig. 4.10 and Fig. 4.11, the node corresponding to $w'''(t)$ is followed by three

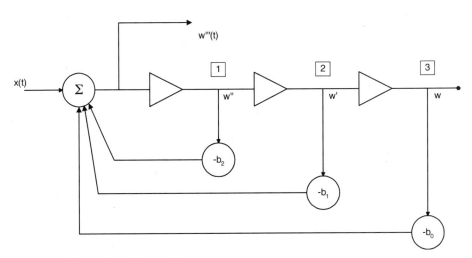

Figure 4.10 Realization of Eq. (4.49).

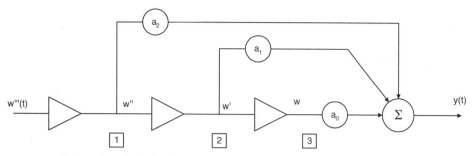

Figure 4.11 Realization of Eq. (4.50).

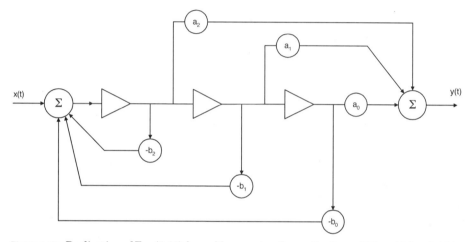

Figure 4.12 Realization of Eq. (4.46) formed by merging the realizations of Figs. 4.10 and 4.11.

integrators in cascade. The node labeled "1" in Fig. 4.10 is equivalent to the node labeled "1" in Fig. 4.11. Likewise for the nodes labeled "2" and "3" in these two figures. What this means is that we can combine the two realizations so that they share a single string of integrators as shown in Fig. 4.12. The filter may now be simulated in software using numerical integration techniques to implement each of these integrators.

Biquadratic form

The approach demonstrated above could be extended to filters with any number of poles and zeros, but it rarely is. It is usually more convenient to implement a high-order filter as a cascade of lower-order sections where each section has at most two poles and two zeros. Because the two polynomials in Eq. (4.44) have real coefficients, it can be shown that the roots of these polynomials either are real or they occur in complex conjugate pairs. Therefore it is possible to factor

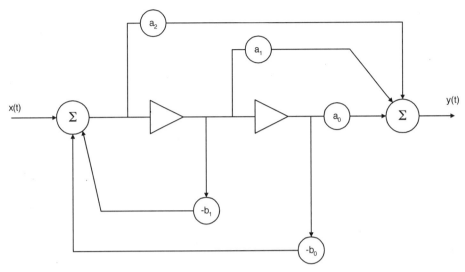

Figure 4.13 Biquad section.

Eq. (4.44) into the form

$$H(s) = H_0 \prod_{k=1}^{N/2} \frac{a_{2k}s^2 + a_{1k}s + a_{0k}}{s^2 + b_{1k}s + b_{0k}} \quad N \text{ even}$$

or

$$H(s) = \frac{H_0}{(s-d)} \prod_{k=1}^{(N-1)/2} \frac{a_{2k}s^2 + a_{1k}s + a_{0k}}{s^2 + b_{1k}s + b_{0k}} \quad N \text{ odd} \quad (4.55)$$

where the roots of each quadratic polynomial form a complex conjugate pair and all of the a, b, and d are real. The form represented by Eq. (4.55) is called the *biquadratic* form of the filter. Each factor in Eq. (4.55) can be implemented using a *biquad section* of the form shown in Fig. 4.13. The complete filter is then implemented by cascading the appropriate number of biquad sections.

Polynomial expansions

For most of the analog filter families, the filter's transfer function $H(s)$ is most easily determined as a ratio of two polynomials in s where the polynomials are expressed as products of binomial terms

$$H(s) = H_0 \frac{\prod_{m=1}^{M}(s - q_m)}{\prod_{n=1}^{N}(s - p_n)} \quad (4.56)$$

where p_n and q_m denote, respectively, the filter's poles and zeros. However, to take advantage of the simulation approaches discussed above, the transfer

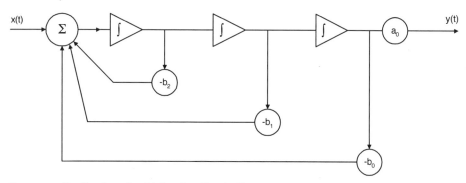

Figure 4.14 Realization of a third-order all-pole filter.

function needs to be in sum-of-powers form as in Eq. (4.44) or in the biquadratic form of Eq. (4.55). Expanding a product of binomials into a sum of powers by hand can be quite a tedious exercise—especially when the p_n and q_m are complex. Fortunately, it is relatively straightforward to create software that can perform this expansion. The class CmplxPolynomial described in Section 4.8 can be used for expanding products of binomials into sum-of-powers form.

Software for simulation of analog filters

A class AnalogPoleZeroFilter which can be used to simulate analog filters is provided in file pzfilt.cpp. The heart of this class is its Run function which uses numeric integration to implement Eqs. (4.53) and (4.54). The specifics of the numeric integration are isolated in a separate class NumericInteg (which is provided in file numinteg.cpp). The constructor for AnalogPoleZeroFilter will instantiate an array of NumericInteg objects, with the number of integrators being equal to the number of poles in the filter's transfer function.

Several common filter types—Butterworth, Chebyshev, and Bessel—are "all pole" filters having transfer functions of the form

$$H(s) = \frac{H_0}{X(s)}$$

An implementation for a third-order all-pole filter is shown in Fig. 4.14. Although AnalogPoleZeroFilter can be used with A_Coef[k] set to zero for all k other than zero, improved execution speed can be achieved by implementing a class specifically designed for simulating all-pole filters. Such a class, AnalogAllPoleFilt, is provided in file polefilt.cpp.

References

1. Chen, C-T. *Linear System Theory and Design*, Holt, Rinehart, and Winston, New York, 1984.
2. Schwartz, R. J. and B. Friedland *Linear Systems*, McGraw-Hill, New York, 1965.
3. Spiegel, M. R. *Laplace Transforms*, Schaum's Outline Series, McGraw-Hill, New York, 1965.
4. Van Valkenburg, M. E. *Network Analysis*, Prentice-Hall, Englewood Cliffs, NJ, 1974.

Chapter 5

Classical Analog Filters

Digital filter designs are often based on common analog filter designs. Therefore, a certain amount of background material concerning analog filters is a necessary foundation for the study of digital filters. This chapter reviews the basics of analog filter specification and then examines in detail the characteristics of the four most widely used types of analog filters—*Butterworth*, *Chebyshev*, *elliptical*, and *Bessel*.

5.1 Filter Fundamentals

Ideal filters would have rectangular magnitude responses, as shown in Fig. 5.1. The desired frequencies would be passed with no attenuation, while the undesired frequencies would be completely blocked. If such filters could be implemented, they would enjoy widespread use. Unfortunately, ideal filters are noncausal and therefore not realizable. However, there are practical filter designs that approximate the ideal filter characteristics, and which are realizable. Each of the major analog filter types—Butterworth, Chebyshev, Elliptical, and Bessel—optimizes a different aspect of the approximation.

Magnitude response features of lowpass filters

The magnitude response of a practical *lowpass filter* (LPF) usually will have one of the four general shapes shown in Figs. 5.2 through 5.5. In all four cases the filter characteristics divide the spectrum into three general regions, as shown. The *passband* extends from zero up to the cutoff frequency ω_c. The *transition band* extends from ω_c up to the beginning of the stopband at ω_1 and the *stopband* extends upward from ω_1 to infinity. The cutoff frequency ω_c is the frequency at which the amplitude response falls to a specified fraction (usually -3 dB, sometimes -1 dB) of the peak passband values. Defining the frequency ω_1, which marks the beginning of the stopband, is not so straightforward.

72   Chapter Five

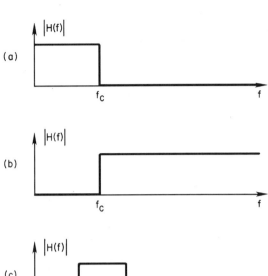

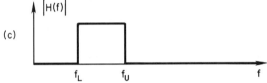

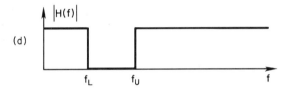

Figure 5.1 Ideal filter responses: (*a*) lowpass, (*b*) highpass, (*c*) bandpass, and (*d*) bandstop.

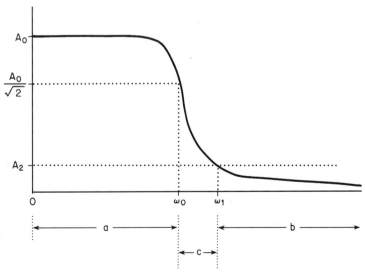

Figure 5.2 Monotonic magnitude response of a practical lowpass filter: (*a*) passband, (*b*) stopband, and (*c*) transition-band.

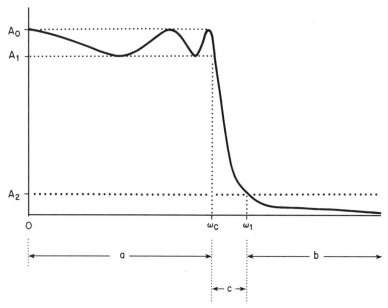

Figure 5.3 Magnitude response of a practical lowpass filter with ripples in the passband: (*a*) passband, (*b*) stopband, and (*c*) transition-band.

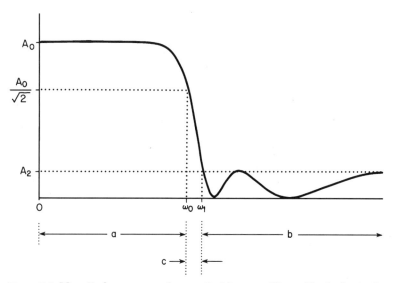

Figure 5.4 Magnitude response of a practical lowpass filter with ripples in the stopband: (*a*) passband, (*b*) stopband, and (*c*) transition-band.

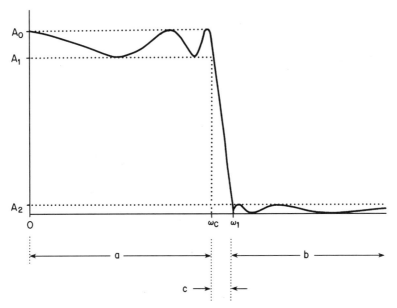

Figure 5.5 Magnitude response of a practical lowpass filter with ripples in the passband and stopband: (a) passband, (b) stopband, and (c) transition-band.

In Fig. 5.2 or 5.3 there is no particular feature that indicates exactly where ω_1 should be located. The usual approach involves specifying a *minimum stopband loss* α_2 (or conversely a *maximum stopband amplitude* A_2) and then defining ω_1 as the lowest frequency at which the loss exceeds and continues to exceed α_2. The width W_T of the transition band is equal to $\omega_c - \omega_1$. The quantity W_T/ω_c is sometimes called the *normalized transition width*. In the case of response shapes like those in Figs. 5.4 and 5.5, the minimum stopband loss is clearly defined by the peaks of the stopband ripples.

Scaling of lowpass filter responses

In plots of practical filter responses, the frequency axes are almost universally plotted on logarithmic scales. Magnitude response curves for LPFs are scaled so that the cutoff frequency occurs at a convenient frequency such as 1 radian per second (rad/s), 1 Hz, or 1 kHz. A single set of such normalized curves then can be denormalized to fit any particular cutoff requirement.

Transfer functions. For common filter types such as Butterworth, Chebyshev, and Bessel, transfer functions are usually presented in a scaled form such that $\omega_c = 1$. Given such a normalized response, we can easily scale the transfer function to yield the corresponding response for $\omega_c = \alpha$. If the normalized response for $\omega_c = 1$ is given by

$$H_N(s) = \frac{K \prod_{i=1}^{m}(s - z_i)}{\prod_{i=1}^{n}(s - p_i)} \tag{5.1}$$

then the corresponding response for $\omega_c = \alpha$ is given by

$$H_\alpha(s) = \frac{K \prod_{i=1}^{m}(s - \alpha z_i)}{\alpha^{(m-n)} \prod_{i=1}^{n}(s - \alpha p_i)} \tag{5.2}$$

Magnitude scaling. The vertical scales of a filter's magnitude response can be presented in several different forms. In theoretical presentations, the magnitude response is often plotted on a linear scale. In practical design situations it is convenient to work with plots of attenuation in decibels, using a high-resolution linear scale in the passband and a lower-resolution linear scale in the stopband. This allows details of the passband response to be shown, as well as large attenuation values deep into the stopband. In nearly all cases, the data are normalized to present 0-dB attenuation at the peak of the passband.

Phase response. The phase response is plotted as a phase angle in degrees or radians versus frequency. By adding or subtracting the appropriate number of full-cycle offsets (that is, 2π radians or $360°$), the phase response can be presented either as a single curve extending over several full cycles (Fig. 5.6) or as an equivalent set of curves, each extending over a single cycle (Fig. 5.7). Phase calculations usually will yield results confined to a single 2π cycle.

Step response. Normalized step response plots are obtained by computing the step response from the normalized transfer function. The inherent scaling of the time axis will thus depend on the transient characteristics of the normalized

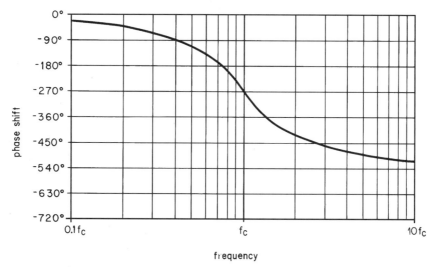

Figure 5.6 Phase response extending over multiple cycles.

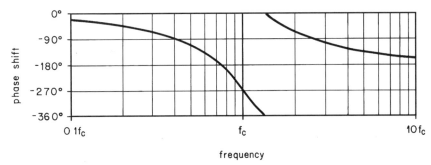

Figure 5.7 Phase response confined to a single-cycle range.

filter. The amplitude axis scaling is not dependent on normalization. The usual lowpass presentation requires that the response be denormalized by dividing the frequency axis by some form of the cutoff frequency.

Impulse response. A normalized impulse response plot is obtained by computing the impulse response from the normalized transfer function. Since an impulse response will always have an area of unity, both the time axis and the amplitude axis will exhibit inherent scaling that depends on the transient characteristics of the normalized filter. The usual lowpass presentation will require that the response be denormalized by multiplying the amplitude by some form of the cutoff frequency and dividing the time axis by the same factor.

Highpass filters

Highpass filters are usually designed via transformation of lowpass designs. Normalized lowpass transfer functions can be converted into corresponding highpass transfer functions simply by replacing each occurrence of s with $1/s$. This causes the magnitude response to be "flipped" around a line at f_c, as shown in Fig. 5.8. (Note that this flip works only when the frequency is plotted on a logarithmic scale.) Rather than trying to draw a flipped response curve, it is much simpler to take the reciprocals of all the important frequencies for the highpass filter in question, then read the appropriate response directly from the lowpass curves.

Bandpass filters

Bandpass filters are classified as wideband or narrowband based on the relative width of their passbands. Different methods are used for obtaining the transfer function for each type.

Wideband bandpass filter. Wideband bandpass filters can be realized by cascading a LPF and a highpass filter. This approach is acceptable as long as the

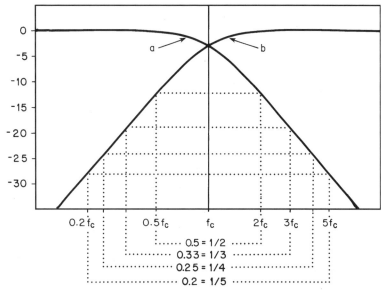

Figure 5.8 Relationship between lowpass and highpass magnitude responses: (a) lowpass response and (b) highpass response.

bandpass filters used exhibit relatively sharp transitions from the passband to cutoff. Relatively narrow bandwidths and/or gradual rolloffs that begin within the passband can cause a significant center-band loss as shown in Fig. 5.9. In situations where such losses are unacceptable, other bandpass filter realizations must be used. A general rule of thumb is to use narrowband techniques for passbands that are an octave or smaller.

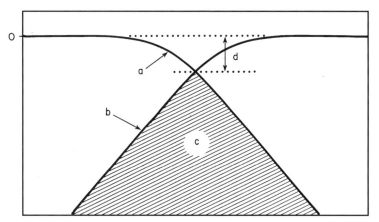

Figure 5.9 Center-band loss in a bandpass filter realized by cascading lowpass and highpass filters: (a) lowpass response, (b) highpass response, (c) passband of BPF, and (d) center-band loss.

78 Chapter Five

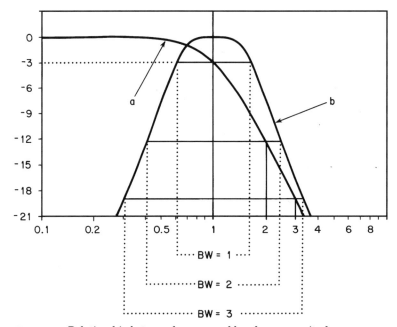

Figure 5.10 Relationship between lowpass and bandpass magnitude responses: (*a*) normalized lowpass response and (*b*) normalized bandpass response.

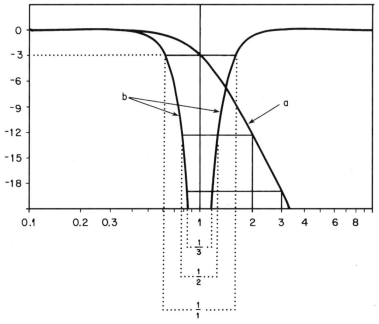

Figure 5.11 Relationship between lowpass and bandstop magnitude responses: (*a*) normalized lowpass response and (*b*) normalized bandstop response.

Narrowband bandpass filters. A normalized LPF can be converted into a normalized narrowband bandpass filter by substituting $[s - (1/s)]$ for s in the lowpass transfer function. The center frequency of the resulting bandpass filter will be at the cutoff frequency of the original LPF, and the passband will be symmetrical about the center frequency when plotted on a logarithmic frequency scale. At any particular attenuation level, the bandwidth of the bandpass filter will equal the frequency at which the lowpass filter exhibits the same attenuation (see Fig. 5.10). This particular bandpass transformation preserves the magnitude response shape of the lowpass prototype but distorts the transient responses.

Bandstop filters. A normalized LPF can be converted into a normalized bandstop filter by substituting $s/(s^2 - 1)$ for s in the lowpass transfer function. The center frequency of the resulting bandstop filter will be at the cutoff frequency of the original LPF, and the stopband will be symmetrical about the center frequency when plotted on a logarithmic frequency scale. At any particular attenuation level, the width of the stopband will be equal to the reciprocal of the frequency at which the LPF exhibits the same attenuation (see Fig. 5.11).

5.2 Butterworth Filters

Butterworth LPFs are designed to have an amplitude response characteristic that is as flat as possible at low frequencies and that decreases monotonically with increasing frequency.

Transfer function

The general expression for the transfer function of an nth-order Butterworth LPF is given by

$$H(s) = \frac{1}{\prod_{i=1}^{n}(s - s_i)} = \frac{1}{(s - s_1)(s - s_2)\cdots(s - s_n)} \quad (5.3)$$

where

$$s_i = e^{j\pi[(2i+n-1)/2n]} = \cos\left(\pi\frac{2i+n-1}{2n}\right) + j\sin\left(\pi\frac{2i+n-1}{2n}\right) \quad (5.4)$$

Example 5.1 Determine the transfer function for a lowpass third-order Butterworth filter. The third-order transfer function will have the form

$$H(s) = \frac{1}{(s - s_1)(s - s_2)(s - s_3)}$$

The values for s_1, s_2, and s_3 are obtained from Eq. 5.4:

$$s_1 = \cos\left(\frac{2\pi}{3}\right) + j\sin\left(\frac{2\pi}{3}\right) = -0.5 + 0.866j$$

$$s_2 = \cos(\pi) + j\sin(\pi) = -1$$

$$s_3 = \cos\left(\frac{4\pi}{3}\right) + j\sin\left(\frac{4\pi}{3}\right) = -0.5 - 0.866j$$

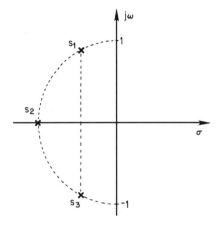

Figure 5.12 Pole locations for a third-order Butterworth LPF.

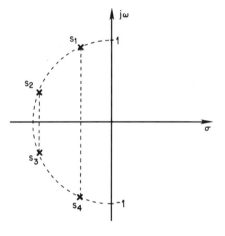

Figure 5.13 Pole locations for a fourth-order Butterworth LPF.

Thus,

$$H(s) = \frac{1}{(s + 0.5 - 0.866j)(s + 1)(s + 0.5 + 0.866j)}$$

$$= \frac{1}{s^3 + 2s^2 + 2s + 1}$$

The form of Eq. (5.3) indicates that an nth-order Butterworth filter will always have n poles and no finite zeros. Also true, but not quite so obvious, is the fact that these plots lie at equally spaced points on the left half of a circle in the s plane. As shown in Fig. 5.12 for the third-order case, any odd-order Butterworth LPF will have one ral pole at $s = -1$, and all remaining poles will occur in complex conjugate pairs. As shown in Fig. 5.13 for the fourth-order case, the poles of any even-order Butterworth LPF will all occur in complex conjugate pairs. Pole values for orders 2 through 8 are listed in Table 5.1. A class ButterworthTransFunc for generating Butterworth transfer functions is provided in file buttfunc.cpp.

TABLE 5.1 Poles of Lowpass Butterworth Filters

n	Pole values
2	$-0.707107 \pm 0.707107j$
3	-1.0
	$-0.5 \pm 0.866025j$
4	$-0.382683 \pm 0.923880j$
	$-0.923880 \pm 0.382683j$
5	-1.0
	$-0.809017 \pm 0.587785j$
	$-0.309017 \pm 0.951057j$
6	$-0.258819 \pm 0.965926j$
	$-0.707107 \pm 0.707107j$
	$-0.965926 \pm 0.258819j$
7	-1.0
	$-0.900969 \pm 0.433884j$
	$-0.623490 \pm 0.781831j$
	$-0.222521 \pm 0.974928j$
8	$-0.195090 \pm 0.980785j$
	$-0.555570 \pm 0.831470j$
	$-0.831470 \pm 0.555570j$
	$-0.980785 \pm 0.195090j$

Frequency response

As shown in the following code fragment, frequency response data for Butterworth filters can be generated by creating an instance of `ButterworthTransFunc` and then using the function `FilterFrequencyResponse` belonging to the base class `FilterTransFunc` in file `filtfunc.cpp`. from which `ButterworthTransFunc` inherits:

```
filter_function = new ButterworthTransFunc(order);
filter_function->LowpassDenorm(passband_edge);
filter_function->FilterFrequencyResponse();
```

Details of the `FilterFrequencyResponse()` function are discussed in Chap. 3. Figures 5.14 through 5.16 show, respectively, the passband magnitude response, the stopband magnitude response, and the phase response for Butterworth filters of various orders. These plots are normalized for a cutoff frequency of 1 Hz. To denormalize them, simply multiply the frequency axis by the desired cutoff frequency f_c.

Example 5.2 Use Figs. 5.15 and 5.16 to determine the magnitude and phase response at 800 Hz of a sixth-order Butterworth LPF having a cutoff frequency of 400 Hz. By setting $f_c = 400$, the $n = 6$ response of Fig. 5.15 is denormalized to obtain the response shown in Fig. 5.17. This plot shows that the magnitude at 800 Hz is approximately -36 dB. The corresponding response calculated by `FilterFrequencyResponse` is -36.12466 dB. Likewise, the $n = 6$ response of Fig. 5.16 is denormalized to obtain

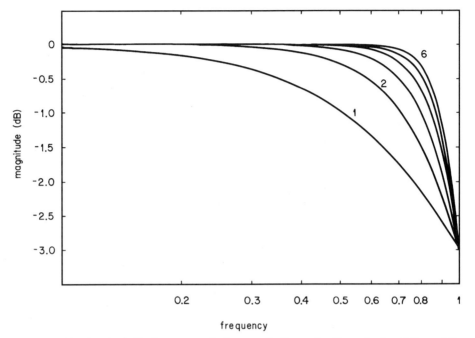

Figure 5.14 Passband amplitude response for lowpass Butterworth filters of orders 1 through 6.

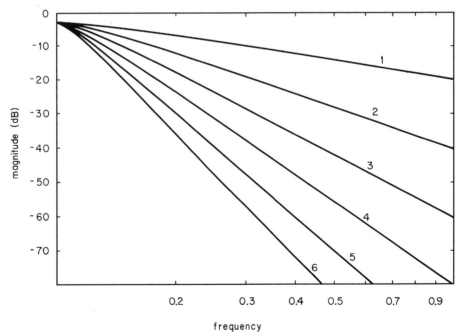

Figure 5.15 Stopband amplitude response for lowpass Butterworth filters of orders 1 through 6.

Classical Analog Filters

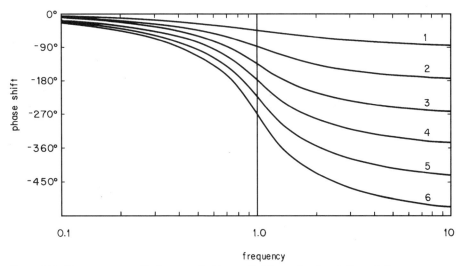

Figure 5.16 Phase response for lowpass Butterworth filters of orders 1 through 6.

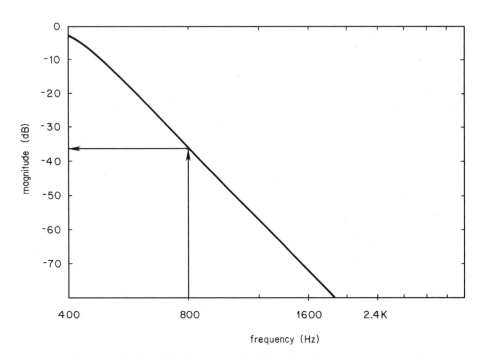

Figure 5.17 Denormalized amplitude response for Example 5.2.

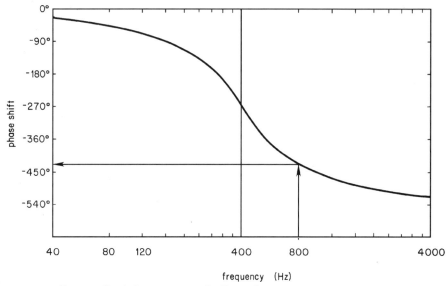

Figure 5.18 Denormalized phase response for Example 5.2.

the response shown in Fig. 5.18. This plot shows that the phase response at 800 Hz is approximately −425°. The corresponding value calculated by FilterFrequencyResponse is −65.475°, which "unwraps" to −425.475°.

Determination of minimum order

Usually in the real world, the order of the desired filter is not given as in Example 5.2, but instead the order must be chosen based on the required performance of the filter. For Butterworth LPFs, the minimum order n that will ensure a magnitude of A_1 or lower at all frequencies ω_1 and above can be obtained by using

$$n = \frac{\log(10^{-A_1/10} - 1)}{2\log(\omega_1/\omega_c)} \tag{5.5}$$

where ω_c = 3-dB frequency
ω_1 = frequency at which the magnitude response first falls below A_1

(Note: The value of A_1 is assumed to be in decibels. The value will be negative, thus canceling the minus sign in the numerator exponent.)

Impulse response

To obtain the impulse response for an nth-order Butterworth filter, we need to take the inverse Laplace transform of the transfer function. Application of the

Heaviside expansion to Eq. (5.3) produces

$$h(t) = \mathcal{L}^{-1}[H(s)] = \sum_{r=1}^{n} K_r e^{s_r t} \tag{5.6}$$

where

$$K_r = \left. \frac{(s - s_r)}{(s - s_1)(s - s_2) \cdots (s - s_n)} \right|_{s=s_r}$$

The values of both K_r and s_r are, in general, complex, but for the Butterworth LPF case all the complex pole values occur in complex conjugate pairs. When the order n is even, this will allow Eq. (5.6) to be put in the form

$$h(t) = \sum_{r=1}^{n/2} \left[2 \operatorname{Re}(K_r) e^{\sigma_r t} \cos(\omega_r t) - 2 \operatorname{Im}(K_r) e^{\sigma_r t} \sin(\omega_r t) \right] \tag{5.7}$$

where $s_r = \sigma_r + j\omega_r$ and the roots s_r are numbered such that for $r = 1, 2, \ldots, n/2$ the s_r lie in the same quadrant of the s plane. [This last restriction prevents two members of the same complex conjugate pair from being used independently in evaluation of Eq. (5.7).] When the order n is odd, Eq. (5.6) can be put into the form

$$h(t) = Ke^{-t} + \sum_{r=1}^{(n-1)/2} \left[2 \operatorname{Re}(K_r) e^{\sigma_r t} \cos(\omega_r t) - 2 \operatorname{Im}(K_r) e^{\sigma_r t} \sin(\omega_r t) \right] \tag{5.8}$$

where no two of the roots s_r, $r = 1, 2, \ldots, (n-1)/2$ form a complex conjugate pair. Equations (5.7) and (5.8) are implemented in the ImpulseResponse class, which is provided in file impresp.cpp. This class was used to generate the impulse responses shown in Figs. 5.19 and 5.20. Calculation of the various K_r values in the constructor for ImpulseResponse has been generalized to support pole-zero filters (such as elliptical filters) in addition to all-pole filters such as Butterworth filters. Therefore, the ImpulseResponse class can be used for any of the analog filter types presented in this book. These responses are normalized for lowpass filters having cutoff frequencies equal to 1 rad/s. To denormalize the response, divide the time axis by the desired cutoff frequency $\omega_c = 2\pi f_c$ and multiply the time axis by the same factor.

Example 5.3 Determine the instantaneous amplitude of the output 1.6 ms after a unit impulse is applied to the input of a fifth-order Butterworth LPF having $f_c = 250$ Hz. The $n = 5$ response of Fig. 5.20 is denormalized as shown in Fig. 5.21. This plot shows that the response amplitude at $t = 1.6$ ms is approximately 378.

Step response

The step response can be obtained by integrating the impulse response. The class StepResponse is provided in file stepresp.cpp. This class makes use of the

86 Chapter Five

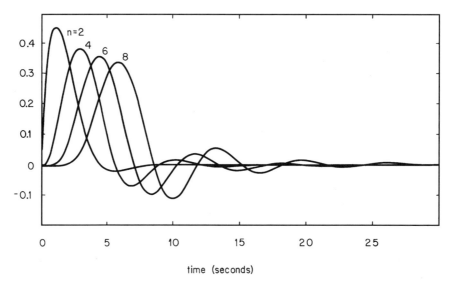

Figure 5.19 Impulse response of even-order Butterworth filters.

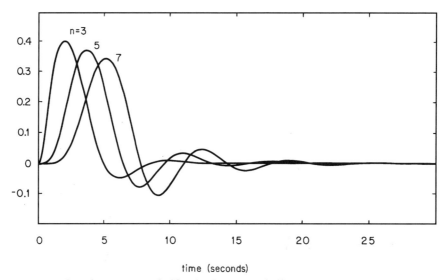

Figure 5.20 Impulse response of odd-order Butterworth filters.

ImpulseResponse class to generate samples of the impulse response, which are then integrated to produce the step response. The StepResponse class can be used with any of the analog filter types presented in this book. Step responses for Butterworth LPFs are shown in Figs. 5.22 and 5.23. These responses are normalized for lowpass filters having a cutoff frequency equal to 1 rad/s. To denormalize the response, divide the time axis by the desired cutoff frequency $\omega_c = 2\pi f_c$.

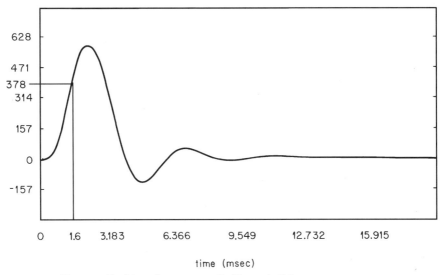

Figure 5.21 Denormalized impulse response for Example 5.3.

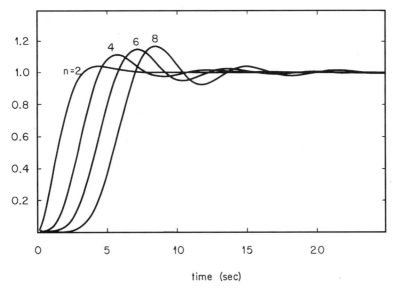

Figure 5.22 Step response of even-order Butterworth filters.

Example 5.4 Determine how long it will take for the step response of a third-order Butterworth LPF with $f_c = 4$ kHz to first reach 100 percent of its final value. By setting $\omega_c = 2\pi f_c = 8000\pi = 25{,}132.7$, the $n = 3$ response of Fig. 5.23 is denormalized to obtain the response shown in Fig. 5.24. This plot indicates that the step response first reaches a value of 1 in approximately 150 μs.

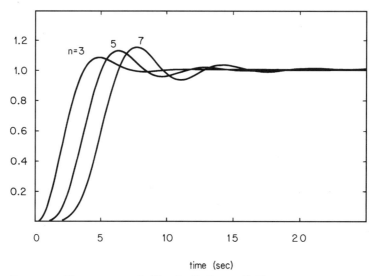

Figure 5.23 Step response of odd-order Butterworth filters.

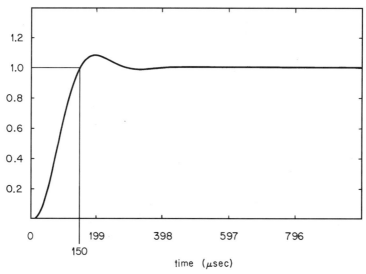

Figure 5.24 Denormalized step response for Example 5.4.

5.3 Chebyshev Filters

Chebyshev filters are designed to have an amplitude response characteristic that has a relatively sharp transition from the passband to the stopband. This sharpness is accomplished at the expense of ripples that are introduced into the response. Specifically, Chebyshev filters are obtained as an equiripple approximation to the passband of an ideal lowpass filter. This results in a filter

TABLE 5.2 Chebyshev Polynomials

n	$T_n(\omega)$
0	1
1	ω
2	$2\omega^2 - 1$
3	$4\omega^3 - 3\omega$
4	$8\omega^4 - 8\omega^2 + 1$
5	$16\omega^5 - 20\omega^3 + 5\omega$
6	$32\omega^6 - 48\omega^4 + 18\omega^2 - 1$
7	$64\omega^7 - 112\omega^5 + 56\omega^3 - 7\omega$
8	$128\omega^8 - 256\omega^6 + 160\omega^4 - 32\omega^2 + 1$
9	$256\omega^9 - 576\omega^7 + 432\omega^5 - 120\omega^3 + 9\omega$
10	$512\omega^{10} - 1280\omega^8 + 1120\omega^6 - 400\omega^4 + 50\omega^2 + 1$

characteristic for which

$$|H(j\omega)|^2 = \frac{1}{1 + \epsilon^2 T_n^2(\omega)} \qquad (5.9)$$

where $\epsilon^2 = 10^{r/10} - 1$
  $T_n(\omega)$ = Chebyshev polynomial of order n
  r = passband ripple, dB

Chebyshev polynomials are listed in Table 5.2.

Transfer function

The general shape of the Chebyshev magnitude response is as shown in Fig. 5.25. This response can be normalized as in Fig. 5.26 so that the ripple bandwidth ω_r is equal to 1, or the response can be normalized as in Fig. 5.27 so that the 3-dB frequency ω_0 is equal to 1. Normalization based on the ripple bandwidth involves simpler calculations, but normalization based on the 3-dB point makes it easier to compare Chebyshev responses to those of other filter types.

The general expression for the transfer function of an nth-order Chebyshev LPF is given by

$$H(s) = \frac{H_0}{\prod_{i=1}^{n}(s - s_i)} = \frac{H_0}{(s - s_1)(s - s_2)\cdots(s - s_n)} \qquad (5.10)$$

90   Chapter Five

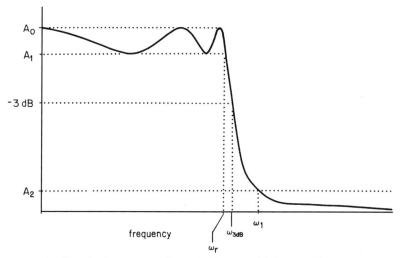

Figure 5.25 Magnitude response of a typical lowpass Chebyshev filter.

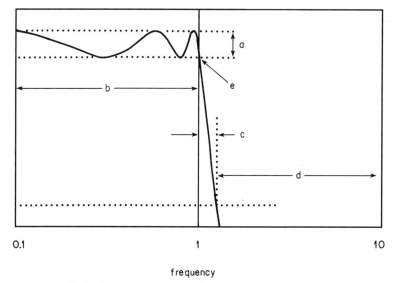

Figure 5.26 Chebyshev response normalized to have passband end at $\omega = 1$ rad/s. Features are: (*a*) ripple limits, (*b*) passband, (*c*) transition-band, (*d*) stopband, and (*e*) intersection of response and lower ripple limit at $\omega = 1$.

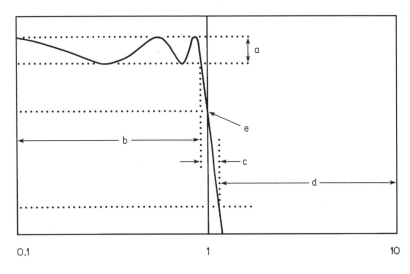

Figure 5.27 Chebyshev response normalized to have 3-dB point at $\omega = 1$ rad/s. Features are: (a) ripple limits, (b) passband, (c) transition-band, (d) stopband, and (e) response that is 3 dB down at $\omega = 1$.

where

$$H_0 = \begin{cases} \prod_{i=1}^{n}(-s_i) & n \text{ odd} \\ 10^{r/20}\prod_{i=1}^{n}(-s_i) & n \text{ even} \end{cases} \quad (5.11)$$

$$s_i = \sigma_i + j\omega_i \quad (5.12)$$

$$\sigma_i = \left[\frac{(1/\gamma) - \gamma}{2}\right]\sin\frac{(2i-1)\pi}{2n} \quad (5.13)$$

$$\omega_i = \left[\frac{(1/\gamma) + \gamma}{2}\right]\cos\frac{(2i-1)\pi}{2n} \quad (5.14)$$

$$\gamma = \left(\frac{1 + \sqrt{1 + \epsilon^2}}{\epsilon}\right)^{1/n} \quad (5.15)$$

$$\epsilon = \sqrt{10^{r/10} - 1} \quad (5.16)$$

The pole formulas are somewhat more complicated than for the Butterworth filter examined in Sec. 5.2, and several parameters—ϵ, γ, and r—must be determined before the pole values can be calculated. Also, all the poles are involved in the calculation of the numerator H_0.

Algorithm 5.1 Determing poles of a Chebyshev filter.

This algorithm computes the poles of an nth-order Chebyshev lowpass filter normalized for a ripple bandwidth of 1 Hz.

1. Determine the maximum amount (in dB) of ripple that can be tolerated in the passband magnitude response. Set r equal to or less than this value.
2. Use Eq. (5.16) to compute ϵ.
3. Select an order n for the filter that will ensure adequate performance.
4. Use Eq. (5.15) to compute γ.
5. For $i = 1, 2, \ldots, n$, use Eqs. (5.13) and (5.14) to compute the real part σ_i and imaginary part ω_i of each pole.
6. Use Eq. (5.11) to compute H_0.
7. Substitute the values of H_0 and s_1 through s_n into Eq. (5.10).

Example 5.5 Consider the case of a third-order Chebyshev filter with 0.5-dB passband ripple. Algorithm 5.1 can be used to determine the transfer-function numerator and poles (normalized for ripple bandwidth equal to 1) with the following results:

$$\epsilon = 0.349311 \qquad \gamma = 1.806477 \qquad s_1 = -0.313228 + 1.021928j$$

$$s_2 = -0.626457 \qquad s_3 = -0.313228 - 1.021928j \qquad H_0\, 0.715695$$

The form of Eq. (5.10) shows that an nth-order Chebyshev filter will always have n poles and no finite zeros. The poles will all lie on the left half of an ellipse in the s plane. The major axis of the ellipse lies on the $j\omega$ axis, and the minor axis lies on the σ axis. The dimensions of the ellipse and the locations of the poles will depend upon the amount of ripple permitted in the passband. Values of passband ripple typically range from 0.1 to 1 dB. The smaller the passband ripple, the wider the transition band will be. In fact, for zero ripple, the Chebyshev filter and Butterworth filter have exactly the same transfer function and response characteristics. Pole locations for third-order Chebyshev filters having different ripple limits are compared in Fig. 5.28. Pole values for ripple limits of 0.1, 0.5, and 1 dB are listed in Tables 5.3, 5.4, and 5.5 for orders 2 through 8.

All of the transfer functions and pole values presented so far are for filters normalized to have a ripple bandwidth of 1. Algorithm 5.2 can be used to renormalize the transfer function to have a 3-dB frequency of 1.

Algorithm 5.2 Renormalizing Chebyshev LPF transfer functions.

This algorithm assumes that ϵ, H_0, and the pole values s_i have been obtained for the transfer function having a ripple bandwidth of 1.

1. Compute A using

$$A = \frac{\cosh^{-1}(1/\epsilon)}{n} = \frac{1}{n}\ln\left(\frac{1+\sqrt{1-\epsilon^2}}{\epsilon}\right)$$

TABLE 5.3 Pole Values for Lowpass Chebyshev Filters with 0.1 dB Passband Ripple

n	Pole values
2	$-1.186178 \pm 1.380948j$
3	-0.969406
	$-0.484703 \pm 1.206155j$
4	$-0.637730 \pm 0.465000j$
	$-0.264156 \pm 1.122610j$
5	-0.538914
	$-0.435991 \pm 0.667707j$
	$-0.166534 \pm 1.080372j$
6	$-0.428041 \pm 0.283093j$
	$-0.313348 \pm 0.773426j$
	$-0.114693 \pm 1.056519j$
7	-0.376778
	$-0.339465 \pm 0.463659j$
	$-0.234917 \pm 0.835485j$
	$-0.083841 \pm 1.041833j$
8	$-0.321650 \pm 0.205314j$
	$-0.272682 \pm 0.584684j$
	$-0.182200 \pm 0.875041j$
	$-0.063980 \pm 1.032181j$

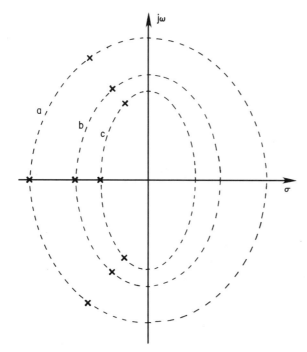

Figure 5.28 Comparison of pole locations for third-order lowpass Chebyshev filters with different amounts of passband ripple: (a) 0.01 dB, (b) 0.1 dB, and (c) 0.5 dB.

TABLE 5.4 Pole Values for Lowpass Chebyshev Filters with 0.5 dB Passband Ripple

n	Pole values
2	$-0.712812 \pm 1.00402j$
3	-0.626457
	$-0.313228 \pm 1.021928j$
4	$-0.423340 \pm 0.420946j$
	-0.175353 ± 1.016253
5	-0.362320
	$-0.293123 \pm 0.625177j$
	$-0.111963 \pm 1.011557j$
6	$-0.289794 \pm 0.270216j$
	$-0.212144 \pm 0.738245j$
	$-0.077650 \pm 1.008461j$
7	-0.256170
	$-0.230801 \pm 0.447894j$
	$-0.159719 \pm 0.807077j$
	$-0.057003 \pm 1.006409j$
8	$-0.219293 \pm 0.199907j$
	$-0.185908 \pm 0.569288j$
	$-0.124219 \pm 0.852000j$
	$-0.043620 \pm 1.005002j$

TABLE 5.5 Pole Values for Lowpass Chebyshev Filters with 1.0 dB Passband Ripple

n	Pole values
2	$-0.548867 \pm 0.895129j$
3	-0.494171
	$-0.247085 \pm 0.965999j$
4	$-0.336870 \pm 0.407329j$
	$-0.139536 \pm 0.983379j$
5	-0.289493
	$-0.234205 \pm 0.611920j$
	$-0.089458 \pm 0.990107j$
6	$-0.232063 \pm 0.266184j$
	$-0.169882 \pm 0.727227j$
	$-0.062181 \pm 0.993411j$
7	-0.205414
	$-0.185072 \pm 0.442943j$
	$-0.128074 \pm 0.798156j$
	$-0.045709 \pm 0.995284j$
8	$-0.175998 \pm 0.198206j$
	$-0.149204 \pm 0.564444j$
	$-0.099695 \pm 0.844751j$
	$-0.035008 \pm 0.996451j$

TABLE 5.6 Factors for Renormalizing Chebyshev Transfer Functions

Ripple	Order						
	2	3	4	5	6	7	8
0.1	1.94322	1.38899	1.21310	1.13472	1.09293	1.06800	1.05193
0.2	1.67427	1.28346	1.15635	1.09915	1.06852	1.05019	1.03835
0.3	1.53936	1.22906	1.12680	1.08055	1.05571	1.04083	1.03121
0.4	1.45249	1.19348	1.10736	1.06828	1.04725	1.03464	1.02649
0.5	1.38974	1.16749	1.09310	1.05926	1.04103	1.03009	1.02301
0.6	1.34127	1.14724	1.08196	1.05220	1.03616	1.02652	1.02028
0.7	1.30214	1.13078	1.07288	1.04644	1.03218	1.02361	1.01806
0.8	1.26955	1.11699	1.06526	1.04160	1.02883	1.02116	1.01618
0.9	1.24176	1.10517	1.05872	1.03745	1.02596	1.01905	1.01457
1.0	1.21763	1.09487	1.05300	1.03381	1.02344	1.01721	1.01316
1.1	1.19637	1.08576	1.04794	1.03060	1.02121	1.01557	1.01191
1.2	1.17741	1.07761	1.04341	1.02771	1.01922	1.01411	1.01079
1.3	1.16035	1.07025	1.03931	1.02510	1.01741	1.01278	1.00978
1.4	1.14486	1.06355	1.03558	1.02272	1.01576	1.01157	1.00886
1.5	1.13069	1.05740	1.03216	1.02054	1.01425	1.01046	1.00801

2. Using the value of A obtained in step 1, compute R as

$$R = \cosh A = \frac{e^A + e^{-A}}{2}$$

(Table 5.6 lists R factors for various orders and ripple limits. If the required combination can be found in this table, steps 1 and 2 can be skipped.)

3. Use R to compute $H_{3dB}(s)$ as

$$H_{3dB}(s) = \frac{H_0/R}{\prod_{i=1}^{n} [s - (s_i/R)]}$$

A class ChebyshevTransFunc, for generating Chebyshev transfer functions, is provided in file chebfunc.cpp. This class inherits from the base class FilterTransFunc which, as discussed in Chap. 3, provides several functions which are common to the transfer functions of all filter types.

Frequency response

As shown in the following code fragment, frequency response data for Chebyshev filters can be generated by creating an instance of ChebyshevTransFunc and then using the function FilterFrequencyResponse belonging to the base class `FilterTransFunc` from which `ChebyshevTransFunc` inherits:

```
filter_function = new ChebyshevTransFunc(order,
                                         passband_ripple,
                                         ripple_bw_norm);
filter_function->LowpassDenorm(passband_edge);
filter_function->FilterFrequencyResponse();
```

The variable `ripple_bw_norm` is set to 1 if the transfer function is to be normalized based on its ripple bandwidth. Otherwise, the filter will be normalized based on a 3-dB bandwidth.

Figures 5.29 through 5.32 show the magnitude and phase responses for Chebyshev filters with passband ripple limits of 0.5 dB. For comparison purposes, Figs. 5.33 and 5.34 show passband responses for ripple limits of 0.1 and 1.0 dB. These plots are normalized for a cutoff frequency of 1 Hz. To denormalize them, simply multiply the frequency axis by the desired cutoff frequency f_c.

Impulse response

Impulse responses for Chebyshev LPFs with 0.5-dB ripple are shown in Fig. 5.35. The class `ImpulseResponse`, used to generate the data for these plots was discussed in Sec. 5.2. These responses are normalized for LPFs having a 3-dB frequency of 1 Hz. To denormalize the response, divide the time axis by

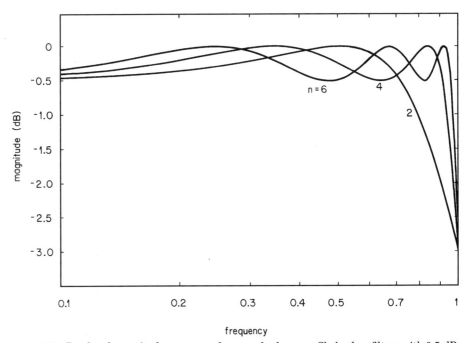

Figure 5.29 Passband magnitude response of even-order lowpass Chebyshev filters with 0.5-dB ripple.

Classical Analog Filters 97

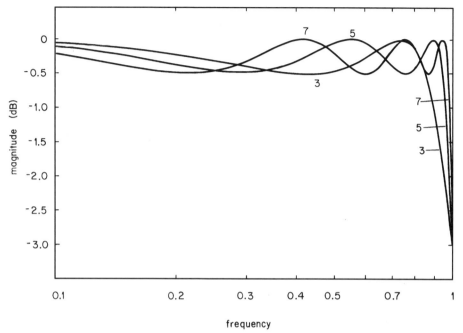

Figure 5.30 Passband magnitude response of odd-order lowpass Chebyshev filters with 0.5-dB ripple.

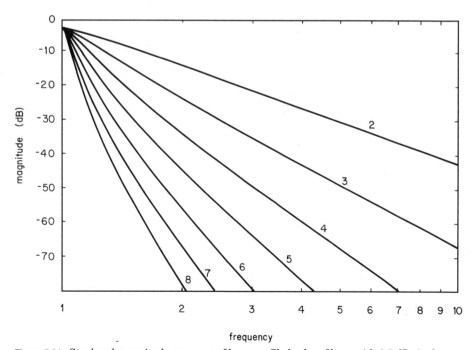

Figure 5.31 Stopband magnitude response of lowpass Chebyshev filters with 0.5-dB ripple.

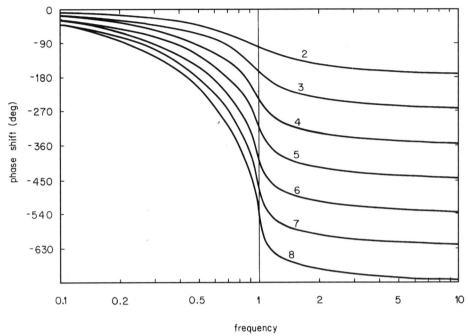

Figure 5.32 Phase response of lowpass Chebyshev filters with 0.5-dB passband ripple.

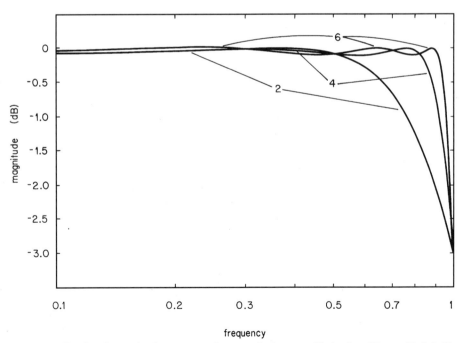

Figure 5.33 Passband magnitude response of even-order lowpass Chebyshev filters with 0.1-dB ripple.

Classical Analog Filters 99

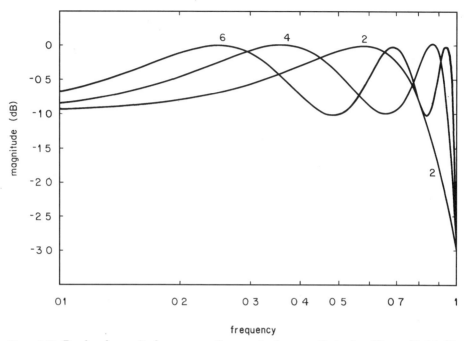

Figure 5.34 Passband magnitude response of even-order lowpass Chebyshev filters with 1.0-dB ripple.

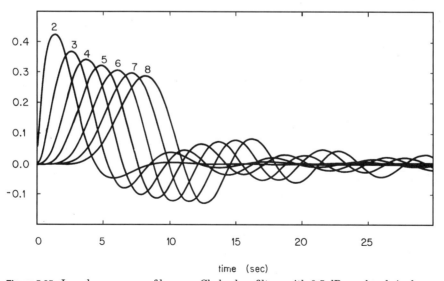

Figure 5.35 Impulse response of lowpass Chebyshev filters with 0.5-dB passband ripple.

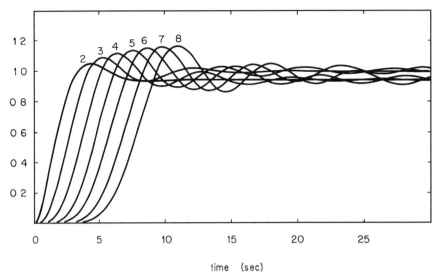

Figure 5.36 Step response of lowpass Chebyshev filters with 0.5-dB passband ripple.

the desired cutoff frequency f_c and multiply the amplitude axis by the same factor.

Step response

The step response can be obtained by integrating the impulse response. Step responses for Chebyshev LPFs with 0.5-dB ripple are shown in Fig. 5.36. These responses are normalized for lowpass filters having a cutoff frequency equal to 1 Hz. To denormalize the response, divide the time axis by the desired cutoff frequency f_c.

5.4 Elliptical Filters

By allowing ripples in the passband, Chebyshev filters obtain better selectivity than Butterworth filters. Elliptical filters improve on the performance of Chebyshev filters by permitting ripples in *both* the passband and stopband. The response of an elliptical filter satisfies

$$|H(j\omega)|^2 = \frac{1}{1 + \epsilon^2 R_n^2(\omega, L)}$$

where $R_n(\omega, L)$ is an nth-order *Chebyshev rational function* with ripple parameter L. Elliptical filters are sometimes called *Cauer filters*.

Parameter specification

As shown in Sec. 5.2, determination of the (amplitude-normalized) transfer function for a Butterworth LPF requires specification of just two parameters—

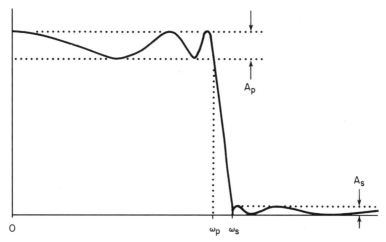

Figure 5.37 Frequency response showing parameters used to specify an elliptical filter.

cutoff frequency ω_c and filter order n. Determination of the transfer function for a Chebyshev filter requires specification of these two parameters plus a third—passband ripple. Determination of the transfer function for an elliptical filter requires specification of the filter order n plus the following four parameters, which are depicted in Fig. 5.37:

A_p = maximum passband loss, dB

A_s = minimum stopband loss, dB

ω_p = passband cutoff frequency

ω_s = stopband cutoff frequency

The design procedures presented in this section assume that the maximum passband amplitude is unity. Therefore, A_p is the size of the passband ripples, and A_s is the size of the stopband ripples. Any four of the five filter parameters can be specified independently, with the fifth being fixed by the nature of the elliptical filter's response. The usual design strategy involves specifying A_p, A_s, ω_p, and ω_s based on requirements of the intended application. Algorithm 5.3 then can be used to compute the minimum value of n for which an elliptical filter can yield the desired performance. Since n must be an integer, not all combinations of A_p, A_s, ω_p, and ω_s can be realized exactly. The design procedure presented in this section can yield a filter that meets the specified A_p, ω_p, and ω_s and that meets or exceeds the specification on A_s.

Algorithm 5.3 Determining the required order for elliptical filters.

1. Based on requirements of the intended application, determine the maximum passband loss A_p and minimum stopband loss A_s in decibels.

2. Based on requirements of the intended application, determine the passband cutoff frequency ω_p and stopband cutoff frequency ω_s.
3. Using ω_p and ω_s, compute the *selectivity factor* k as $k = \omega_p/\omega_s$.
4. Using the selectivity factor computed in step 3, compute the *modular constant* q using

$$q = u + 2u^5 + 15u^9 + 150u^{13} \tag{5.17}$$

where

$$u = \frac{1 - \sqrt[4]{1-k^2}}{2(1 + \sqrt[4]{1-k^2})} \tag{5.18}$$

5. Using the values of A_p and A_s determined in step 1, compute the *discrimination factor* D as

$$D = \frac{10^{A_s/10} - 1}{10^{A_p/10} - 1} \tag{5.19}$$

6. Using the value of D from step 5 and the value of q from step 4, compute the minimum required order n as

$$n = \left\lceil \frac{\log 16D}{\log(1/q)} \right\rceil \tag{5.20}$$

where $\lceil x \rceil$ denotes the smallest integer equal to or greater than x.

The actual minimum stopband loss provided by any given combination of A_p, ω_p, ω_s, and n is given by

$$A_s = 10\log\left(1 + \frac{10^{A_p 10} - 1}{16q^n}\right) \tag{5.21}$$

where q is the modular constant given by Eq. (5.17).

Example 5.6 Using Algorithm 5.3 to determine the minimum order for an elliptical filter for which $A_p = 1$, $A_s \geq 50.0$, $\omega_p = 3000.0$, and $\omega_s = 3200.0$, we obtain:

$$k = \frac{300}{3200} = 0.9375$$

$$u = 0.12897$$

$$q = 0.12904$$

$$D = \frac{10^5 - 1}{10^{0.01} - 1} = 4{,}293{,}093.82$$

$$n = \lceil 8.81267 \rceil = 9$$

Normalized transfer function

The design of elliptical filters is greatly simplified by designing a frequency-normalized filter having the appropriate response characteristics, and then frequency-scaling this design to the desired operating frequency. The simplification results from the particular type of normalizing that is performed. Instead of normalizing so that either a 3-dB bandwidth or the ripple bandwidth equals unity, an elliptical filter is normalized so that

$$\sqrt{\omega_{pN}\omega_{sN}} = 1 \tag{5.22}$$

where ω_{pN} and ω_{sN} are, respectively, the normalized passband cutoff frequency and the normalized stopband cutoff frequency. If we let α represent the frequency-scaling factor such that

$$\omega_{pN} = \frac{\omega_p}{\alpha} \qquad \omega_{sN} = \frac{\omega_s}{\alpha} \tag{5.23}$$

then we can solve for the value of α by substituting Eq. (5.23) into Eq. (5.22) to obtain

$$\sqrt{\frac{\omega_p \omega_s}{\alpha^2}} = 1$$

$$\alpha = \sqrt{\omega_p \omega_s} \tag{5.24}$$

As it turns out, the only way that the frequencies ω_{pN} and ω_{sN} enter into the design procedure (given by Algorithm 5.4) is via the selectivity factor k that is given by

$$k = \frac{\omega_{pN}}{\omega_{sN}} = \frac{\omega_p/\alpha}{\omega_s/\alpha} = \frac{\omega_p}{\omega_s} \tag{5.25}$$

Since Eq. (5.25) indicates that k can be obtained directly from the desired ω_p and ω_s, we can design a *normalized* filter without having to determine the normalized frequencies ω_{pN} and ω_{sN}. However, once a normalized design is obtained, the frequency-scaling factor α as given by Eq. (5.24) will be needed to frequency-scale the design to the desired operating frequency.

Algorithm 5.4 Generating normalized transfer functions for elliptical filters.

1. Use Algorithm 5.3 or any other equivalent method to determine viable combination of values for A_p, A_s, ω_p, ω_s, and n.
2. Using ω_p and ω_s, compute the *selectivity factor* k as $k = \omega_p/\omega_s$.
3. Using the selectivity factor computed in step 3, compute the *modular constant* q using

$$q = u + 2u^5 + 15u^9 + 150u^{13} \tag{5.26}$$

where

$$u = \frac{1 - \sqrt[4]{1-k^2}}{2(1 + \sqrt[4]{1-k^2})} \tag{5.27}$$

4. Using the values of A_p and n from step 1, compute V as

$$V = \frac{1}{2n} \ln\left(\frac{10^{A_p/20} + 1}{10^{A_p/20} - 1}\right) \tag{5.28}$$

5. Using the value of q from step 3 and the value of V from step 4, compute p_0 as

$$p_0 = \left|\frac{q^{1/4} \sum_{m=0}^{\infty} (-1)^m q^{m(m+1)} \sinh[(2m+1)V]}{0.5 + \sum_{m=1}^{\infty} (-1)^m q^{m^2} \cosh 2mV}\right| \tag{5.29}$$

6. Using the value of k from step 2 and the value of p_0 from step 5, compute W as

$$W = \left[\left(1 + \frac{p_0^2}{k}\right)(1 + kp_0^2)\right]^{1/2} \tag{5.30}$$

7. Determine r, the number of quadratic sections in the filter, as $r = n/2$ for even n, and $r = (n-1)/2$ for odd n.
8. For $i = 1, 2, \ldots, r$, compute X_i as

$$X_i = \frac{2q^{1/4} \sum_{m=0}^{\infty} (-1)^m q^{m(m+1)} \sin[(2m+1)\mu\pi/n]}{1 + 2\sum_{m=1}^{\infty} (-1)^m q^{m^2} \cos(2m\mu\pi/n)} \tag{5.31}$$

where

$$\mu = \begin{cases} i & n \text{ odd} \\ i - \frac{1}{2} & n \text{ even} \end{cases}$$

9. For $i = 1, 2, \ldots, r$, compute Y_i as

$$Y_i = \left[\left(1 - \frac{X_i^2}{k}\right)(1 - kX_i^2)\right]^{1/2} \tag{5.32}$$

10. For $i = 1, 2, \ldots, r$, use the W, X_i, and Y_i from steps 6, 8, 9; compute the coefficients a_i, b_i, and c_i as

$$a_i = \frac{1}{X_i^2} \tag{5.33}$$

$$b_i = \frac{2p_0 Y_i}{1 + p_0^2 X_i^2} \tag{5.34}$$

$$c_i = \frac{(p_0 Y_i)^2 + (X_i W)^2}{(1 + p_0^2 X_i^2)^2} \tag{5.35}$$

11. Using a_i and c_i, compute H_0 as

$$H_0 = \begin{cases} p_0 \prod_{i=1}^{r} \frac{c_i}{a_i} & n \text{ odd} \\ 10^{-A_p/20} \prod_{i=1}^{r} \frac{c_i}{a_i} & n \text{ even} \end{cases} \quad (5.36)$$

12. Finally, compute the normalized transfer function $H_N(s)$ as

$$H_N(s) = \frac{H_0}{d} \prod_{i=1}^{r} \frac{s^2 + a_i}{s^2 + b_i s + c_i} \quad (5.37)$$

where

$$d = \begin{cases} s + p_0 & n \text{ odd} \\ 1 & n \text{ even} \end{cases}$$

A C++ class EllipticalTransFunc, which implements steps 1 through 11 of Algorithm 5.4, is provided in file elipfunc.cpp. Step 12 is implemented separately by creating an instance of EllipticalTransFunc and then using the function FilterFrequencyResponse belonging to the base class FilterTransFunc from which EllipticalTransFunc inherits:

```
filter_function = new EllipticalTransFunc(order,
                                          passband_ripple,
                                          stopband_ripple,
                                          passband_edge,
                                          stopband_edge,
                                          upper_summation_limit);
filter_function->FilterFrequencyResponse();
```

The variable upper_summation_limit is the number of terms to be included in the evaluations of the infinite summations in Eqs. (5.29) and (5.31).

Example 5.7 Use Algorithm 5.4 to obtain the coefficients of the normalized transfer function for the ninth-order elliptical filter having $A_p = 0.1$ dB, $\omega_p = 3000$ rad/s, and $\omega_s = 3200$ rad/s. Determine the actual minimum stopband loss.

solution Using the formulas from Algorithm 5.4 plus Eq. (5.21), we obtain

$$q = 0.129041 \quad V = 0.286525 \quad p_0 = 0.470218$$
$$W = 1.221482 \quad r = 4 \quad A_s = 51.665651$$

The coefficients X_i, Y_i, a_i, b_i, and c_i obtained via steps 8 through 10 for $i = 1, 2, 3, 4$ are listed in Table 5.7. Using Eq. (5.36), we obtain $H_0 = 0.015317$. The normalized frequency response of this filter is shown in Figs. 5.38, 5.39, and 5.40. (The phase response shown in Fig. 5.40 may seem a bit peculiar. At first glance, the discontinuities in the phase response might be taken for jumps of 2π caused by $+\pi$ to $-\pi$ "wrap-around" of the arctangent operation. However, this is not the case. The discontinuities in Fig. 5.40 are jumps of π that coincide with the nulls in the magnitude response.

TABLE 5.7 Coefficients for Example 5.7

i	X_i	Y_i	a_i	b_i	c_i
1	0.4894103	0.7598211	4.174973	0.6786235	0.4374598
2	0.7889940	0.3740371	1.606396	0.3091997	0.7415493
3	0.9196814	0.1422994	1.182293	0.1127396	0.8988261
4	0.9636668	0.0349416	1.076828	0.0272625	0.9538953

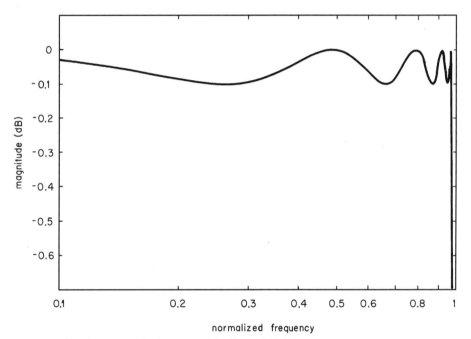

Figure 5.38 Passband magnitude response for Example 5.7.

Denormalized transfer function

As noted in Sec. 3.9, if we have a response normalized for $\omega_{cN} = 1$, we can frequency-scale the transfer function to yield an identical response for $\omega_c = \alpha$ by multiplying each pole and each zero by α and dividing the overall transfer function by $\alpha^{(n_z - n_p)}$ where n_z is the number of zeros and n_p is the number of poles. An elliptical filter has a transfer function of the form given by Eq. (5.37). For odd n, there is a real pole at $s = p_0$ and r complex conjugate pairs of poles that are roots of

$$s^2 + b_i s + c_i = 0 \quad i = 1, 2, \ldots, r$$

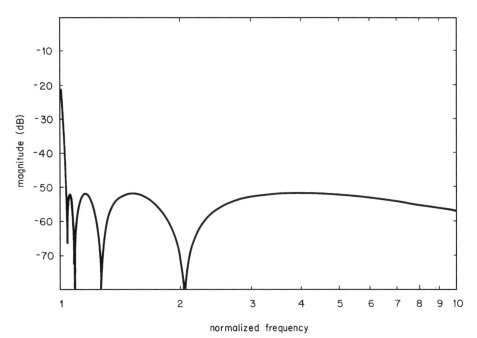

Figure 5.39 Stopband magnitude response for Example 5.7.

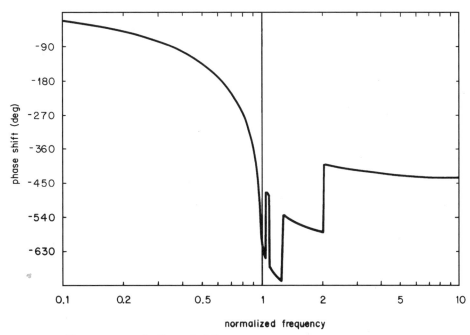

Figure 5.40 Phase response for Example 5.7.

Using the quadratic formula, the ith pair of complex pole values can be expressed as

$$p_i = \frac{-b_i \pm \sqrt{b_i^2 - 4c_i}}{2}$$

The zeros of the normalized transfer function occur at $s = \pm j\sqrt{a_i}, i = 1, 2, \ldots, r$. For even n, the number of poles equals the number of zeros so $\alpha^{(n_z - n_p)} = 1$. For odd n, $n_z - n_p = -1$, so the transfer function must be divided by $1/\alpha$ or multiplied by α. If we multiply the poles and zeros by α and multiply the overall transfer function by 1 or α as appropriate, we obtain the frequency-scaled transfer function $H(s)$ as

$$H(s) = K \prod_{i=1}^{r} \frac{s^2 + \alpha^2 a_i}{s^2 + \alpha b_i s + \alpha^2 c_i} \tag{5.38}$$

where

$$K = \begin{cases} \frac{H_0 \alpha}{s + \alpha p_0} & n \text{ odd} \\ H_0 & n \text{ even} \end{cases}$$

Comparison of Eqs. (5.37) and (5.38) reveals that the frequency rescaling consists of making the following substitutions in Eq. (5.37):

$\alpha^2 a_i$ replaces a_i

$\alpha^2 c_i$ replaces c_i

αb_i replaces b_i

αH_0 replaces H_0 (n odd)

αp_0 replaces p_0 (n odd)

5.5 Bessel Filters

Bessel filters are designed to have maximally flat group delay characteristics. As a consequence, there is no ringing in the impulse and step responses.

Transfer function

The general expression for the transfer function of an nth-order Bessel lowpass filter is given by

$$H(s) = \frac{b_0}{q_n(s)} \tag{5.39}$$

where $q_n(s) = \sum_{k=1}^{n} b_k s^k$

$b_k = \frac{(2n-k)!}{2^{n-k} k! (n-k)!}$

TABLE 5.8 Denominator Polynomials for Transfer Functions of Bessel Filters Normalized to Have Unit Delay at $\omega = 0$

n	$q_n(s)$
2	$s^2 + 3s + 3$
3	$s^3 + 6s^2 + 15s + 15$
4	$s^4 + 10s^3 + 45s^2 + 105s + 105$
5	$s^5 + 15s^4 + 105s^3 + 420s^2 + 945s + 945$
6	$s^6 + 21s^5 + 210s^4 + 1260s^3 + 4725s^2 + 10{,}395s + 10{,}395$
7	$s^7 + 28s^6 + 378s^5 + 3150s^4 + 17{,}325s^3 + 62{,}370s^2 + 135{,}135s + 135{,}135$
8	$s^8 + 36s^7 + 630s^6 + 6930s^5 + 9450s^4 + 270{,}270s^3 + 945{,}945s^2 + 2{,}027{,}025s + 2{,}027{,}025$

The following recursion can be used to determine $q_n(s)$ from $q_{n-1}(s)$ and $q_{n-2}(s)$:

$$q_n = (2n-1)q_{n-1} + s^2 q_{n-2} \qquad (5.40)$$

Table 5.8 lists $q_n(s)$ for $n = 2$ through $n = 8$.

Unlike the transfer functions for Butterworth and Chebyshev filters, Eq. (5.39) does not provide an explicit expression for the poles of the Bessel filter. The numerator of Eq. (5.39) will be a polynomial in s, upon which numerical root-finding methods (such as Algorithm 4.1) must be used to determine the pole locations for $H(s)$. Table 5.9 lists approximate pole locations for $n = 2$ through $n = 8$.

The transfer functions given by Eq. (5.39) are for Bessel filters normalized to have unit delay at $\omega = 0$. The poles p_k and denominator coefficients b_k can be renormalized for a 3-dB frequency of $\omega = 1$ using

$$p'_k = A p_k \qquad b'_k = A^{n-k} b_k \qquad (5.41)$$

where the value of A appropriate for n is selected from Table 5.10.

A C++ class `BesselTransFunc`, for generating Bessel filter transfer functions, is provided in file `bessfunc.cpp`. The constructor for this class uses the recursion given in Eq. (5.40) to generate the coefficients for the transfer function's denominator polynomials. If the third input parameter is set TRUE, the transfer function will be normalized to have unit delay at zero frequency. If the third input parameter is set FALSE, the transfer function will be normalized to have a 3-dB attenuation at the passband edge. `BesselTransFunc` uses the Laguerre method to find the roots of the denominator polynomial and thereby determine the filter's pole locations.

Frequency response

As shown in the following code fragment, frequency response data for Bessel filters can be generated by creating an instance of `BesselTransFunc` and then

TABLE 5.9 Poles of Bessel Filters Normalized to Have Unit Delay at $\omega = 0$

n	Pole values
2	$-1.5 \pm 0.8660j$
3	-2.3222
	$-1.8390 \pm 1.7543j$
4	$-2.1039 \pm 2.6575j$
	$-2.8961 \pm 0.8672j$
5	-3.6467
	$-2.3247 \pm 3.5710j$
	$-3.3520 \pm 1.7427j$
6	$-2.5158 \pm 4.4927j$
	$-3.7357 \pm 2.6263j$
	$-4.2484 \pm 0.8675j$
7	-4.9716
	$-2.6857 \pm 5.4206j$
	$-4.0701 \pm 3.5173j$
	$-4.7584 \pm 1.7939j$
8	$-5.2049 \pm 2.6162j$
	$-4.3683 \pm 4.4146j$
	$-2.8388 \pm 6.3540j$
	$-5.5878 \pm 0.8676j$

TABLE 5.10 Factors for Renormalizing Bessel Filter Poles from Unit Delay at $\omega = 0$ to 3-dB Attenuation at $\omega = 1$

n	A
2	1.35994
3	1.74993
4	2.13011
5	2.42003
6	2.69996
7	2.95000
8	3.17002

using the function FilterFrequencyResponse belonging to the base class FilterTransFunc from which BesselTransFunc inherits:

```
filter_function = new BesselTransFunc(order,
                                      passband_edge,
                                      norm_for_unit_delay);
filter_function->LowpassDenorm(passband_edge);
filter_function->FilterFrequencyResponse();
```

Classical Analog Filters

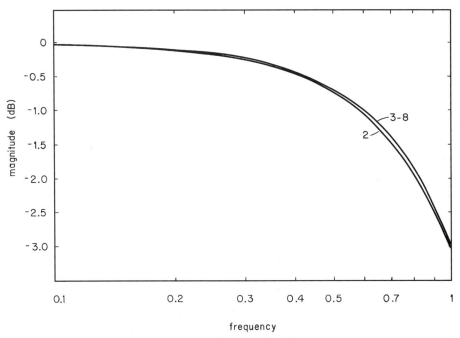

Figure 5.41 Passband magnitude response of lowpass Bessel filters.

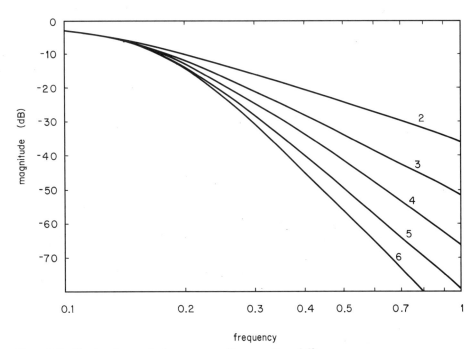

Figure 5.42 Stopband magnitude response of lowpass Bessel filters.

112   Chapter Five

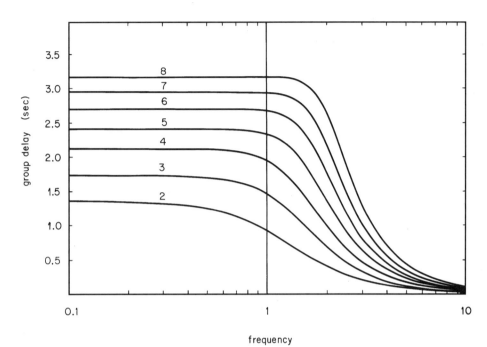

Figure 5.43 Group delay response of lowpass Bessel filters.

Figures 5.41 and 5.42 show the magnitude responses for Bessel filters of several different orders.

Group delay

Group delay for lowpass Bessel filters of several different orders are plotted in Fig. 5.43. The data for these plots was generated by the FilterFrequencyResponse function, which performs numerical differentiation of the phase response to evaluate the group delay.

Chapter 6

Foundations of Digital Signal Processing

Digital signal processing (DSP) is based on the fact that an analog signal can be digitized and input to a general-purpose digital computer or special-purpose digital processor. Once this is accomplished, we are free to perform all sorts of mathematical operations on the sequence of digital data samples inside the processor. Some of these operations are simply digital versions of classical analog techniques, while others have no counterpart in analog circuit devices or processing methods. This chapter covers *digitization* and introduces the fundamental types of processing that can be performed on the sequence of digital values once they are inside the processor. The remainder of the book is devoted to more detailed explorations of the myriad ways in which these fundamental processing operations can be combined to achieve a desired result.

6.1 Digitization

Digitization is the process of converting an analog signal such as a time-varying voltage or current into a sequence of digital values. Digitization actually involves two distinct parts—*sampling* and *quantization*—which are usually analyzed separately. Three basic types of sampling, shown in Fig. 6.1, are *ideal*, *instantaneous*, and *natural*. From the illustration we can see that the sampling process converts a signal that is defined over a continuous time interval into a signal that has nonzero amplitude values only at discrete instants of time (as in ideal sampling) or over a number of discretely separate but internally continuous subintervals of time (as in instantaneous and natural sampling). The signal that results from a sampling process is called a *sampled-data signal*. The signals resulting from ideal sampling are also referred to as *discrete-time signals*.

Each of the three basic sampling types occurs at different places within a DSP system. The output from a sample-and-hold amplifier or a *digital-to-analog*

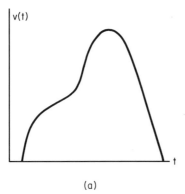 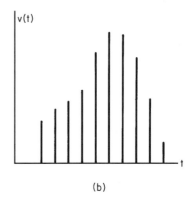

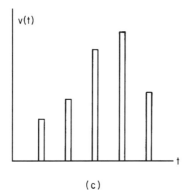 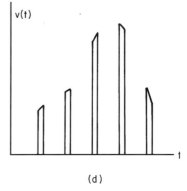

Figure 6.1 An analog signal (a) and three different types of sampling: (b) ideal, (c) instantaneous, and (d) natural.

converter (DAC) is an instantaneously sampled signal. In the output of a practical *analog-to-digital converter* (ADC) used to sample a signal, each sample will, of course, exist for some nonzero interval of time. However, within the software of the digital processor, these values still can be interpreted as the amplitudes for a sequence of ideal samples. In fact, this is almost always the best approach, since the ideal sampling model results in the simplest processing for most applications. Natural sampling is encountered in the analysis of the analog multiplexing that is often performed prior to A/D conversion in multiple-signal systems. In all three of the sampling approaches presented, the sample values are free to assume any appropriate value from the continuum of possible analog signal values.

Quantization is the part of digitization that is concerned with converting the amplitudes of an analog signal into values that can be represented by binary numbers having some finite number of bits. A quantized, or *discrete-valued*, signal is shown in Fig. 6.2. The sampling and quantization processes will

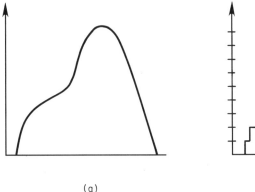
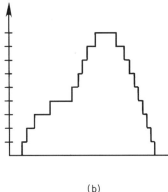

Figure 6.2 An analog signal (*a*) and the corresponding quantized signal (*b*).

introduce some significant changes in the spectrum of a digitized signal. The details of the changes will depend on both the precision of the quantization operation and the particular sampling model that most aptly fits the actual situation.

Ideal sampling

In *ideal sampling*, the sampled-data signal, as shown in Fig. 6.3, comprises a sequence of uniformly spaced impulses, with the weight of each impulse equal to the amplitude of the analog signal at the corresponding instant in time. Although not mathematically rigorous, it is convenient to think of the sampled-data signal as the result of multiplying the analog signal $x(t)$ by a periodic train of unit impulses:

$$x_s(\cdot) = x(t) \sum_{n=-\infty}^{\infty} \delta(t - nT)$$

Based on Property 11 from Table 2.2, this means that the spectrum of the sampled-data signal could be obtained by convolving the spectrum of the analog

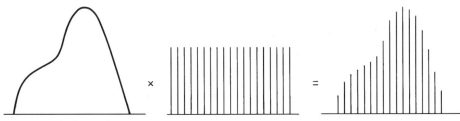

Figure 6.3 Ideal sampling.

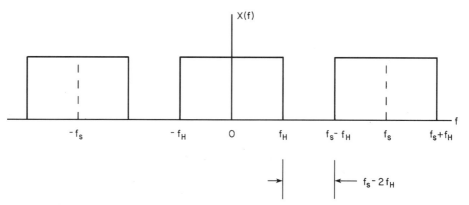

Figure 6.4 Spectrum of an ideally sampled signal.

signal with the spectrum of the impulse train:

$$\mathcal{F}\left[x(t)\sum_{n=-\infty}^{\infty}\delta(t-nT)\right] = X(f) \otimes \left[f_s \sum_{m=-\infty}^{\infty}\delta(t-mf_s)\right]$$

As illustrated in Fig. 6.4, this convolution produces copies, or *images*, of the original spectrum that are periodically repeated along the frequency axis. Each of the images is an exact (to within a scaling factor) copy of the original spectrum. The center-to-center spacing is equal to $f_s - 2f_H$. As long as f_s is greater than 2 times f_H, the original signal can be recovered by a lowpass filtering operation that removes the extra images introduced by the sampling.

Sampling rate selection

If f_s is less than $2f_H$, the images will overlap, or *alias*, as shown in Fig. 6.5, and recovery of the original signal will not be possible. The minimum alias-free sampling rate of $2f_H$ is called the *Nyquist rate*. A signal sampled exactly at its Nyquist rate is said to be *critically sampled*.

Uniform sampling theorem. If the spectrum $X(f)$ of a function $x(t)$ vanishes beyond an upper frequency of f_H Hz or ω_H rad/s, then $x(t)$ can be completely determined by its values at uniform intervals of less than $1/(2f_H)$ or π/ω. If sampled within these constraints, the original function $x(t)$ can be reconstructed from the samples by

$$x(t) = \sum_{n=-\infty}^{\infty} x(nT) \frac{\sin[2f_s(t-nT)]}{2f_s(t-nT)}$$

where T is the sampling interval.

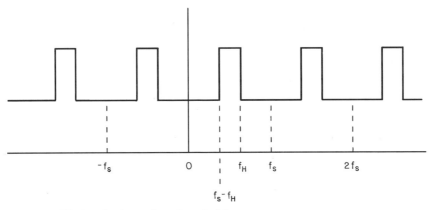

Figure 6.5 Aliasing due to overlap of spectral images.

Since practical signals cannot be strictly band-limited, sampling of a real-world signal must be performed at a rate greater than $2f_H$ where the signal is known to have negligible (that is, typically less than 1 percent) spectral energy above the frequency of f_H. When designing a signal processing system, we will rarely, if ever, have reliable information concerning the exact spectral occupancy of the noisy real-world signals that our system will eventually face. Consequently, in most practical design situations, a value is selected for f_H based on the requirements of the particular application, and then the signal is lowpass filtered prior to sampling. Filters used for this purpose are called *antialiasing filters* or *guard filters*. The sample-rate selection and guard filter design are coordinated so that the filter provides attenuation of 40 dB or more for all frequencies above $f_s/2$. The spectrum of an ideally sampled practical signal is shown in Fig. 6.6. Although some aliasing does occur, the aliased components are suppressed at least 40 dB below the desired components. Antialias filtering must be performed prior to sampling. In general, there is no way to eliminate aliasing once a signal has been improperly sampled. The particular type (Butterworth, Chebyshev, elliptical, etc.) and order of the filter should be chosen to provide the necessary stopband attenuation while preserving the passband characteristics most important to the intended application.

Instantaneous sampling

In instantaneous sampling, each sample has a nonzero width and a flat top. As shown in Fig. 6.7, the sampled-data signal resulting from instantaneous sampling can be viewed as the result of convolving a sample pulse $p(t)$ with an ideally sampled version of the analog signal. The resulting sampled-data signal can thus be expressed as

$$x_s(\cdot) = p(t) \otimes \left[x(t) \sum_{n=-\infty}^{\infty} \delta(t - nT) \right]$$

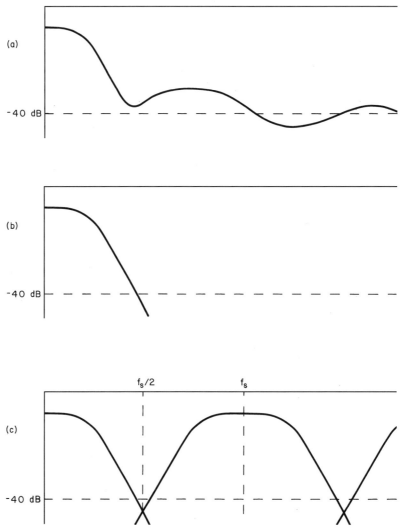

Figure 6.6 Spectrum of an ideally sampled practical signal: (a) spectrum of raw analog signal, (b) spectrum after lowpass filtering, and (c) spectrum after sampling.

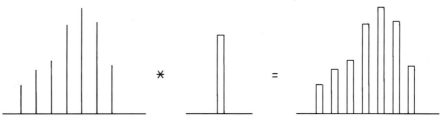

Figure 6.7 Instantaneous sampling.

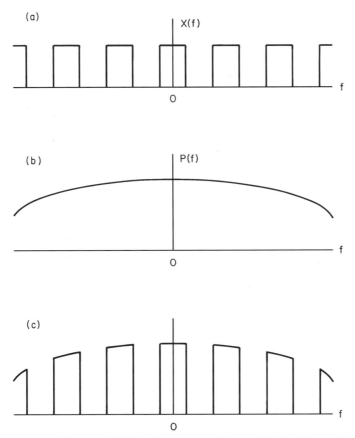

Figure 6.8 Spectrum (a) of an instantaneously sampled signal is equal to the spectrum (b) of an ideally sampled signal multiplied by the spectrum (c) of a single sampling pulse.

where $p(t)$ is a single rectangular sampling pulse and $x(t)$ is the original analog signal. Based on Property 10 from Table 2.2, this means that the spectrum of the instantaneous sampled-data signal can be obtained by multiplying the spectrum of the sample pulse with the spectrum of the ideally sampled signal:

$$\mathcal{F}\left\{p(t) \otimes \left[x(t) \sum_{n=-\infty}^{\infty} \delta(t - nT)\right]\right\} = P(f) \cdot \left\{X(f) \otimes \left[f_s \sum_{m=-\infty}^{\infty} \delta(f - mf_s)\right]\right\}$$

As shown in Fig. 6.8, the resulting spectrum is similar to the spectrum produced by ideal sampling. The only difference is the amplitude distortion introduced by the spectrum of the sampling pulse. This distortion is sometimes called the *aperture effect*. Notice that distortion is present in all the images, including the one at baseband. The distortion will be less severe for narrow sampling pulses. As the pulses become extremely narrow, instantaneous sampling begins

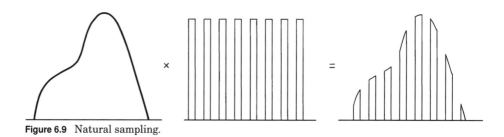

Figure 6.9 Natural sampling.

to look just like ideal sampling, and distortion due to the aperture effect all but disappears.

Natural sampling

In natural sampling, each sample's amplitude follows the analog signal's amplitude throughout the sample's duration. As shown in Fig. 6.9, this is mathematically equivalent to multiplying the analog signal by a periodic train of rectangular pulses:

$$x_s(\cdot) = x(t) \cdot \left\{ p(t) \otimes \left[\sum_{n=-\infty}^{\infty} \delta(t - nT) \right] \right\}$$

The spectrum of a naturally sampled signal is found by convolving the spectrum of the analog signal with the spectrum of the sampling pulse train:

$$\mathcal{F}[x_s(\cdot)] = X(f) \otimes \left[P(f) \, f_s \sum_{m=-\infty}^{\infty} \delta(f - mf_s) \right]$$

As shown in Fig. 6.10, the resulting spectrum is similar to the spectrum produced by instantaneous sampling. In instantaneous sampling, all frequencies of the sampled signal's spectrum are attenuated by the spectrum of the sampling pulse, while in natural sampling each image of the basic spectrum will be attenuated by a factor that is equal to the value of the sampling pulse's spectrum at the center frequency of the image. In communications theory, natural sampling is called *shaped-top pulse amplitude modulation*.

Discrete-time signals

In the discussion so far, weighted impulses have been used to represent individual sample values in a *discrete-time signal*. This was necessary in order to use continuous mathematics to connect continuous-time analog signal representations with their corresponding discrete-time digital representations. However, once we are operating strictly within the digital or discrete-time realms, we can dispense with the Dirac delta impulse and adopt in its place the unit sample

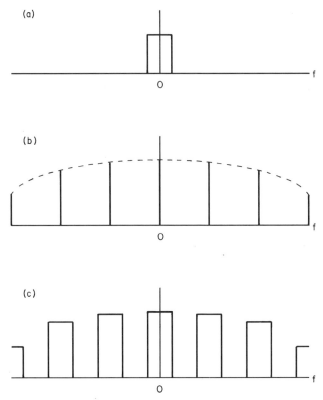

Figure 6.10 Spectrum (a) of a naturally sampled signal is equal to the spectrum (b) of the analog signal convolved with the spectrum (c) of the sampling pulse train.

function, which is much easier to work with. The unit sample function is also referred to as a *Kronecker delta impulse* [1]. Figure 6.11 shows both the Dirac delta and Kronecker delta representations for a typical signal. In the function sampled using a Dirac impulse train, the independent variable is continuous time t, and integer multiples of the sampling interval T are used to define the discrete sampling instants. On the other hand, the Kronecker delta notation assumes uniform sampling with an implicitly defined sampling interval. The independent variable is the integer-valued index n whose values correspond to the discrete instants at which samples can occur. In most theoretical work, the implicitly defined sampling interval is dispensed with completely by treating all the discrete-time functions as though they have been normalized by setting $T = 1$.

Notation

Writers in the field of digital signal processing are faced with the problem of finding a convenient notational way to distinguish between continuous-time

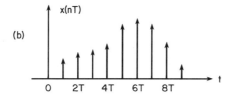
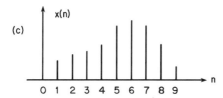

Figure 6.11 Sampling with Dirac and Kronecker impulses: (*a*) continuous signal, (*b*) sampling with Dirac impulses, and (*c*) sampling with Kronecker impulses.

functions and discrete-time functions. Since the early 1970s, a number of different approaches have appeared in the literature, but none of the schemes advanced so far have been perfectly suited for all situations. In fact, some authors use two or more different notational schemes within different parts of the same book. In keeping with long-established mathematical practice, functions of a continuous variable are almost universally denoted with the independent variable enclosed in parentheses: $x(t)$, $H(e^{j\omega})$, $\phi(f)$ and so on. Many authors, such as Oppenheim and Schafer [2], Rabiner and Gold [3], and Roberts and Mullis [4], make no real notational distinction between functions of continuous variables and functions of discrete variables, instead relying on context to convey the distinction. This approach, while easy for the writer, can be confusing for the reader. Another approach involves using subscripts for functions of a discrete variable:

$$x_k \stackrel{\triangle}{=} x(kT)$$

$$H_n \stackrel{\triangle}{=} H(e^{jn\theta})$$

$$\phi_m \stackrel{\triangle}{=} \phi(mF)$$

This approach quickly becomes typographically unwieldy when the independent variable is represented by a complicated expression. A more recent practice [5] uses parentheses to enclose the independent variable of continuous-variable functions and brackets to enclose the independent variable of discrete-variable functions:

$$x[k] = x(kT)$$
$$H[n] = H(e^{jn\theta})$$
$$\phi[m] = \phi(mF)$$

For the remainder of this book, we will adopt this practice and remind ourselves to be careful in situations where the bracket notation for discrete-variable functions could be confused with the bracket notation used for arrays in C++.

6.2 Discrete-Time Fourier Transform

The Fourier series given by Eq. (2.13) can be rewritten to make use of the discrete sequence notation that was introduced in the previous section:

$$x(t) = \sum_{n=-\infty}^{\infty} X[n] e^{j2\pi nFt}$$

where $F = \frac{1}{t_0}$ = sample spacing in the frequency domain
t_0 = period of $x(t)$

Likewise, Eq. (2.14) can be written as

$$X[n] = \frac{1}{t_0} \int_{t_0} x(t) e^{-jn2\pi Ft}\, dt$$

The fact that the signal $x(t)$ and sequence $F[n]$ form a Fourier series pair with a frequency domain sampling interval of F can be indicated as

$$x(t) \overset{FS;F}{\longleftrightarrow} X[n]$$

In Sec. 6.1 the results concerning the impact of sampling on a signal's spectrum were obtained using the *continuous-time Fourier transform* in conjunction with a periodic train of Dirac impulses to model the sampling of the continuous-time signal $x(t)$. Once we have defined a discrete-time sequence $x[n]$, the *discrete-time Fourier transform* (DTFT) can be used to obtain the corresponding spectrum directly from the sequence without having to resort to impulses and continuous-time Fourier analysis.

The DTFT, which links the discrete-time and continuous-frequency domains, is defined by

$$X(e^{j\omega T}) = \sum_{n=-\infty}^{\infty} x[n] e^{-j\omega nT} \qquad (6.1)$$

and the corresponding inverse is given by

$$x[n] = \frac{1}{2\pi} \int_{-\pi}^{\pi} X(e^{j\omega}) e^{j\omega nT} d\omega \tag{6.2}$$

If Eqs. (6.1) and (6.2) are compared to the DTFT definitions given by certain texts [2, 3, 5], an apparent disagreement will be found. The cited texts define the DTFT and its inverse as

$$X(e^{j\omega}) = \sum_{n=-\infty}^{\infty} x[n] e^{-j\omega n} \tag{6.3}$$

$$x[n] = \frac{1}{2\pi} \int_{-\pi}^{\pi} X(e^{j\omega}) e^{j\omega nT} d\omega \tag{6.4}$$

The disagreement is due to the notation used by these texts, in which ω is used to denote the *digital* frequency given by

$$\omega = \frac{\Omega}{F_s} = \Omega T$$

where Ω = analog frequency
F_s = sampling frequency
T = sampling interval

In most DSP books other than the three previously cited, the analog frequency is denoted by ω rather than by Ω. Whether ω or Ω is the "natural" choice for denoting analog frequency depends on the overall approach taken in developing Fourier analysis of sequences. Books that begin with sequences, then proceed to Fourier analysis of sequences, and finally tie sequences to analog signals via sampling tend to use ω for the first frequency variable encountered, which is *digital* frequency. Other books that begin with analog theory and then move to sampling and sequences tend to use ω for the first frequency variable encountered, which in this case is *analog* frequency. In this book, we will adopt the convention used by Peled and Liu [6], denoting analog frequency by ω and digital frequency by $\lambda = \omega T$. The function $X(e^{j\omega T})$ is periodic with a period of $\omega_p = 2\pi/T$, and $X(e^{j\lambda})$ is periodic with a period of $\lambda_p = 2\pi$.

Independent of the ω versus Ω controversy, the notations $X(e^{j\omega T})$ or $X(e^{j\lambda})$ are commonly used rather than $X(\omega)$ or $X(\lambda)$ so that the form of Eq. (6.1) remains similar to the form of the z transform given in Chap. 7, which is

$$X(z) = \sum_{n=-\infty}^{\infty} x[n] z^{-n} \tag{6.5}$$

If $e^{j\omega T}$ is substituted for z in Eq. (6.5), the result is identical to Eq. (6.1). This indicates that the DTFT is nothing more than the z transform evaluated on the unit circle. [Note: $e^{j\omega} = \cos\omega + j\sin\omega$, $0 \leq \omega \leq 2\pi$, does in fact define the unit circle in the z plane since $|e^{j\omega}| = (\cos^2\omega + \sin^2\omega)^{1/2} \equiv 1$.]

Convergence conditions

If the time sequence $x[n]$ satisfies

$$\sum_{n=-\infty}^{\infty} |x[n]| < \infty$$

then $X(e^{j\omega T})$ exists and the series in Eq. (6.1) converges uniformly to $X(e^{j\omega T})$. If $x[n]$ satisfies

$$\sum_{n=-\infty}^{\infty} |x[n]|^2 < \infty$$

then $X(e^{j\omega T})$ exists and the series in Eq. (6.1) converges in a mean-square sense to $X(e^{j\omega T})$, that is,

$$\lim_{M\to\infty} \int_{-\pi}^{\pi} |X(e^{j\omega T}) - X_M(e^{j\omega T})|^2 \, d\omega = 0$$

where

$$X_M(e^{j\omega T}) = \sum_{n=-M}^{M} x[n] e^{-j\omega n T}$$

The function $X_M(e^{j\omega T})$ is a form of the Dirichlet kernel discussed in Sec. 10.1.

Relationship to Fourier series

Since the Fourier series represents a periodic continuous-time function in terms of a discrete-frequency function, and the DTFT represents a discrete-time function in terms of a periodic continuous-frequency function, we might suspect that some sort of duality exists between the Fourier series and DTFT. It turns out that such a duality does indeed exist. Specifically if

$$f[k] \stackrel{DTFT}{\longleftrightarrow} F(e^{j\omega T})$$

and we set $\omega_0 = T$

$$x(t) = F(e^{j\omega T})|_{\omega = T}$$
$$X[n] = f[k]|_{k=-n}$$

then $x(t) \stackrel{FS;\omega_0}{\longleftrightarrow} X[n]$

6.3 Discrete-Time Systems

In Chap. 4 we saw how continuous-time systems such as filters and amplifiers can accept analog input signals and operate upon them to produce different analog output signals. *Discrete-time systems* perform essentially the same role for digital or discrete-time signals.

Difference equations

Although Chap. 4 deliberately avoided discussing differential equations and their accompanying headaches in the analysis of analog systems, *difference equations* are much easier to work with, and they play an important role in the analysis and synthesis of discrete-time systems. A *discrete-time, linear, time-invariant* (DTLTI) system, which accepts as an input sequence $x[n]$ and produces an output sequence $y[n]$, can be described by a linear difference equation of the form

$$y[n] + a_1 y[n-1] + a_2 y[n-2] + \cdots + a_k y[n-k]$$
$$= b_0 x[n] + b_1 x[n-1] + b_2 x[n-2] + \cdots + b_k x[n-k] \quad (6.6)$$

Such a difference equation can describe a DTLTI system having any initial conditions as long as they are specified. This is in contrast to the discrete-convolution and discrete-transfer functions that are limited to describing digital filters that are initially relaxed (that is, all inputs and outputs are initially zero). In general, the computation of the output $y[n]$ at point n using Eq. (6.6) will involve previous outputs $y[n-1], y[n-2], y[n-3]$ and so on. However, in some filters all of the coefficients $a_1, a_2, \ldots, a_k$ are equal to zero, thus yielding

$$y[n] = b_0 x[n] + b_1 x[n-1] + b_2 x[n-2] + \cdots + b_k x[n-k] \quad (6.7)$$

in which the computation of $y[n]$ does not involve previous output values. Difference equations involving previous output values are called *recursive difference equations*, and equations in the form of (6.7) are called *nonrecursive difference equations*.

Example 6.1 Determine a nonrecursive difference equation for a simple moving-average lowpass filter in which the output at $n = i$ is equal to the arithmetic average of the five inputs from $n = i - 4$ through $n = i$.

solution The desired difference equation is given by

$$y[n] = \frac{x[n] + x[n-1] + x[n-2] + x[n-3] + x[n-4]}{5}$$

$$= 0.2x[n] + 0.2x[n-1] + 0.2x[n-2] + 0.2x[n-3] + 0.2x[n-4]$$

Discrete convolution

A discrete-time system's impulse response is the output response produced when a unit sample function is applied to the input of the previously relaxed system. As we might expect from our experiences with continuous systems, we can obtain the output $y[n]$ due to any input by performing a *discrete convolution* of the input signal $x[n]$ and the impulse response $h[n]$. This discrete convolution is given by

$$y[n] = \sum_{m=0}^{\infty} h[m]x[n-m]$$

If the impulse response has nonzero values at an infinite number of points along the discrete-time axis, the corresponding digital filter is called an *infinite impulse response* (IIR) filter. On the other hand, if $h[m] = 0$ for all $m \geq M$, the filter is called a *finite impulse response* (FIR) filter, and the convolution summation can be rewritten as

$$y[n] = \sum_{m=0}^{M-1} h[m]x[n-m]$$

FIR filters are also called *transversal filters*. The following summation identities will often prove useful in the evaluation of convolution summations:

$$\sum_{n=0}^{N} \alpha^n = \frac{1-\alpha^{N+1}}{1-\alpha} \quad \alpha \neq 1 \tag{6.8}$$

$$\sum_{n=0}^{N} n\alpha^n = \frac{\alpha}{(1-\alpha)^2}(1-\alpha^N - N\alpha^N + N\alpha^{N+1}) \quad \alpha \neq 1 \tag{6.9}$$

$$\sum_{n=0}^{N} n^2\alpha^n = \frac{\alpha}{(1-\alpha)^3}[(1+\alpha)(1-\alpha^N) - 2(1-\alpha)N\alpha^N - (1-\alpha)^2 N^2 \alpha^N] \quad \alpha \neq 1$$

$$\tag{6.10}$$

6.4 Diagramming Discrete-Time Systems

Block diagrams

As they are for continuous-time systems, block diagrams are useful in the design and analysis of discrete-time systems. Construction of block diagrams for discrete-time systems involves three basic building blocks: the *unit-delay element*, *multiplier*, and *summer*.

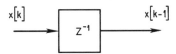

Figure 6.12 Block diagram representation of a unit-delay element.

Unit-delay element. As its name implies, a unit-delay element generates an output that is identical to its input delayed by one sample interval:

$$y[k] = x[k-1]$$

The unit-delay element usually is drawn as shown in Fig. 6.12. The term z^{-1} is used to denote a unit delay because delaying a discrete-time signal by one sample time multiplies the signal's z transform by z^{-1}. (See Property 5 in Table 7.4.) Delays of p sample times may be depicted as p unit delays in series or as a box enclosing z^{-p}.

Multiplier. A multiplier generates as output the product of a fixed constant and the input signal

$$y[k] = ax[k]$$

A multiplier can be drawn in any of the ways shown in Fig. 6.13. The form shown in Fig. 6.13c usually is reserved for adaptive filters and other situations where the factor a is not constant. [Note that a system containing multiplication by a nonconstant factor would not be a *linear time-invariant* (LTI) system!]

a)

b)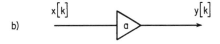

c)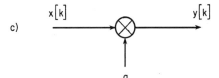

Figure 6.13 Block diagram representations of a multiplier.

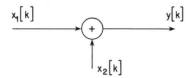

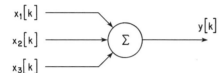

Figure 6.14 Block diagram representations of a summer.

Summer. A summer adds two or more discrete-time signals to generate the discrete-time output signal:

$$y[k] = x_1[k] + x_2[k] + \cdots + x_n[k]$$

A summer is depicted using one of the forms shown in Fig. 6.14. A negative sign can be placed next to a summer's input paths as required to indicate a signal that is to be subtracted rather than added.

Example 6.2 Draw a block diagram for a simple moving-average lowpass filter in which the output at $k = i$ is equal to the arithmetic average of the three inputs for $k = i - 2$ through $k = i$.

solution The difference equation for the desired filter is

$$y[k] = \frac{1}{3}x[k] + \frac{1}{3}x[k-1] + \frac{1}{3}x[k-2]$$

The block diagram for this filter will be as shown in Fig. 6.15. It should be noted that block diagram representations in general are not unique and that a given system can be represented in several different ways.

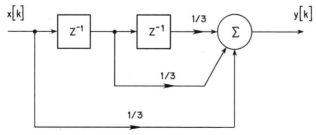

Figure 6.15 Block diagram for Example 6.2.

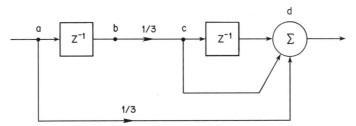

Figure 6.16 Block diagram of a discrete-time system.

Signal flow graphs

A modified form of a directed graph, called a *signal flow graph* (SFG), can be used to depict all the same information as a block diagram but in a more compact form. Consider the block diagram in Fig. 6.16, which has some labeled points added for ease of reference. The *oriented graph*, or *directed graph*, for this system is obtained by replacing each multiplier, each connecting branch, and each delay element with a directed line segment called an *edge*. Furthermore, each branching point and each adder is replaced by a point called a *node*. The resulting graph is shown in Fig. 6.17. A signal flow graph is obtained by associating a signal with each node and a linear operation with each edge of the directed graph. The node weights correspond to signals present within the discrete-time system. Associated with each edge is the linear operation (delay or constant gain) that must be performed on the signal associated with the edge's *from* node in order to obtain the signal associated with the edge's *to* node. For a node which is the *to* node for two or more edges, the signal associated with the node is the sum of all signals produced by the incoming edges. For the graph shown in Fig. 6.17, the following correspondences can be identified:

Node a	$x[k]$
Node b	$x[k-1]$
Node c	$\frac{1}{3}x[k-1]$
Node d	Summer producing $y[k]$
Edge (a, b)	First delay element
Edge (c, d)	Second delay element

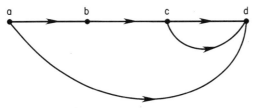

Figure 6.17 Directed graph corresponding to the block diagram of Fig. 6.16.

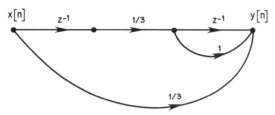

Figure 6.18 Signal flow graph derived from the directed graph of Fig. 6.17.

Edge (a, d) Bottom multiplier

Edge (b, c) Top multiplier

Edge (c, d) Unity gain connection from point c to summer

The resulting signal flow graph is shown in Fig. 6.18. It is customary to use multiplication by z^{-1} as a shorthand notation for unit delay, even though the signals in an SFG are time domain signals, and multiplication by z^{-1} is a frequency domain operation.

6.5 Quantization

Although floating-point formats are used in some digital filters, cost and speed considerations will often dictate the use of fixed-point formats having a relatively short word length. Such formats will force some precision to be lost in representations of the signal samples, filter coefficients, and computation results. A digital filter designed under an infinite-precision assumption will not perform up to design expectations if implemented with short word-length, fixed-point arithmetic. In many cases, the degradations can be so severe as to make the filter unusable. This section examines the various types of degradations caused by finite-precision implementations and explores what can be done to achieve acceptable performance in spite of the degradations.

Fixed-point numeric formats

Binary fixed-point representation of numbers enjoys widespread use in DSP applications, where there is usually some control over the range of values that must be represented. Typically, all of the coefficients $h[n]$ for a digital filter will be scaled such that

$$|h[n]| \leq 1.0 \quad \text{for } n = 1, 2, \ldots, N \tag{6.11}$$

Once scaled in this way, each coefficient can be expressed as

$$h = b_0 2^0 + b_1 2^{-1} + b_2 2^{-2} + \cdots \tag{6.12}$$

where each of the b_n is a single bit; that is, $b_n \in \{0, 1\}$. If we limit our representation to a length of $L + 1$ bits, the coefficients can be represented as a fixed-point

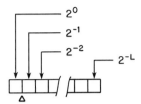

Figure 6.19 Fixed-point binary number format.

binary number of the form shown in Fig. 6.19. As shown in the figure, a small triangle is often used to represent the binary point so that it cannot be easily confused with a decimal point. The expansion of Eq. (6.12) can then be written as

$$h = \sum_{h=0}^{L} b_k 2^{-k} \qquad (6.13)$$

The bit shown to the left of the binary point in Fig. 6.19 is necessary to represent coefficients for which the equality in Eq. (6.11) holds, but its presence complicates the implementation of arithmetic operations. If we eliminate the need to exactly represent coefficients that equal unity, we can use the fixed-point fractional format shown in Fig. 6.20. Using this scheme, some values are easy to write:

$$\frac{1}{2} = {}_\triangle 1000$$

$$\frac{3}{8} = {}_\triangle 01100$$

$$\frac{5}{64} = {}_\triangle 000101$$

Some other values are not easy. Consider the case of 1/10, which expands as

$$\frac{1}{10} = 2^{-4} + 2^{-5} + 2^{-8} + 2^{-9} + 2^{-12} + 2^{-13} + \cdots$$

$$= \sum_{k=1}^{\infty} (2^{-4k} + 2^{-4k-1})$$

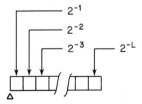

Figure 6.20 Alternative fixed-point binary number format.

The corresponding fixed-point binary representation is a repeating fraction given by

$$\frac{1}{10} = {}_\triangle 0001\overline{10011}\cdots$$

If we are limited to a 16-bit fixed-point binary representation, we can truncate the fraction after 16 bits to obtain

$$\frac{1}{10} = {}_\triangle 0001100110011001\cdots$$

The actual value of this 16-bit representation is

$$2^{-4} + 2^{-5} + 2^{-8} + 2^{-9} + 2^{-12} + 2^{-13} + 2^{-16} = \frac{6553}{65{,}536} \cong 0.099990845$$

Thus, the value represented in 16 bits is too small by approximately 9.155×10^{-6}.

Instead of truncating, we could use a rounding approach. Rounding a binary value is simple—just add 1 to the first (leftmost) bit that is not being retained in the rounded format. In the current example we add 1 to bit 16. This generates a carry into b_{15}, which propagates into b_{14} to yield

$${}_\triangle 0001100110011010 = \frac{6554}{65{,}536} \cong 0.100006104$$

This value is too large by approximately 6.1×10^{-6}.

In many DSP applications where design simplicity, low cost, or high speed is important, the word length may be significantly shorter than 16 bits, and the error introduced by either truncating or rounding the coefficients can be quite severe.

Floating-point numeric formats

A fixed-point fractional format has little use in a general-purpose computer, where there is little or no *a priori* control over the range of values that may need to be represented. Clearly, any time a value equals or exceeds 1.0, it cannot be represented in the format of Fig. 6.20. Floating-point formats remove this limitation by effectively allowing the binary point to shift position as needed. For floating-point representations, a number is typically expanded in the form

$$h = 2^a \sum_{k=0}^{L} b_k 2^{-k}$$

A typical floating-point numeric format is shown in Fig. 6.21. The fields denoted *i* and *f* contain a fixed-point value of the form shown in Fig. 6.19 where the binary

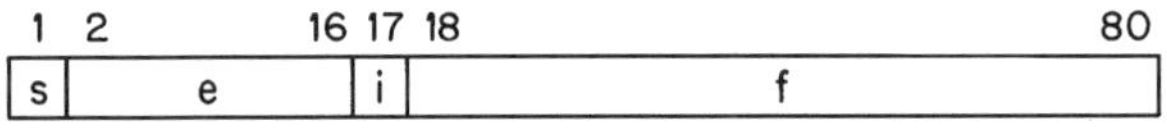

```
  1 2         16 17 18                              80
 ┌──┬─────────┬──┬─────────────────────────────────────┐
 │s │    e    │i │                 f                   │
 └──┴─────────┴──┴─────────────────────────────────────┘
```

Figure 6.21 A typical floating-point numeric format.

point is assumed to lie between i and the most significant bit of f. This fixed-point value is referred to as the *mantissa*. If the bits in field f are designated from left to right as $f_1, f_2, \ldots, f_{63}$, the value of the mantissa is given by

$$m = i + \sum_{k=1}^{63} f_k 2^{-k}$$

The field denoted as e in a 15-bit integer value is used to indicate the power of 2 by which the numerator must be multiplied in order to obtain the value being represented. This can be a positive or negative power of 2, but rather than using a sign in conjunction with the exponent, most floating-point formats use an offset. A 15-bit binary field can have values ranging from 0 to 32,767. Values from 0 to 16,382 are interpreted as negative powers of 2, and values from 16,384 to 32,766 are interpreted as positive powers of 2. The value 16,383 is interpreted as $2^0 = 1$, and the value 32,767 is reserved for representing infinity and specialized values called NaN (*not-a-number*). The sign bit denoted by s is the sign of the overall number. The value represented by a floating-point number in the format of Fig. 6.21 can be obtained as

$$v = (-1)^s 2^{(e-16,383)} \left(i + \sum_{k=1}^{63} f_k 2^{-k} \right)$$

provided $e \neq 32,767$.

Suppose we wish to represent 1/10 in the floating-point format of Fig. 6.21. One way to accomplish this is to set the mantissa equal to a 64-bit fixed-point representation of 1/10 and set $e = 16,383$ to indicate a multiplier of unity. Using the hexadecimal notation discussed previously, we can write the results of such an approach as

$$s = 0$$
$$e = \text{0x3fff}$$
$$i = 0$$
$$f = \text{0x0ccccccccccccccc}$$

With the various fields packed together, the resulting 80-bit floating-point representation of 1/10 is $W = \text{0x3fff0ccccccccccccccc}$. Slightly more precision can be squeezed into the representation if we shift f four places to the left and modify e to indicate multiplication by 2^{-4}. Such an approach yields

$$W = \text{0x3ffbcccccccccccccccc}$$

Numbers greater than 1.0 present no problem for this format. The value 57 is represented as

$$s = 0$$
$$e = \text{0x4004} \quad (\text{that is, } 2^5)$$
$$i = 1$$
$$f = \text{0x6400000000000000}$$
$$W = \text{0x4004e400000000000000}$$

In other words, this representation stores 57 by making use of the fact that

$$57 = 2^5(2^0 + 2^{-1} + 2^{-2} + 2^{-5})$$

Quantized coefficients

When the coefficients of a digital filter are quantized, the filter becomes a different filter. The resulting filter is still a discrete-time linear time-invariant system—it's just not the system we set out to design. Consider the 21-tap LPF having the coefficients listed in Table 6.1. The values given in the table, having 15 decimal digits in the fractional part, will be used as the baseline approximation to the coefficients' infinite-precision values. Let us force the coefficient values into a fixed-point fractional format having a 16-bit magnitude plus one sign bit. After truncating the bits in excess of 16, the coefficient values listed in Table 6.2 are obtained. The magnitude response of a filter using such coefficients is virtually identical to the response obtained using the floating-point coefficients of Table 6.1. If the coefficients are further truncated to 14- or 12-bit magnitudes, slight degradations in stopband attenuation can be observed.

TABLE 6.1 Coefficients for a 21-tap Lowpass Filter

n	$h[n]$
0,20	0.000
1,20	−0.000823149720361
2,20	−0.002233281959082
3,20	0.005508892585759
4,20	0.017431813641454
5,20	−0.000000000000050
6,20	−0.049534952531101
7,20	−0.049511869643024
8,20	0.084615800641299
9,20	0.295322344140975
10	0.40

TABLE 6.2 Truncated 16-bit Coefficients for a 21-tap Lowpass Filter

n	Sign	Hex value	Decimal value
0,20	+	0000	0.0
1,20	−	0035	−0.000808715820312
2,20	−	0092	−0.002227783203125
3,20	+	0169	0.005508422851562
4,20	+	0476	0.017425537109375
5,20	+	0000	0.0
6,20	−	0cae	−0.049530029296875
7,20	−	0cac	−0.049499511718750
8,20	+	15a9	0.084609985351562
9,20	+	4b9a	0.295318603515625
10	+	6666	0.399993896484375

TABLE 6.3 Truncated 10-bit Coefficients for a 21-tap Lowpass Filter

n	Sign	Hex value	Decimal value
0,20	+	000	0.0
1,20	−	000	−0.0
2,20	−	008	−0.001953125
3,20	+	014	0.0048828125
4,20	+	044	0.0166015625
5,20	+	000	0.0
6,20	−	0c8	−0.048828125
7,20	−	0c8	−0.048828125
8,20	+	158	0.083984375
9,20	+	4b8	0.294921875
10	+	664	0.3994140625

The degradations in filter response are quite significant for the 10-bit coefficients listed in Table 6.3. As shown in Fig. 6.22, the fourth sidelobe is narrowed, and the fifth sidelobe peaks at −50.7 dB—a value significantly worse than the −68.2 dB of the baseline case. The filter responses for 8- and 6-bit coefficients are shown in Figs. 6.23 and 6.24, respectively.

Quantization noise

The finite digital word lengths used to represent numeric values within a digital filter limit the precision of other quantities besides the filter coefficients. Each sample of the input and output, as well as all intermediate results of

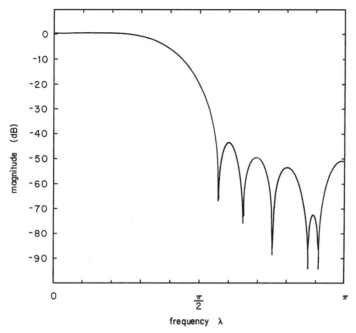

Figure 6.22 Magnitude response for 21-tap lowpass filter with coefficients quantized to 10 bits plus sign.

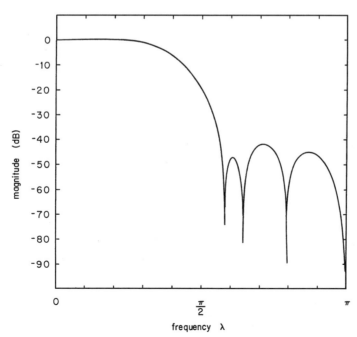

Figure 6.23 Magnitude response for 21-tap lowpass filter with coefficients quantized to 8 bits plus sign.

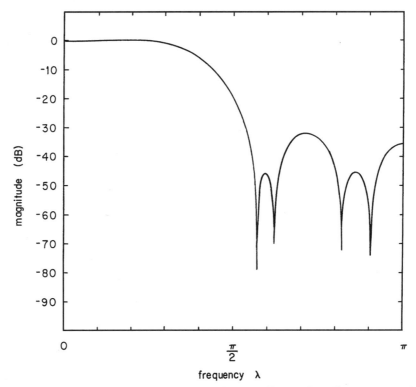

Figure 6.24 Magnitude response for 21-tap lowpass filter with coefficients quantized to 6 bits plus sign.

mathematical operations, must be represented with finite precision. As shown in the previous section, the effects of coefficient quantization are straightforward and easy to characterize. The effects of signal quantization are somewhat different.

Typically, an *analog-to-digital converter* (ADC) is used to sample and quantize an analog signal that can be thought of as a continuous amplitude function of continuous time. The ADC can be viewed as a sampler and quantizer in cascade. The transfer characteristic of a typical quantizer is shown in Fig. 6.25. This particular quantizer *rounds* the analog value to the nearest "legal" quantized value. The resulting sequence of quantized signal values $y[n]$ can be viewed as the sampled continuous-time signal $x[n]$ plus an error sequence $e[n]$ whose values are equal to the errors introduced by the quantizer:

$$y[n] = x[n] + e[n]$$

A typical discrete-time signal along with the corresponding quantized sequence and error sequence are shown in Fig. 6.26. Because the quantizer rounds to the nearest quantizer level, the magnitude of the error will never exceed $Q/2$, where Q is the increment between two consecutive legal quantizer output levels,

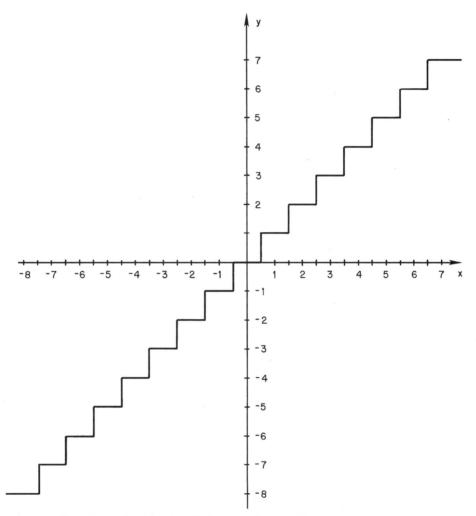

Figure 6.25 Typical transfer characteristic for a rounding quantizer.

that is,

$$\frac{-Q}{2} \leq e(t) \leq \frac{Q}{2} \quad \text{for all } t$$

The error usually is assumed to be uniformly distributed between $-Q/2$ and $Q/2$ and, consequently, to have a mean and variance of 0 and $Q^2/12$, respectively. For most practical applications, this assumption is reasonable. The quantization interval Q can be related to the number of bits in the digital word. Assume a word length of $L+1$ bits with 1 bit used for the sign and L bits for the magnitude. For the fixed-point format of Fig. 6.20, the relationship between Q and L is then given by $Q = 2^{-L}$.

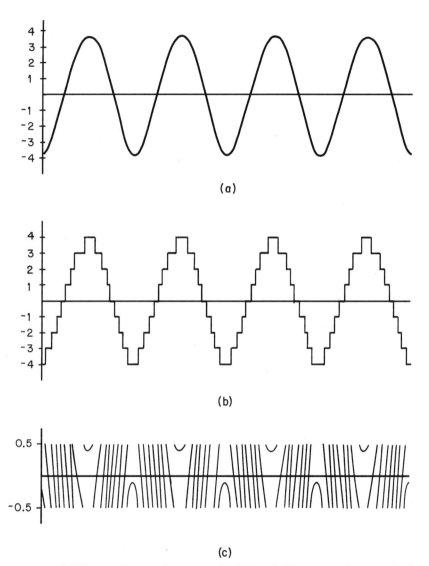

Figure 6.26 (a) Discrete-time continuous-amplitude signal, (b) corresponding quantized signal, and (c) error sequence.

It is often useful to characterize the quantization noise by means of a *signal-to-noise ratio* (SNR). In order to accomplish this characterization, the following additional assumptions are usually made:

1. The error sequence is assumed to be a sample sequence of a stationary random process; that is, the statistical properties of the error sequence do not change over time.

2. The error is a white noise process; or, equivalently, the error signal is uncorrelated.
3. The error sequence $e[n]$ is uncorrelated with the sequence of unquantized samples $x[n]$.

Based on these assumptions, the power of the quantization noise is equal to the error variance that was given previously as

$$\sigma_e^2 = \frac{Q^2}{12} = \frac{2^{-2L}}{12}$$

If we let σ_x^2 denote the signal power, then the SNR is given by

$$\frac{\sigma_x^2}{\sigma_e^2} = \frac{\sigma_x^2}{2^{-2L}/12} = (12 \cdot 2^{2L})\sigma_x^2$$

Expressed in decibels, this SNR is

$$10 \log\left(\frac{\sigma_x^2}{\sigma_e^2}\right) = 10 \log 12 + 20L \log 2 + 10 \log \sigma_x^2$$

$$= 10.792 + 6.021L + 10 \log \sigma_x^2 \qquad (6.14)$$

The major insight to be gained from Eq. (6.14) is that the SNR improves by 6.02 dB for each bit added to the digital word format. We are not yet in a position to compute an SNR using Eq. (6.14), because the term σ_x^2 needs some further examination. How do we go about obtaining a value for σ_x^2? Whatever the value of σ_x^2 may be originally, we must realize that in practical systems, the input signal is subjected to some amplification prior to digitization. For a constant amplifier gain of A, the unquantized signal becomes $Ax[n]$, the signal power becomes $A^2\sigma_x^2$, and the corresponding SNR is given by

$$\text{SNR} = 10 \log\left(\frac{A^2\sigma_x^2}{\sigma_e^2}\right) = 10.792 + 6.021L + 10 \log(A^2\sigma_x^2) \qquad (6.15)$$

A general rule of thumb often used in practical DSP applications is to set A so that $A\sigma$ is equal to 25 percent of the ADC full-scale value. Since we have been treating full scale as being normalized to unity, this indicates a value of A such that

$$A\sigma_x = 0.25 \quad \text{or} \quad A = \frac{1}{4\sigma_x}$$

Substituting this value of A into Eq. (6.15) yields

$$\text{SNR} = 10.79 + 6.02L + 10\log\frac{1}{16}$$
$$= 6.02L - 1.249 \text{ dB}$$

Using a value of $A = 1/(4\sigma_x)$ means that the ADC will introduce clipping any time the unquantized input signal exceeds $4\sigma_x$. Increasing A improves the SNR but decreases the *dynamic range*, that is, the range of signal values that can be accomodated without clipping. Thus, for a fixed word length, we can improve the SNR at the expense of degraded dynamic range. Conversely, by decreasing A, we can improve dynamic range at the expense of degraded SNR. The only way to simultaneously improve both dynamic range and quantization SNR is to increase the number of bits in the digital word.

References

1. Cadzow, J. A. *Discrete-Time Systems*, Prentice-Hall, Englewood Cliffs, NJ, 1973.
2. Oppenheim, A. V. and R. W. Schafer. *Digital Signal Processing*, Parentice-Hall, Englewood Cliffs, NJ, 1975.
3. Rabiner, L. R. and B. Gold. *Theory and Application of Digital Signal Processing*, Prentice-Hall, Englewood Cliffs, NJ, 1975.
4. Roberts, A. A. and C. T. Mullis. *Digital Signal Processing*, Addison-Wesley, Reading, MA, 1987.
5. Oppenheim, A. V. and R. W. Schafer. *Discrete-Time Signal Processing*, Prentice-Hall, Englewood Cliffs, NJ, 1989.
6. Peled, A. and B. Liu. *Digital Signal Processing*, Wiley, New York, 1976.

Chapter 7

Transform Analysis of Discrete-Time Systems

The *two-sided*, or *bilateral*, z transform of a discrete-time sequence $x[n]$ is defined by

$$X(z) = \sum_{n=-\infty}^{\infty} x[n]z^{-n} \qquad (7.1)$$

and the *one-sided*, or *unilateral*, z transform is defined by

$$X(z) = \sum_{n=0}^{\infty} x[n]z^{-n} \qquad (7.2)$$

Some authors (for example, [1]) use the unqualified term *z transform* to refer to Eq. (7.1), while others (for example, [2]) use the two-sided transform, and explicitly identify the one-sided transform. For causal sequences (that is, $x[n] = 0$ for $n < 0$) the one-sided and two-sided transforms are equivalent.

7.1 Region of Convergence

For some values of z, the series in Eq. (7.1) does not converge to a finite value. The portion of the z plane for which the series does converge is called the *region of convergence* (ROC). Whether Eq. (7.1) converges depends on the magnitude of z rather than a specific complex value of z. In other words, for a given sequence $x[n]$, if the series in Eq. (7.1) converges for a value of $z = z_1$, then the series will converge for all values of z for which $|z| = |z_1|$. Conversely, if the series diverges for $z = z_2$, then the series will diverge for all values of z for which $|z| = |z_2|$. Because convergence depends on the magnitude of z, the region of convergence will always be *bounded* by circles centered at the origin of the z plane. This is not to say that the region of convergence will always be a circle—it can be the interior of a circle, the exterior of a circle, an annulus,

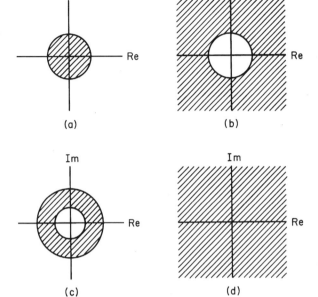

Figure 7.1 Possible configuration of the region of convergence for the z transform: (a) interior of a circle, (b) exterior of a circle, (c) an annulus, and (d) the entire plane.

or the entire z plane as shown in Fig. 7.1. Each of these four cases can be loosely viewed as an annulus—a circle's interior being an annulus with an inner radius of zero and a finite outer radius, a circle's exterior being an annulus with nonzero inner radius and infinite outer radius, and the entire z plane being an annulus with an inner radius of zero and an infinite outer radius. In some cases the ROC has an inner radius of zero, but the origin itself is not part of the region.

By definition, the ROC cannot contain any poles, since the series becomes infinite at the poles. The ROC for a z transform will always be a single contiguous region in the z plane. If we assume that the sequence $x[n]$ has a finite magnitude for all finite values of n, the nature of the ROC can be related to the nature of the sequence in several ways, as discussed in the paragraphs that follow and as summarized in Table 7.1.

Finite-duration sequences

If $x[n]$ is nonzero over only a finite range of n, then the z transform can be rewritten as

$$X(z) = \sum_{n=N_1}^{N_2} x[n]z^{-n}$$

This series will converge provided that $|x[n]| < \infty$ for $N_1 \leq n \leq N_2$ and $|z^{-n}| < \infty$

TABLE 7.1 Properties of the Region of Convergence for the z Transform

$x[n]$	ROC for $X(z)$
All	Includes no poles
All	Single contiguous region
Single sample at $n = 0$	Entire z plane
Finite-duration, causal, $x[n] = 0$ for all $n < 0$, $x[n] \neq 0$ for some $n > 0$	z plane except for $z = 0$
Finite-duration, causal, $x[n] \neq 0$ for some $n < 0$, $x[n] = 0$ for all $n > 0$	z plane except for $z = \infty$
Finite-duration, causal, $x[n] \neq 0$ for some $n < 0$, $x[n] \neq 0$ for some $n > 0$	z plane except for $z = 0$ and $z = \infty$
Right-sided, $x[n] = 0$ for all $n < 0$	Outward from outermost pole
Right-sided, $x[n] \neq 0$ for some $n < 0$	Outward from outermost pole, $z = \infty$ is excluded
Left-sided, $x[n] = 0$ for all $n > 0$	Inward from innermost pole
Left-sided, $x[n] \neq 0$ for some $n > 0$	Inward from innermost pole, $z = 0$ is excluded
Two-sided	Annulus

for $N_1 \leq n \leq N_2$. For negative values of n, $|z^{-n}|$ will be infinite for $z = \infty$; and for positive values of n, $|z^{-n}|$ will be infinite for $z = 0$. Therefore, a sequence having nonzero values only for $n = N_1$ through $n = N_2$ will have a z transform that converges everywhere in the z plane except for $z = \infty$ when $N_1 < 0$ and $z = 0$ when $N_2 > 0$. Note that a single sample at $n = 0$ is the only finite-duration sequence defined over the entire z plane.

Infinite-duration sequences

The sequence $x[n]$ is a *right-sided sequence* if $x[n]$ is zero for all n less than some finite value N_1. It can be shown (see [3] and [4]) that the z transform $X(z)$ of a right-sided sequence will have an ROC that extends outward from the outermost finite pole of $X(z)$. In other words, the ROC will be the area outside a circle whose radius equals the magnitude of the pole of $X(z)$ having the largest magnitude (see Fig. 7.2). If $N_1 < 0$, this ROC will not include $z = \infty$.

The sequence $x[n]$ is a *left-sided sequence* if $x[n]$ is zero for all n greater than some finite value N_2. The z transform $X(z)$ of a left-sided sequence will have an ROC that extends inward from the innermost pole of $X(z)$. The ROC will be the interior of a circle whose radius equals the magnitude of the pole of $X(z)$ having the smallest magnitude (see Fig. 7.3). If $N_2 > 0$, this ROC will not include $z = 0$.

The sequence $x[n]$ is a *two-sided sequence* if $x[n]$ has a nonzero values extending to both $-\infty$ and $+\infty$. The ROC for the z transform of a two-sided sequence will be an annulus.

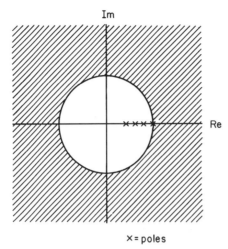

Figure 7.2 Region of convergence for the z transform of a right-sided sequence.

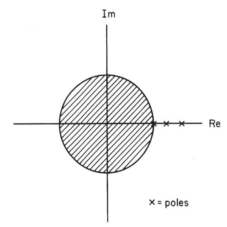

Figure 7.3 Region of convergence for the z transform of a left-sided sequence.

Convergence of the unilateral z transform

Note that all of the properties discussed in the preceding are for the two-sided z transform defined by Eq. (7.1). Since the one-sided z transform is equivalent to the two-sided transform when $x[n] = 0$ for $n < 0$, the ROC for a one-sided transform will always look like the ROC for the two-sided transform of either a causal finite-duration sequence or a causal right-sided sequence. For all causal systems, the ROC for the bilateral transform always consists of the area outside a circle of radius $R \geq 0$. Therefore, for two-sided transforms of causal sequences and for all one-sided transforms, the ROC can be (and frequently is) specified in terms of a *radius of convergence* R such that the transform converges for $|z| > R$.

7.2 Relationship Between the Laplace and z Transforms

The z transform can be related to both the Laplace and Fourier transforms. A sequence can be obtained by sampling a function of continuous time. Specifically, for a causal sequence

$$x[n] = \sum_{n=0}^{\infty} x_a(nT)\delta(t - nT) \qquad (7.3)$$

the Laplace transform is given by

$$X(s) = \sum_{n=0}^{\infty} x_a(nT) e^{-nT_s} \qquad (7.4)$$

Let $X_a(s)$ denote the Laplace transform of $x_a(t)$. The pole-zero pattern for $X(s)$ consists of the pole-zero pattern for $X_a(s)$ replicated at intervals of $\omega_s = 2\pi/T$ along the $j\omega$ axis in the s plane. If we modify Eq. (7.4) by substituting

$$z = e^{sT} \qquad (7.5)$$

$$x[n] = x_a(nT) \qquad (7.6)$$

we obtain the z transform defined by Eq. (7.1).

Relationships between features in the s plane and features in the z plane can be established using Eq. (7.5). Since $s = \sigma + j\omega$ with σ and ω real, we can expand Eq. (7.5) as

$$z = e^{sT} = e^{\sigma T} e^{j\omega T} = e^{\sigma T}(\cos \omega T + j \sin \omega T)$$

Because $|e^{j\omega T}| = (\cos^2 \omega T + \sin^2 \omega T)^{1/2} = 1$, and $T > 0$, we can conclude that $|z| < 1$ for $\sigma < 0$. Or, in other words, the left half of the s plane maps into the interior of the unit circle in the z plane. Likewise, $|z| = 1$ for $\sigma = 0$, so the $j\omega$ axis of the s plane maps onto the unit circle in the z plane. The "extra" replicated copies of the pole-zero pattern for $X(s)$ will all map into a single pole-zero pattern in the z plane. When evaluated around the unit circle (that is, $z = e^{j\lambda}$), the z transform yields the discrete-time Fourier transform (DTFT) (see Sec. 6.2).

7.3 System Functions

Given the relationships between the Laplace transform and the z transform that were noted in the previous section, we might suspect that the z transform of a discrete-time system's unit sample response (that is, digital impulse response) plays a major role in the analysis of the system in much the same way that the Laplace transform of a continuous-time system's impulse response yields the system's transfer function. This suspicion is indeed correct. The z transform of a discrete-time system's unit sample response is called the *system function*, or *transfer function*, of the system and is denoted by $H(z)$.

The system function can also be derived from the linear difference equation that describes the filter. As discussed in Sec. 6.3, a DTLTI system can be described by a linear difference equation of the form

$$y[n] + a_1 y[n-1] + a_2 y[n-2] + \cdots + a_k y[n-k]$$
$$= b_0 x[n] + b_1 x[n-1] + b_2 x[n-2] + \cdots + b_k x[n-k] \quad (7.7)$$

If we take the z transform of each term in this equation, we obtain

$$Y(z) + a_1 z^{-1} Y(z) + a_2 z^{-2} Y(z) + \cdots + a_k z^{-k} Y(z)$$
$$= b_0 X(z) + b_1 z^{-1} X(z) + b_2 z^{-2} X(z) + \cdots + b_m z^{-m} X(z) \quad (7.8)$$

Factoring out $Y(z)$ and $X(z)$ and then solving for $H(z) = Y(z)/X(z)$ yields

$$H(z) = \frac{Y(z)}{X(z)} = \frac{b_0 + b_1 z^{-1} + b_2 z^{-2} + \cdots + b_m z^{-m}}{1 + a_1 z^{-1} + a_2 z^{-2} + \cdots + a_k z^{-k}}$$

Both the numerator and denominator of $H(z)$ can be factored to yield

$$H(z) = \frac{b_0 (z - q_1)(z - q_2) \cdots (z - q_m)}{(z - p_1)(z - p_2) \cdots (z - p_k)}$$

The poles of $H(z)$ are $p_1, p_2, \ldots, p_k$, and the zeros are $q_1, q_2, \ldots, q_m$.

7.4 Common z-Transform Pairs and Properties

The use of the unilateral z transform by some authors and the use of the bilateral transform by others do not present as many problems as we might expect, because in DSP, most of the sequences of interest are causal sequences, or sequences that can easily be made causal. As noted previously, for causal sequences the one-sided and two-sided transforms are equivalent. It comes down to a matter of being careful about definitions. An author using the unilateral default (that is, "z transform" means "unilateral z transform") might say that the z transform of $x[n] = a^n$ is given by

$$X(z) = \frac{z}{z - a} \quad \text{for } |z| > |a| \quad (7.9)$$

On the other hand, an author using the bilateral default might say that

TABLE 7.2 Common Bilateral z-Transform Pairs

$x[n]$	$X(z)$	ROC
$\delta[n]$	1	All z
$\delta[n-m], m > 0$	z^{-m}	$z \neq 0$
$\delta[n-m], m < 0$	z^{-m}	$z \neq \infty$
$u[n]$	$\dfrac{z}{z-1}$	$|z| > 1$
$-u[-n-1]$	$\dfrac{z}{z-1}$	$|z| < 1$
$-a^n u[-n-1]$	$\dfrac{z}{z-a}$	$|z| < |a|$
$-na^n u[-n-1]$	$\dfrac{az}{(z-a)^2}$	$|z| < |a|$

Eq. (7.9) represents the z transform of $x[n] = a^n u[n]$ where $u[n]$ is the unit step sequence. Neither author is concerned with the values of a^n for $n < 0$: the first author is eliminating these values by the way the transform is defined, and the second author is eliminating these values by multiplying them with a unit step sequence that is zero for $n < 0$. There are a few useful bilateral transform pairs that consider values of $x[n]$ for $n < 0$. These pairs are listed in Table 7.2. However, the majority of the commonly used z-transform pairs involve values of $x[n]$ only for $n \geq 0$. These pairs are most conveniently tabulated as unilateral transforms with the understanding that any unilateral transform pair can be converted into a bilateral transform pair by replacing $x[n]$ with $x[n]u[n]$. Some common unilateral z-transform pairs are listed in Table 7.3. Some useful properties exhibited by both the unilateral and bilateral z transforms are listed in Table 7.4.

7.5 Inverse z Transform

The inverse z transform is given by the contour integral

$$x[n] = \frac{1}{2\pi j} \oint_C X(z) z^{n-1} \, dz \qquad (7.10)$$

where the integral notation indicates a counterclockwise closed contour that encircles the origin of the z plane and that lies within the region of convergence for $X(z)$. If $X(z)$ is rational, the residue theorem can be used to evaluate Eq. (7.10). However, direct evaluation of the inversion integral is rarely performed in actual practice. In practical situations, inversion of the z transform is usually performed indirectly, using established transform pairs and transform properties.

TABLE 7.3 Common Unilateral z-transform Pairs

$x[n]$	$X(z)$	R	
1	$\dfrac{z}{z-1}$	1	
$u_1[n]$	$\dfrac{z}{z-1}$	1	
$\delta[n]$	1	0	($z=0$ included)
nT	$\dfrac{Tz}{(z-1)^2}$	1	
$(nT)^2$	$\dfrac{T^2 z(z+1)}{(z-1)^3}$	1	
$(nT)^3$	$\dfrac{T^3 z(z^2+4z+1)}{(z-1)^4}$	1	
a^n	$\dfrac{z}{z-a}$	$|a|$	
$(n+1)a^n$	$\dfrac{z^2}{(z-a)^2}$	$|a|$	
$\dfrac{(n+1)(n+2)}{2!}a^n$	$\dfrac{z^3}{(z-a)^3}$	$|a|$	
$\dfrac{(n+1)(n+2)(n+3)}{3!}a^n$	$\dfrac{z^4}{(z-a)^4}$	$|a|$	
$\dfrac{(n+1)(n+2)(n+3)(n+4)}{4!}a^n$	$\dfrac{z^5}{(z-a)^5}$	$|a|$	
na^n	$\dfrac{az}{(z-a)^2}$	$|a|$	
$n^2 a^n$	$\dfrac{az(z+a)}{(z-a)^3}$	$|a|$	
$n^3 a^n$	$\dfrac{az(z^2+4az+a^2)}{(z-a)^4}$	$|a|$	
$\dfrac{a^n}{n!}$	$e^{a/z}$	0	
e^{-anT}	$\dfrac{z}{z-e^{-aT}}$	$|e^{-aT}|$	
$a^a \sin n\omega T$	$\dfrac{az \sin \omega T}{z^2 - 2az \cos \omega T + a^2}$	$|a|$	
$a^a \cos n\omega T$	$\dfrac{z^2 - za \cos \omega T}{z^2 - 2az \cos \omega T + a^2}$	$|a|$	
$e^{-anT} \sin \omega_0 nT$	$\dfrac{ze^{-aT} \sin \omega_0 T}{z^2 - 2ze^{-aT} \cos \omega_0 T + e^{-2aT}}$	$|e^{-aT}|$	
$e^{-anT} \cos \omega_0 nT$	$\dfrac{z^2 - ze^{-aT} \cos \omega_0 T}{z^2 - 2ze^{-aT} \cos \omega_0 T + e^{-2aT}}$	$|e^{-aT}|$	

Transform Analysis of Discrete-Time Systems

TABLE 7.4 Properties of the z Transform

Property no.	Time function	Transform
	$x[n]$	$X(z)$
	$y[n]$	$Y(z)$
1	$ax[n]$	$aX(z)$
2	$x[n] + y[n]$	$X(z) + Y(z)$
3	e^{-anT}	$X(e^{at}z)$
4	$a^n x[n]$	$X\left(\dfrac{z}{a}\right)$
5	$x[n-m]$	$z^{-m}X(z)$
6	$x[n] \otimes y[n]$	$X(z)Y(z)$
7	$nx[n]$	$-z\dfrac{d}{dz}X(z)$
8	$x[-n]$	$X(z^{-1})$
9	$x^*[n]$	$X^*(z^*)$

Inverse z transform via partial fraction expansion

Consider a system function of the general form given by

$$H(z) = \frac{b_0 z^m + b_1 z^{m-1} + \cdots + b_{m-1} z^1 + b_m}{z^k + a_1 z^{k-1} + \cdots + a_{k-1} z^1 + a_k}$$

Such a system function can be expanded into a sum of simpler terms that can be more easily inverse-transformed. Linearity of the z transform allows us then to sum the simpler inverse transforms to obtain the inverse of the original system function. The method for generating the expansion differs slightly depending upon whether the system function's poles are all distinct or some are multiple poles. Since most practical filter designs involve system functions with distinct poles, the more complicated multiple-pole procedure is not presented. For a discussion of the multiple-pole case, see [2].

Algorithm 7.1 Partial fraction expansion for $H(z)$ having simple poles.

1. Factor the denominator of $H(z)$ to produce

$$H(z) = \frac{b_0 z^m + b_1 z^{m-1} + \cdots + b_{m-1} z^1 + b_m}{(z - p_1)(z - p_2)(z - p_3) \cdots (z - p_k)}$$

2. Compute c_0 as given by

$$c_0 = H(z)|_{z=0} = \frac{b_m}{(-p_1)(-p_2)(-p_3) \cdots (-p_k)}$$

3. Compute c_i for $1 \leq i \leq k$ using

$$c_i = \left. \frac{z - p_i}{z} H(z) \right|_{z=p_i}$$

4. Formulate the discrete-time function $h[n]$ as given by

$$h[n] = c_0 \delta(n) + c_1 (p_1)^n + c_2 (p_2)^n + \cdots + c_k (p_k)^n$$

The function $h[n]$ is the inverse z transform of $H(z)$.

Example 7.1 Use the partial fraction expansion to determine the inverse z transform of

$$H(z) = \frac{z^2}{z^2 + z - 2}$$

solution

1. Factor the denominator of $H(z)$ to produce

$$H(z) = \frac{z^2}{(z-1)(z+2)}$$

2. Compute c_0 as

$$c_0 = H(z)|_{z=0} = 0$$

3. Compute c_1, c_2 as

$$c_1 = \left[\frac{(z-1)}{z} \frac{z^2}{(z-1)(z+2)} \right]\bigg|_{z=1} = \left. \frac{z^2}{z^2 + 2z} \right|_{z=1} = \frac{1}{3}$$

$$c_2 = \left[\frac{(z+2)}{z} \frac{z^2}{(z-1)(z+2)} \right]\bigg|_{z=-2} = \left. \frac{z^2}{z^2 - z} \right|_{z=-2} = \frac{2}{3}$$

4. The inverse transform $h[n]$ is given by

$$h[n] = \frac{1}{3}(1)^n + \frac{2}{3}(-2)^n$$

$$= \frac{1 + 2(-2)^n}{3} \quad n = 0, 1, 2, \ldots$$

References

1. Rabiner, L. R. and B. Gold. *Theory and Application of Digital Signal Processing*, Prentice-Hall, Englewood Cliffs, NJ, 1975.
2. Cadzow, J. A. *Discrete-Time Systems*, Prentice-Hall, Englewood Cliffs, NJ, 1973.
3. Oppenheim, A. V. and R. W. Schafer. *Digital Signal Processing*, Prentice-Hall, Englewood Cliffs, NJ, 1975.
4. Oppenheim, A. V. and R. W. Schafer. *Discrete-Time Signal Processing*, Prentice-Hall, Englewood Cliffs, NJ, 1989.

Chapter 8

Discrete Fourier Transform (DFT)

The *Fourier series* (FS), introduced in Chap. 2, links the continuous-time domain to the discrete-frequency domain; and the *Fourier transform* (FT) links the continuous-time domain to the continuous-frequency domain. The *discrete-time Fourier transform* (DTFT), introduced in Sec. 6.2, links the discrete-time domain to the continuous-frequency domain. In this chapter, we examine the *discrete Fourier transform* (DFT) which links the discrete-time and discrete-frequency domains.

8.1 Discrete Fourier Transform

The DFT and its inverse are given by

$$X[m] = \sum_{n=0}^{N-1} x[n]\, e^{-j2\pi mnFT} \quad m = 0, 1, \ldots, N-1 \tag{8.1a}$$

$$= \sum_{n=0}^{N-1} x[n] \cos(2\pi mnFT) - j \sum_{n=0}^{N-1} x[n] \sin(2\pi mnFT) \tag{8.1b}$$

$$x[n] = \sum_{m=0}^{N-1} X[m]\, e^{j2\pi mnFT} \quad n = 0, 1, \ldots, N-1 \tag{8.2a}$$

$$= \sum_{m=0}^{N-1} X[m] \cos(2\pi mnFT) + j \sum_{m=0}^{N-1} X[m] \sin(2\pi mnFT) \tag{8.2b}$$

It is a common practice in the DSP literature to "bury the details" of Eqs. (8.1) and (8.2) by defining $W_N = e^{j2\pi/N} = e^{j2\pi FT}$ and rewriting Eqs. (8.1a) and (8.2a) as

$$X[m] = \sum_{n=0}^{N-1} x[n]\, W_N^{-mn} \qquad (8.3)$$

$$x[n] = \sum_{m=0}^{N-1} X[m]\, W_N^{mn} \qquad (8.4)$$

Because the exponents in Eqs. (8.3) and (8.4) differ only in sign, another common practice in writing DFT software is to write only a single routine that can evaluate either equation depending on the value of an input flag being equal to +1 or −1. In the early days of DSP, when memory and disk space were expensive, this was a problem; now, with memory resources having become relatively inexpensive writing software using two separate routines usually will pay for itself in terms of clarity, execution speed, and simplified calling sequences.

Parameter selection

In designing a DFT for a particular application, values must be chosen for the parameters N, T, and F. N is the number of time sequence values $x[n]$ over which the DFT summation is performed to compute each frequency-sequence value. It is also the total number of frequency-sequence values $X[m]$ produced by the DFT. For convenience, the complete set of N consecutive time-sequence values is called the *input record*, and the complete set of N consecutive frequency-sequence values is called the *output record*. T is the time interval between two consecutive samples of the time sequence, and F is the frequency interval between two consecutive samples of the frequency sequence. The selection of values for N, F, and T is subject to the following constraints, which are consequences of the sampling theorem and the inherent properties of the DFT:

1. The inherent properties of the DFT require that $FNT = 1$.
2. The sampling theorem requires that $T \leq 1/(2 f_H)$, where f_H is the highest significant frequency component in the continuous-time signal.
3. The record length in time is equal to NT or $1/F$.
4. Many fast DFT algorithms (see Chap. 9) require that N be an integer power of 2.

Example 8.1 Choose values of N, F, and T given that F must be 5 Hz or less, N must be an integer power of 2, and the bandwidth of the input signal is 300 Hz. For the values chosen, determine the longest signal that can fit into a single input record.

solution From constraint number 2, $T \leq 1/(2 f_H)$. Since $f_H = 300$ Hz, $T \leq 1.66$ ms. If we select $F = 5$ and $T = 0.0016$, then $N \geq 125$. Because N must be an integer power of 2, we choose $N = 128 = 2^7$, and F becomes 4.883 Hz. Using these values, the input record will span $NT = 204.8$ ms.

Example 8.2 Assuming that $N = 256$ and that F must be 5 Hz or less, determine the highest input-signal bandwidth that can be accommodated without aliasing.

solution Since $FNT = 1$, then $T \geq (FN)^{-1}$ or $T \geq 781.25$ μs. This sampling interval corresponds to a maximum f_H of 640 Hz.

Periodicity

A periodic function of time will have a discrete-frequency spectrum, and a discrete-time function will have a spectrum that is periodic. Since the DFT relates a discrete-time function to a corresponding discrete-frequency function, this implies that both the time function and frequency function are periodic as well as discrete. This means that some care must be exercised in selecting DFT parameters and in interpreting DFT results, but it does not mean that the DFT can be used only on periodic signals. Based on the DFT's inherent periodicity, it is a common practice to regard the points from $n = 1$ through $n = N/2$ as positive and the points from $n = N/2$ through $n = N - 1$ as negative. Since both the time and frequency sequences are periodic, the values at points $n = N/2$ through $n = -1$ are in fact equal to the values at point $n = -N/2$ through $n = -1$. Under this convention, it is convenient to redefine the concept of even and odd sequences: If $x[N - n] = x[n]$, then $x[n]$ is even symmetric; and if $x[N - n] = -x[n]$, then $x[n]$ is odd symmetric.

8.2 Properties of the DFT

The DFT exhibits a number of useful properties and operational relationships that are similar to the properties of the continuous Fourier transform.

Linearity

The DFT relating $x[n]$ and $X[m]$:

$$x[n] \underset{IDFT}{\overset{DFT}{\longleftrightarrow}} X[m]$$

is homogeneous

$$ax[n] \underset{IDFT}{\overset{DFT}{\longleftrightarrow}} aX[m]$$

additive

$$x[n] + y[n] \underset{IDFT}{\overset{DFT}{\longleftrightarrow}} X[m] + Y[m]$$

and therefore linear

$$ax[n] + by[n] \underset{IDFT}{\overset{DFT}{\longleftrightarrow}} aX[m] + bY[m]$$

Time shifting

A time sequence $x[n]$ can be shifted in time by subtracting an integer from n. Shifting the time sequence will cause the corresponding frequency sequence to be phase-shifted. Specifically, given

$$x[n] \underset{IDFT}{\overset{DFT}{\longleftrightarrow}} X[m]$$

then

$$x[n-k] \underset{IDFT}{\overset{DFT}{\longleftrightarrow}} X[m]e^{-j2\pi mk/N}$$

Frequency shifting

Time sequence modulation is accomplished by multiplying the time sequence by an imaginary exponential term $e^{j2\pi mk/N}$. This will cause a frequency shift of the corresponding spectrum. Specifically, given

$$x[n] \underset{IDFT}{\overset{DFT}{\longleftrightarrow}} X[m]$$

then

$$x[n]e^{j2\pi mk/N} \underset{IDFT}{\overset{DFT}{\longleftrightarrow}} X[m-k]$$

Even and odd symmetry

Consider a time sequence $x[n]$ and the corresponding frequency sequence $X[m] = X_R[m] + jX_I[m]$, where $X_R[m]$ and $X_I[m]$ are real valued. If $x[n]$ is even, then $X[m]$ is real valued and even:

$$x[-n] = x[n] \Leftrightarrow X[m] = X_R[m] = X_R[-m]$$

If $x[n]$ is odd, then $X[m]$ is imaginary and odd:

$$x[-n] = -x[n] \Leftrightarrow X[m] = jX_I[m] = -jX_I[-m]$$

Real and imaginary properties

In general, the DFT of a real-valued time sequence will have an even real component and an odd imaginary component. Conversely, an imaginary-valued time sequence will have an odd real component and an even imaginary component.

Given a time sequence $x[n] = x_R[n] + x_I[n]$ and the corresponding frequency sequence $X[m] = X_R[m] + jX_I[m]$, then

$$x[n] = x_R[n] \Leftrightarrow X_R[m] = X_R[-m] \quad \text{and} \quad X_I[m] = -X_I[-m]$$

$$x[n] = jx_I[n] \Leftrightarrow X_R[m] = -X_R[-m] \quad \text{and} \quad X_I[m] = X_I[-m]$$

8.3 Applying the DFT

Short time-limited signals

Consider the time-limited continuous-time signal and its continuous spectrum shown in Figs. 8.1a and 8.1b. (Remember that a signal cannot be both strictly time limited and strictly band limited.) We can sample this signal to produce the time sequence shown in Fig. 8.1c for input to a DFT. If the input record length N is chosen to be longer than the length of the input time sequence, the entire sequence can fit within the input record as shown. As discussed in Sec. 8.2, the

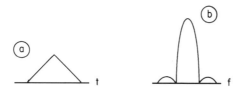

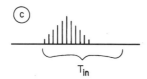

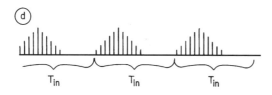

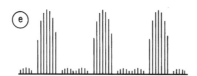

Figure 8.1 Signals and sequences for the DFT of a short time-limited signal: (a) the continuous signal, (b) its continuous spectrum, (c) sampled sequence for input to DFT, (d) periodic sequence the DFT will treat it as, and (e) the resulting periodic discrete-frequency spectrum.

DFT will treat the input sequence as though it is the periodic sequence shown in Fig. 8.1d. This will result in a periodic discrete-frequency spectrum as shown in Fig. 8.1e. The actual output produced by the DFT algorithm will be the sequence of values from $m = 0$ to $m = N - 1$. Of course, there will be some aliasing due to the time-limited nature (and consequently unlimited bandwidth) of the input-signal pulse.

Periodic signals

Consider the band-limited and periodic continuous-time signal and its spectrum shown in Fig. 8.2. We can sample this signal to produce the time sequence shown in Fig. 8.2c for input to the DFT. If the input record length N of the DFT is chosen to be exactly equal to the length of one period of this sequence, the periodic assumption implicit in the DFT will cause the DFT to treat the single input record as though it were the complete sequence. The corresponding periodic discrete-frequency spectrum is shown in Fig. 8.2d. The DFT output sequence will actually consist of just one period that matches exactly the spectrum of Fig. 8.2b. We could not hope for or find a more convenient situation. Unfortunately, this relationship exists only in an N-point DFT where the input signal is both band limited and periodic with a period of exactly N.

Long aperiodic signals

So far we have covered the use of the DFT under relatively favorable conditions that are not likely to exist in many important signal-processing applications.

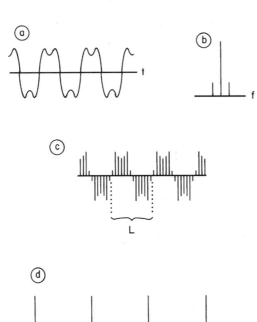

Figure 8.2 Signals and sequences for the DFT of a periodic signal. The length L of the DFT input record equals the period of the signal. (a) The band-limited signal. (b) The signal's spectrum. (c) The resulting time sequence for DFT input. (d) The corresponding discrete-frequency spectrum.

Discrete Fourier Transform (DFT)

Often the signal to be analyzed will be neither periodic nor reasonably time-limited. The corresponding sequence of digitized-signal values will be longer than the DFT input record and will therefore have to be truncated to just N samples before the DFT can be applied. The periodic nature of the DFT will cause the truncated sequence of Fig. 8.3b to be interpreted as though it were the sequence shown in Fig. 8.3c. Notice that in this sequence there is a large discontinuity in the signal at the points corresponding to the ends of the input record. This will introduce additional high-frequency components into the spectrum produced by the DFT. This phenomenon is called *leakage*. To reduce the leakage effects, a common practice is to multiply the truncated input sequence by a tapering window prior to application of the DFT. A good window shape will taper off at the ends of the input record but still have a reasonably compact and narrow spectrum. This is important, since multiplying the time sequence by the window will cause the corresponding frequency sequence to be convolved with the spectrum of the window. A narrow window spectrum will cause minimum smearing of the signal spectrum. Several popular windowing functions and their spectra are treated at length in Chap. 10.

Software notes

The function dft() shown in Listing 8.1 is the "brute-force" implementation of Eq. (8.1). This function is an example of grossly inefficient code. The sine and cosine functions are each evaluated N^2 times to compute an N-point DFT. Since

$$\exp\left(\frac{-2\pi jk}{N}\right) = \exp\left[\frac{-2\pi j(k \bmod N)}{N}\right]$$

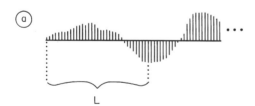

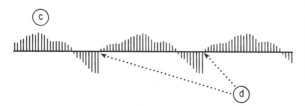

Figure 8.3 Discontinuities caused by truncating the input sequence of a DFT: (a) long input sequence, (b) truncated input sequence, (c) input sequence as interpreted by the DFT, and (d) resulting discontinuities.

it follows that there are only N different values of phi that need to be computed in dft(). We can trade space for speed by precomputing and storing the values of sin(phi) and cos(phi) for phi $= 2\pi k/N, k = 0, 1, \ldots, N-1$. The resulting modified function dft2() is shown in Listing 8.2.

Listing 8.1

```
const double TWO_PI = 6.283185308;

void dft( complex *x_in,
         complex *y_out,
         int num_samps)
{
 int n, m;
 int k;
 double phi;
 double sum_real, sum_imag;

 for( m =0; m<num_samps; m++)
   {
     sum_real = 0.0;
     sum_imag = 0.0;

     for( n =0; n<num_samps; n++)
       {
         phi = TWO_PI * m * n /num_samps;
         sum_real += (real(x_in[n]) * cos(phi) - imag(x_in[k]) * sin(phi));
         sum_real += (imag(x_in[n]) * cos(phi) + real(x_in[k]) * sin(phi));
       }
     y_out[m] = complex(sum_real, sum_imag);
   }
 return;
}
```

Listing 8.2

```
const double TWO_PI = 6.283185308;

void dft2( complex *x_in,
          complex *y_out,
          int num_sumps)
{
 int n, m;
 int k;
 double *cos_val, *sin_val;
 double w_factor;
 double sum_real, sum_imag;

 cos_val = new double[num_samps];
```

```
  sin_val = new double[num_samps];
  w_factor = TWO_PI/num_samps;

for( k=0; k<num_samps; k++)
  {
    cos_val[k] = (double) cos(w_factor * k);
    sin_val[k] = (double) sin(w_factor * k);
  }

for( m =0; m<num_samps; m++)
    {
      sum_real = 0.0;
      sum_imag = 0.0;

      for( n =0; n<num_samps; n++)
         {
           k = (m*n) % num_samps;
           sum_real += (real(x_in[n]) * cos_val[k] + imag(x_in[n]) * sin_val[k]);
           sum_imag += (imag(x_in[n]) * cos_val[k] - real(x_in[n]) * sin_val[k]);
         }
      y_out[m] = complex(sum_real, sum_imag);
    }
 delete [] cos_val;
 delete [] sin_val;
 return;
}
```

Chapter

9

Fast Fourier Transforms

For all but the smallest transforms, the direct computation of the DFT in the form presented in Chapter 8 is prohibitively expensive in terms of required computer operations. Fortunately, a number of "fast" transforms have been developed that are mathematically equivalent to the DFT, but which require significantly fewer computer operations for their implementation. This chapter will examine some of the more useful of these "fast" algorithms.

9.1 Computational Complexity of the DFT

Consider the basic form of the DFT given by

$$X[m] = \sum_{n=0}^{N-1} x[n]\, e^{-j2\pi mn/N}$$

It is readily apparent that computation of $X[m]$ for any single value of m will require (in general) N complex multiplications and N complex additions. Therefore, computing a complete set of N values for $X[m]$ will entail N^2 complex multiplications and N^2 complex additions. (This total includes some trivial multiplications by 1 as well as some nearly as trival multiplications by j.) Furthermore, values of $e^{-j2\pi mn/N}$ need to be computed for various combinations of m and n.

9.2 Decimation-in-Time Algorithms

Start with the "usual" DFT for an N-point sequence

$$X[m] = \sum_{n=0}^{N-1} x[n] W_N^{mn} \quad m = 0, 1, \ldots N-1$$

where
$$W_N = \exp\left(\frac{-2\pi j}{N}\right)$$

Then break the summation into two separate summations—one for the even-indexed samples of $x[n]$ and one for the odd-indexed samples of $x[n]$,

$$X[m] = \sum_{\substack{n=0 \\ n \text{ even}}}^{N-1} x[n] W_N^{nm} + \sum_{\substack{n=0 \\ n \text{ odd}}}^{N-1} x[n] W_N^{nm}$$

$$= \sum_{n=0}^{N/2-1} x[2n] W_N^{2nm} + \sum_{n=0}^{N/2-1} x[2n+1] W_N^{(2n+1)m}$$

$$= \sum_{n=0}^{N/2-1} x[2n] W_N^{2nm} + W_N^m \sum_{n=0}^{N/2-1} x[2n+1] W_N^{2nm} \qquad (9.1)$$

The factor W_N^{2mn} is equal to $W_{N/2}^{mn}$. Therefore we can represent Eq. (9.1) as

$$X[m] = \sum_{n=0}^{N/2-1} x[2n] W_{N/2}^{nm} + W_N^m \sum_{n=0}^{N/2-1} x[2n+1] W_{N/2}^{nm} \qquad (9.2)$$

Each of the summations in Eq. (9.2) has the form of an ($\frac{N}{2}$)-point DFT. Therefore, we can define x_{even} and x_{odd} as

$$x_{\text{even}}[n] = x[2n] \qquad n = 0, 1, \ldots N/2 - 1$$
$$x_{\text{odd}}[n] = x[2n+1] \qquad n = 0, 1, \ldots N/2 - 1$$

and rewrite Eq. (9.2) as

$$X[m] = X_{\text{even}}[m] + W_N^m X_{\text{odd}}[m]$$

where $X_{\text{even}}[m]$ and $X_{\text{odd}}[m]$ are the ($\frac{N}{2}$)-point DFTs of $x_{\text{even}}[n]$ and $x_{\text{odd}}[n]$ respectively. What all this means is that any N-point DFT (where N is even) can be broken into two ($\frac{N}{2}$)-point DFTs. In turn, if $N/2$ is even, each of these ($\frac{N}{2}$)-point DFTs can be broken into two ($\frac{N}{4}$)-point DFTs, allowing us to express the original N-point DFT in terms of four ($\frac{N}{4}$)-point DFTs. If N is an integer power of 2 (i.e., $N = 2^\nu$) the process of breaking each DFT into two smaller DFTs can be repeated until the original DFT can be computed as a combination of 2^ν 1-point DFTs. Computing the DFT in this indirect manner results in a significant computational savings.

Example 9.1 Let us do the complete breakdown for the case of an 8-point DFT originally defined by

$$X[m] = \sum_{n=0}^{7} x[n] W^{nm} \quad m = 0, 1, \ldots 7$$

(The subscript N has been omitted from W for convenience.) Splitting this into separate DFTs for even and odd n as in Eq. (9.1) yields

$$X[m] = \sum_{n=0}^{3} x[2n]W^{2mn} + W^m \sum_{n=0}^{3} x[2n+1]W^{2mn} \qquad (9.3)$$

If we let $A[m]$ denote the DFT of the even-indexed samples and $B[m]$ denote the DFT of the odd-indexed samples, then we can rewrite Eq. (9.3) as

$$X[m] = A[m] + W^m B[m] \qquad (9.4)$$

$X[m]$ is an 8-point frequency sequence, but $A[m]$ and $B[m]$ are only 4-point frequency sequences. What happens for $m = 4, 5, 6,$ and 7? As discussed in Chap. 9, an N-point DFT is periodic with a period of N samples, i.e., $A[4] = A[0]$, $B[4] = B[0]$, $A[5] = A[1]$, $B[5] = B[1]$, etc. Therefore

$$X[4] = A[0] + W^4 B[0]$$
$$X[5] = A[1] + W^5 B[1]$$
$$X[6] = A[2] + W^6 B[2]$$
$$X[7] = A[3] + W^7 B[3]$$

The operations represented by Eq. (9.4) are depicted in the signal flow graph (SFG) of Fig. 9.1. In an SFG there are nodes and edges. Each node represents a signal that is obtained by summing together all of the signals represented by the edges directed into the node. Each edge represents the multiplication of a weight times the signal that is represented by the edge's source node. An edge's weight is indicated by an annotation near the arrowhead used to indicate the edge's direction. An edge weight of one is assumed if no weight is indicated. For example, the node marked $X[5]$ in the figure has two incident edges. The upper incident edge, (coming from $A[1]$) has no weight indicated. The lower incident edge (coming from $B[1]$) has a weight of W^5. All of this is interpreted to mean that $X[5] = A[1] + W^5 B[1]$.

Each of the two 4-point DFTs in Eq. (9.4) can be broken into two 2-point DFTs as follows:

$$A[m] = \sum_{n=0}^{1} x[4n]W^{4nm} + \sum_{n=0}^{1} x[4n+2]W^{(4n+2)m}$$
$$= C[m] + W^{2m} D[m] \qquad (9.5)$$

$$W^m B[m] = \sum_{n=0}^{1} x[4n+1]W^{(4n+1)m} + \sum_{n=0}^{1} x[4n+3]W^{(4n+3)m}$$
$$= W^m (E[m] + W^{2m} F[m]) \qquad (9.6)$$

The operations represented by Eqs. (9.5) and (9.6) are depicted in the *signal flow graph* (SFG) of Fig. 9.2. Notice that in going from the second term of Eq. (9.4) to Eq. (9.6), the multiplier W^m has not been distributed over the terms inside the parentheses of (9.6). If W^m were distributed over the terms within the parentheses to yield

$$W^m B[m] = W^m E[m] + W^{3m} F[m]$$

166   Chapter Nine

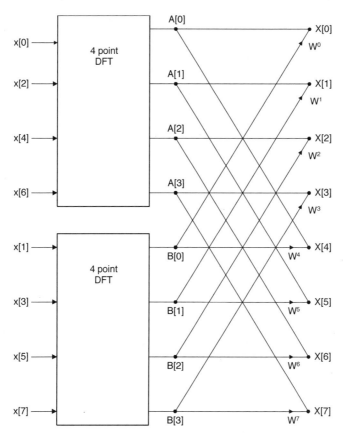

Figure 9.1 Signal flow graph illustrating how an 8-point DFT can be formed in terms of two 4-point DFTs.

it would not be possible to simply extend the SFG of Fig. 9.1 to obtain the SFG of Fig. 9.2 without having to change some of the existing edges between nodes for $B[m]$ and nodes for $X[m]$. Keeping the various multipliers factored is the key to maximizing the reuse of interim results and thereby minimizing the total computational burden.

Finally, each of the 2-point DFTs $C[m]$, $D[m]$, $E[m]$, and $F[m]$ can be broken into two single-point DFTs

$$C[m] = \sum_{n=0}^{0} x[8n]W^{8nm} + \sum_{n=0}^{0} x[8n+4]W^{(8n+4)m}$$

$$= x[0]W^0 + x[4]W^{4m} \tag{9.7}$$

$$W^{2m}D[m] = \sum_{n=0}^{0} x[8n+2]W^{(8n+2)m} + \sum_{n=0}^{0} x[8n+6]W^{(8n+6)m}$$

$$= x[2]W^{2m} + x[6]W^{6m}$$

$$= W^{2m}(x[2] + x[6]W^{4m}) \tag{9.8}$$

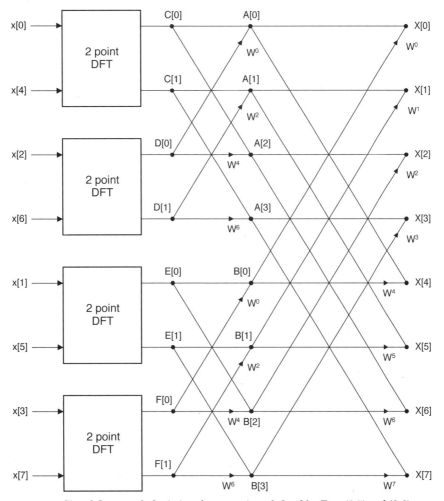

Figure 9.2 Signal flow graph depicting the operations defined by Eqs. (9.5) and (9.6).

$$W^m E[m] = \sum_{n=0}^{0} x[8n+1] W^{(8n+1)m} + \sum_{n=0}^{0} x[8n+5] W^{(8n+5)m}$$

$$= x[1] W^m + x[5] W^{5m}$$

$$= W^m (x[1] + x[5] W^{4m}) \qquad (9.9)$$

$$W^m W^{2m} F[m] = \sum_{n=0}^{0} x[8n+3] W^{(8n+3)m} + \sum_{n=0}^{0} x[8n+7] W^{(8n+7)m}$$

$$= x[3] W^{3m} + x[7] W^{7m}$$

$$= W^m W^{2m} (x[3] + x[7] W^{4m}) \qquad (9.10)$$

168 Chapter Nine

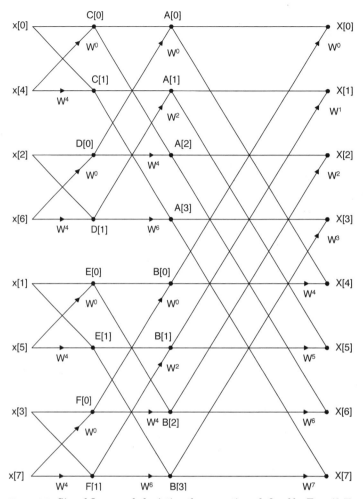

Figure 9.3 Signal flow graph depicting the operations defined by Eqs. (9.7) through (9.10).

The operations represented by Eqs. (9.7) through (9.10) are depicted in the SFG of Fig. 9.3.

How do we go about writing software to implement the 8-point FFT algorithm represented by the SFG of Fig. 9.3? More importantly, how do we extract the salient features of the algorithm so that we can devise software that will work for any value of N that is a power of 2?

Let us take a closer look at Fig. 9.3. The nodes in the leftmost column of the graph are samples of the input sequence. Moving down the column, these samples are not in the "natural" order $x_0, x_1, x_2, \ldots$ These inputs are in "bit-reversed" order—so named because of the way in which naturally-ordered indices are mapped into scrambled indices. For a 2^ν-point FFT, the natural index of each sample is represented as a ν-bit binary value. For any particular

TABLE 9.1 Bit-reversed Index Values for an 8-point FFT

Natural index	Binary value	Reversed binary	Scrambled index
0	000	000	0
1	001	100	4
2	010	010	2
3	011	110	6
4	100	001	1
5	101	101	5
6	110	011	3
7	111	111	7

natural index, the corresponding scrambled index is obtained simply by reversing the order of the v-bit representation. The mapping for an 8-point FFT is given in Table 9.1. File cbitrev.cpp contains a function ComplexBitReverse() that can be used for "in place" scrambling of a naturally-ordered complex-valued array. (As we will see shortly, this same function also can be used for in-place descrambling of an array that is already in bit-reverse scrambled order.)

The input nodes in the first column are in pairs, with each pair being used to compute exactly one pair of nodes in the second column. For example, x_0 and x_4 both are used to compute C_0 and C_1, and neither x_0 nor x_4 is used in the calculation of any other second column nodes. Once x_0 and x_4 have been used to calculate C_0 and C_1, they are not needed again. This means that the calculation can be done "in place," with the resulting C_0 and C_1 being written into the memory locations that originally held x_0 and x_4.

The set of calculations that produces a pair of column $k + 1$ nodes from a pair of column k nodes is sometimes called a *butterfly* because of the shape that the four nodes and the edges between them form in the SFG. We will refer to the butterflies between the first column of nodes and the second column of nodes as *first stage butterflies*.

In going from the second column of nodes to the third column of nodes, the second stage butterflies will each involves two pairs of nodes, but the pairing is different than for the first stage butterflies. For example, C_0 and C_1 were produced by the same first stage butterfly, but in the second stage they are split up, with C_0 being paired with D_0 to produce A_0 and A_2, while C_1 is paired with D_1 to produce A_1 and A_3. Despite the changes in pairing, the in-place relationship still holds—node A_0 is in the same row of the SFG as node C_0, so A_0 can be written into the memory location that held C_0. Likewise, A_2 can be written into the memory location that held D_0.

In the third stage of butterflies, the pair (A_0, A_2) is split up with A_0 being paired with B_0 to produce X_0 and X_4, while A_2 is paired with B_2 to produce X_2 and X_6.

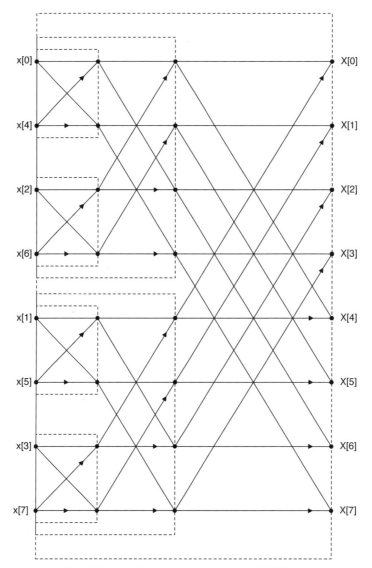

Figure 9.4 Signal flow graph redrawn to show nested DFTs.

Let us redraw Fig. 9.3 to show the boundaries of the 2-point and 4-point DFTs that are nested within the overall 8-point DFT. As shown in Fig. 9.4, the third stage butterflies operate on outputs of 4-point DFTs to their left to produce outputs of the 8-point DFT to their right. The second stage butterflies operate on outputs of 2-point DFTs to their left to produce outputs of the 4-point DFTs to their right. The first stage butterflies operate on outputs of 1-point DFTs to their left to produce outputs of 2-point DFTs to their right. In general, for an N-point DFT with $N = 2^M$, the stage k butterflies will operate on outputs of 2^{k-1}-point DFTs to their left to produce outputs of the 2^k-point DFTs to their right. Notice that every butterfly straddles the gap between two adjacent lefthand DFTs, but none of the butterflies straddle the gap between two righthand DFTs.

The top edge of each butterfly in stage k begins at one of the 2^{k-1} outputs from the top DFT in the lefthand pair. The bottom edge of each butterfly begins at the corresponding output from the bottom DFT in this pair. (Because for stage k each lefthand DFT has 2^{k-1} outputs, the top and bottom edges of each butterfly will always be exactly 2^{k-1} locations apart in the array used to store the results of stage $k-1$.)

In Fig. 9.4, there are 4 second stage butterflies. Two of these butterflies operate on the C and D nodes to produce the A nodes, and the other two operate on the E and F nodes to produce the B nodes. Corresponding butterflies in these two sets involve the same multipliers on corresponding edges. For example, the upper edges into both A_1 and B_1 have a multiplier of unity, and the lower edges into both A_1 and B_1 have a multiplier of W^2. Because of this correspondence, it is more efficient to compute the butterfly beginning at C_0, then compute the butterfly beginning at E_0 before computing the butterfly beginning at C_1. Computing the corresponding butterflies across all groups before computing a different butterfly within the same group avoids the need for recomputing identical values of W^p for each of the $N/2^k$ DFTs to the right of the butterflies. Hence, in function `FftDitSino` (provided in file `dit_sino.cpp`), the inner loop over values of `top_node` is exhausted for a given value of `bfly_pos`, before `bfly_pos` is incremented in the middle loop.

Other variations of decimation-in-time FFT algorithms

The nodes of an SFG can be reordered, and so long as the connections between nodes are preserved, the algorithm represented by the reordered SFG is equivalent to the algorithm represented by the original SFG. Specifically, we can reorder the input nodes in the SFG of Fig. 9.3 to obtain the SFG of Fig. 9.5, which now has naturally-ordered inputs and scrambled outputs. A function for implementing this form of FFT is provided in file `dit_niso.cpp`. Notice that the sequence in which the powers of W are used has been changed. Therefore, a different looping strategy is needed to minimize repeated calculation of the same powers of W.

9.3 Decimation-in-Frequency Algorithms

Start with the "usual" DFT for an N-point sequence

$$X[m] = \sum_{n=0}^{N-1} x[n] W_N^{mn} \quad m = 0, 1, \ldots N-1 \qquad (9.11)$$

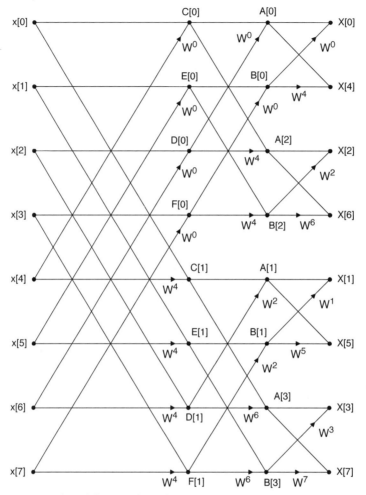

Figure 9.5 Signal flow graph for decimation-in-time FFT with naturally-ordered inputs and scrambled outputs.

where

$$W_N = \exp\left(\frac{-2\pi j}{N}\right)$$

Now specialize Eq. (9.11) into two different forms—one form tailored for computing the even-numbered frequency samples and one form tailored for computing the odd-numbered frequency samples. First let us consider the specialization for even-numbered samples

$$X[2r] = \sum_{n=0}^{N-1} x[n] W_N^{2nr} \quad r = 0, 1, \ldots (N/2) - 1$$

This summation can be split into two pieces

$$X[2r] = \sum_{n=0}^{(N/2)-1} x[n] W_N^{2nr} + \sum_{n=N/2}^{N-1} x[n] W_N^{2nr}$$

$$= \sum_{n=0}^{(N/2)-1} x[n] W_N^{2nr} + \sum_{n=0}^{(N/2)-1} x\left[n+\frac{N}{2}\right] W_N^{2r[n+(N/2)]}$$

The factor $W_N^{2r[n+(N/2)]}$ is equal to $W_{N/2}^{rn}$. Therefore

$$X[2r] = \sum_{n=0}^{(N/2)-1} x[n] W_N^{2nr} + \sum_{n=0}^{(N/2)-1} x\left[n+\frac{N}{2}\right] W_{N/2}^{nr}$$

$$= \sum_{n=0}^{(N/2)-1} \left(x[n] + x\left[n+\frac{N}{2}\right]\right) W_{N/2}^{nr} \quad r = 0, 1, \ldots (N/2) - 1 \quad (9.12)$$

Let us set this result aside for a moment and consider the specialization for odd-numbered frequency samples.

$$X[2r+1] = \sum_{n=0}^{N-1} x[n] W_N^{n(2r+1)} \quad r = 0, 1, \ldots (N/2) - 1 \quad (9.13)$$

Using manipulations similar to the even-sample case, Eq. (9.13) can be reduced to

$$X[2r+1] = \sum_{n=0}^{(N/2)-1} \left(x[n] - x\left[n+\frac{N}{2}\right]\right) W_N^n W_{N/2}^{nr} \quad (9.14)$$

We can define two new ($\frac{N}{2}$)-point time sequences $a[n]$ and $b[n]$

$$a[n] = x[n] + x\left[n+\frac{N}{2}\right] \quad n = 0, 1, \ldots (N/2) - 1$$

$$b[n] = x[n] - x\left[n+\frac{N}{2}\right] \quad n = 0, 1, \ldots (N/2) - 1$$

Then Eq. (9.12) can be seen as the ($\frac{N}{2}$)-point DFT of the sequence $a[n]$

$$X[2r] = \sum_{n=0}^{(N/2)-1} a[n] W_{N/2}^{nr} \quad r = 0, 1, \ldots (N/2) - 1$$

and Eq. (9.14) can be seen as the $(\frac{N}{2})$-point DFT of the sequence $W_N^n b[n]$

$$X[2r+1] = \sum_{n=0}^{(N/2)-1} b[n] W_N^n W_{N/2}^{nr} \quad r = 0, 1, \ldots (N/2) - 1$$

If N is a power of 2, this decomposition process can be continued until the original N-point DFT has been decomposed into N 1-point DFTs.

Example 9.2 Let us do the complete decimation-in-frequency decomposition for the case of an 8-point DFT originally defined by

$$X[m] = \sum_{n=0}^{7} x[n] W^{nm} \quad m = 0, 1, \ldots 7$$

Splitting this into separate DFTs for even and odd m as in Eqs. (9.12) and (9.14) yields

$$X[2r] = \sum_{n=0}^{3} (x[n] + x[n+4]) W_N^{2rn} \tag{9.15}$$

$$X[2r+1] = \sum_{n=0}^{3} (x[n] - x[n+4]) W_N^{(2r+1)n} \tag{9.16}$$

The operations represented by Eqs. (9.15) and (9.16) are depicted in the SFG of Fig. 9.6.

Each of the two 4-point DFTs can be decomposed into two 2-point DFTs as follows:

$$X[4r] = \sum_{n=0}^{1} (a[n] + a[n+2]) W_N^{4rn} \tag{9.17}$$

$$X[4r+2] = \sum_{n=0}^{1} (a[n] - a[n+2]) W_N^{(4r+2)n}$$

$$= \sum_{n=0}^{1} (a[n] - a[n+2]) W_N^{2n} W_N^{4rn} \tag{9.18}$$

$$X[4r+1] = \sum_{n=0}^{1} (b[n] + b[n+2]) W_N^{(4r+1)n}$$

$$= \sum_{n=0}^{1} (b[n] + b[n+2]) W_N^n W_N^{4rn} \tag{9.19}$$

Fast Forward Transforms

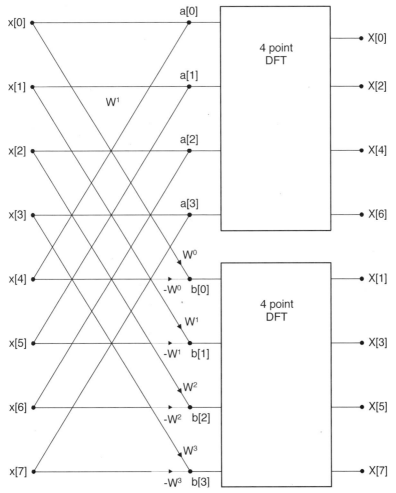

Figure 9.6 Signal flow graph depicting the operations defined by Eqs. (9.15) and (9.16).

$$X[4r+3] = \sum_{n=0}^{1}(b[n] - b[n+2])W_N^{(4r+3)n}$$

$$= \sum_{n=0}^{1}(b[n] - b[n+2])W_N^{3n}W_N^{4rn} \qquad (9.20)$$

Each of the four 2-point DFTs then can be decomposed into two 1-point DFTs. If we let $c[n] = a[n] + a[n+2]$ then Eq. (9.18) becomes

$$X[4r] = \sum_{n=0}^{1} c[n]W_N^{4rn} \quad r = 0, 1 \qquad (9.21)$$

The frequency sequence produced by Eq. (9.21) consists of just two samples, $X[0]$ and $X[4]$, but we can still perform an even-odd decomposition to obtain

$$X[8r] = \sum_{n=0}^{0}(c[n]+c[n+1])W_N^{8rn}$$

$$X[8r+4] = \sum_{n=0}^{0}(c[n]-c[n+1])W_N^{(8r+4)n}$$

Since each of these two equations holds only for $r = 0$, we can simplify them immediately to obtain

$$X[0] = c[0] + c[1]$$

$$X[4] = c[0] - c[1]$$

In a similar fashion we can let

$$d[n] = a[n] - a[n+2]$$

$$f[n] = b[n] + b[n+2]$$

$$g[n] = b[n] - b[n+2]$$

and decompose Eqs. (9.19), (9.19), and (9.20) to obtain

$$X[2] = d[0] + d[1]$$

$$X[6] = d[0] - d[1]$$

$$X[1] = f[0] + f[1]$$

$$X[5] = f[0] - f[1]$$

$$X[3] = g[0] + g[1]$$

$$X[7] = g[0] - g[1]$$

The SFG in Fig. 9.7 has been annotated to indicate which nodes correspond to the various intermediate working sequences $a[n]$, $b[n]$, etc.

As shown in Fig. 9.7, the result of this decimation-in-frequency decomposition is a NISO FFT algorithm with naturally ordered inputs and scrambled order outputs. A function for implementing this form of the FFT is provided in file dif_niso.cpp. Just as for the decimation-in-time case, it is possible to re-order the nodes of the SFG while preserving the connections between the nodes to produce a SINO form of the decimation-in-frequency FFT in which the inputs are in scrambled order and the outputs are in natural order. A function for implementing a DIF-SINO FFT is provided in file dif_sino.cpp.

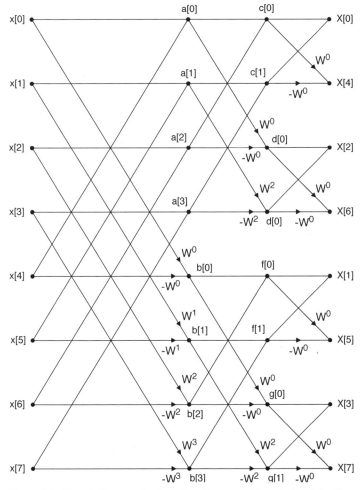

Figure 9.7 Signal flow graph depicting the operations defined by Eqs. (9.17) through (9.21)

9.4 Prime Factor Algorithm

The *prime factor algorithm* (PFA) for computing the DFT can be viewed as a generalization of the radix 2 FFT algorithms developed in prior sections. In a radix 2 FFT, a DFT containing $N = 2^m$ points is computed as an m-stage combination of 2-point DFTs. If, instead of being a power of 2, the size of the DFT is some other composite number $N = N_1 N_2$ [where $\text{GCF}(N_1, N_2) = 1$], it is possible to express the original N-point DFT in the form of an N_1 by N_2 two-dimensional DFT.

Consider a DFT of length $N = N_1 N_2$ with $\text{GCF}(N_1, N_2) = 1$. The one-dimensional input sequence $x[n]$ must be mapped into a two-dimensional sequence

$\hat{x}[n_1, n_2]$. This mapping can be accomplished via the index transformation

$$n \equiv (N_1 n_2 + N_2 n_1) \text{ modulo } N \qquad (9.22)$$

where $n_1 = 0, 1, \ldots N_1 - 1$
$n_2 = 0, 1, \ldots N_2 - 1$

The sequence $\hat{x}[n_1, n_2]$ can be used as the input to a two-dimensional DFT defined by

$$\hat{X}[k_1, k_2] = \sum_{n_2=0}^{N_2-1} W_2^{n_2 k_2} \sum_{n_1=0}^{N_1-1} \hat{x}[n_1, n_2] W_1^{n_1 k_1}$$

where $W_1 = \exp(-j2\pi/N_1)$
$W_2 = \exp(-j2\pi/N_2)$

The two-dimensional result $\hat{X}[k_1, k_2]$ is then mapped into the one-dimensional result $X[k]$ using the index transformation

$$k \equiv (N_1 t_1 k_2 + N_2 t_2 k_1) \text{ modulo } N \qquad (9.23)$$

The values t_1 and t_2 must be chosen such that

$$N_2 t_2 \equiv 1 \text{ modulo } N_1 \qquad (9.24)$$

and
$$N_1 t_1 \equiv 1 \text{ modulo } N_2 \qquad (9.25)$$

Example 9.3 Determine the index mappings needed to apply the PFA to compute a 63-point DFT with $N_1 = 7$ and $N_2 = 9$.

TABLE 9.2 Input Index Mapping for Example 9.3

		\multicolumn{7}{c}{n_1}
		0	1	2	3	4	5	6
	0	0	9	18	27	36	45	54
	1	7	16	25	34	43	52	61
	2	14	23	32	41	50	59	5
	3	21	30	39	48	57	3	12
n_2	4	28	37	46	55	1	10	19
	5	35	44	53	62	8	17	26
	6	42	51	60	6	15	24	33
	7	49	58	4	13	22	31	40
	8	56	2	11	20	29	38	47

TABLE 9.3 Output Index Mapping for Example 9.3

		\multicolumn{7}{c}{k_1}
		0	1	2	3	4	5	6
	0	0	36	9	45	18	54	27
	1	28	1	37	10	46	19	55
	2	56	29	2	38	11	47	20
	3	21	57	30	3	39	12	48
k_2	4	49	22	58	31	4	40	13
	5	14	50	23	59	32	5	41
	6	42	15	51	24	60	33	6
	7	7	43	16	52	25	61	34
	8	35	8	44	17	53	26	62

solution The index transformation for the input sequence is readily enumerated simply by evaluating Eq. (9.22) for all possible combinations of n_1 and n_2. The results are listed in Table 9.2. Before we can use Eq. (9.23) to enumerate the index transformation for the output sequence, we need to use Eqs. (9.24) and (9.25) to determine values for t_2 and t_1. Equation (9.24) is satisfied for $t_2 = 4$, and Eq. (9.25) is satisfied for $t_1 = 4$. Thus Eq. (9.23) becomes

$$k \equiv (28k_2 + 36k_1) \, \text{modulo} \, 63 \qquad (9.26)$$

The index mapping produced by Eq. (9.26) is listed in Table 9.3.

Chapter 10

Windows for Filtering and Spectral Analysis

As discussed in Sec. 8.3, when the input to a DFT is abruptly truncated, large discontinuities may be introduced into the signal. These discontinuities will cause a *leakage* phenomena in which additional high-frequency components appear in the signal spectrum. To reduce these leakage effects, it is a common practice to multiply the truncated input sequence by a tapering window prior to the application of the DFT. As we will discover in Chap. 12, windows are also useful for combating overshoot and ripple in the response of FIR filters designed using the Fourier series method.

The basic idea of windowing is straightforward, and most of the effort in this area has been directed toward finding *good* window functions. A good window shape will taper off at the ends of the input record but still have a reasonably compact and narrow spectrum. This is important, because multiplying the time sequence by the window will cause the corresponding spectrum to be convolved with the spectrum of the window.

10.1 Rectangular Window

Truncating a continuous-time signal can be thought of as multiplying the original signal by a rectangular window such as the one shown in Fig. 10.1. This window has a value of unity for all values of t at which the signal is to be preserved, and a value of zero at all values of t at which the signal is to be eliminated:

$$w(t) = \begin{cases} 1 & |t| < \frac{\tau}{2} \\ 0 & \text{otherwise} \end{cases} \qquad (10.1)$$

The rectangular window's Fourier transform is given by

$$W(f) = \frac{\tau \sin \pi f \tau}{\pi f \tau} \qquad (10.2)$$

182 Chapter Ten

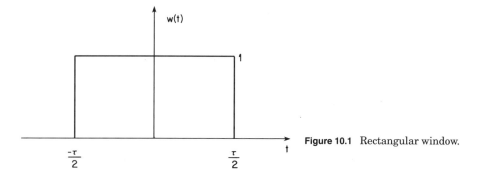

Figure 10.1 Rectangular window.

The magnitude of Eq. (10.2) is plotted in Fig. 10.2. The peaks of the first through ninth sidelobes are attenuated by 13.3, 17.8, 20.8, 23.0, 24.7, 26.2, 27.4, 28.5, and 29.5 dB respectively.

The rectangular window's response will serve primarily as a benchmark to which the responses of other windows can be compared. (Note: By omitting further explanation, some texts such as [1] imply that Eq. (10.2) also applies to the discrete-time version of the rectangular window. However, as we will discover in the following, the Fourier transforms of the continuous-time and discrete-time windows differ significantly. A similar situation exists with respect to the triangular window.)

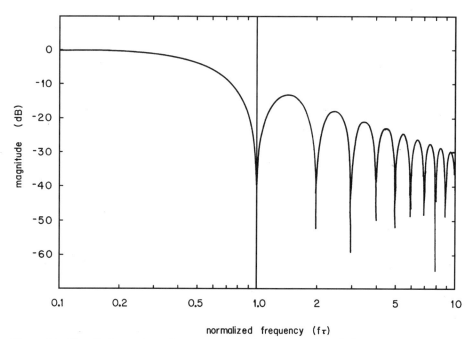

Figure 10.2 Magnitude spectrum for a continuous-time rectangular window.

Software notes

The data for Fig. 10.2 was generated using the class `ContRectangularMagResp` which is provided in file `con_rect.cpp`. This class by itself is simple, consisting of just one constructor and no data members. All of the necessary data members, as well as the code to support user interactions, are provided by the base class `ContinWindowResponse` from which `ContinRectangularMagResp` inherits. Because we will examine several more windows in this chapter, it makes sense to separate out those actions which will be common to all window types and perform these actions in a base class from which each specific window response class can be derived. The class definition for `ContinWindowResponse` is provided in file `con_resp.cpp`. Within this class, all user outputs are routed to the stream `uout` and all user inputs are extracted from the stream `uin`. In most applications, the constructor will be called with arguments `&cin` and `&cout`, but the class has been designed to allow references to other appropriate streams to be used as well. The file `prog_10a.cpp` contains a main program that provides some user interface support and allows any of the windowing functions in this chapter to be conveniently exercised.

Discrete-time window

Since FIR filter coefficients exist only for integer values of n or discrete values of $t = nT$, it is convenient to work with a window function that is defined in terms of n rather than t. If the function defined by Eq. (10.1) is sampled using $N = 2M + 1$ samples with one sample at $t = 0$ and samples at nT for $n = \pm 1, \pm 2, \ldots, \pm M$, the sampled window function becomes

$$w[n] = \begin{cases} 1 & -M \leq n \leq M \\ 0 & \text{otherwise} \end{cases} \qquad (10.3)$$

For an even number of samples, the rectangular window can be defined as either

$$w[n] = 1 \quad -(M-1) \leq n \leq M \qquad (10.4)$$

or
$$w[n] = 1 \quad -M \leq n \leq (M-1) \qquad (10.5)$$

The window specified by Eq. (10.4) will be centered around a point midway between $n = 0$ and $n = 1$, and the window specified by Eq. (10.5) will be centered around a point midway between $n = -1$ and $n = 0$. In many applications (especially in languages such as C that use zero-origin indexing), it is convenient to have $w[n]$ defined for $0 \leq n \leq (N-1)$:

$$w[n] = 1 \quad 0 \leq n \leq (N-1) \qquad (10.6)$$

To emphasize the difference between windows such as Eq. (10.3), which are defined over positive and negative frequencies, and windows such as Eq. (10.6) which are defined over nonnegative frequencies, DSP borrows terminology from the closely related field of time series analysis. Using this borrowed terminology, windows such as Eq. (10.3) are called *lag windows*, and windows such as Eq. (10.6) are called *data windows*. Data windows are also referred to as *tapering windows* and occasionally *tapers* or *faders*. To avoid having to deal with windows centered around $\frac{1}{2}$ or $-\frac{1}{2}$, many authors state that N must be odd for lag windows. However, even-length *data windows* are widely used for leakage reduction in FFT applications.

Frequency windows and spectral windows

The discrete-time Fourier transform (DTFT) of the lag window Eq. (10.3) is given by

$$W(f) = \frac{\sin[\pi f(2M+1)]}{\sin(\pi f)} \qquad (10.7)$$

The form of Eq. (10.7) is closely related to the so-called Dirichlet kernel $D_n(\cdot)$ which is variously defined as

$$D_n(\theta) \triangleq \frac{1}{2\pi} \sum_{k=-n}^{n} \cos k\theta = \frac{\sin\{[n+(1/2)]\theta\}}{\sin(\theta/2)} \qquad \text{per [2]}$$

$$D_n(x) \triangleq \sum_{k=-n}^{n} \exp(2\pi jkx) = \frac{\sin[(2n+1)\pi x]}{\sin(\pi x)} \qquad \text{per [3]}$$

$$D_n(x) \triangleq \frac{1}{2} \sum_{k=-n}^{n} \cos(kx) = \frac{\sin\{[n+(1/2)]x\}}{2\sin(x/2)} \qquad \text{per [4]}$$

The magnitude of Eq. (10.7) is plotted in Fig. 10.3 for $N = 11$ and Fig. 10.4 for $N = 21$. As indicated by these two cases, when the number of points in the window increases, the width of the DTFT sidelobes decreases. The side-lobes in Fig. 10.3 are attenuated by 13.0, 17.1, 19.3, 20.5, and 20.8 dB; and the side lobes in Fig. 10.4 are attenuated by 13.2, 17.6, 20.4, 22.3, 23.7, 24.8, 25.5, 26.1, and 26.3.

The DTFT of the data window Eq. (10.6) is given by

$$W(f) = \exp[-j\pi f(N-1)]\frac{\sin(N\pi f)}{\sin(\pi f)} \qquad (10.8)$$

A function such as Eq. (10.7), which is the Fourier transform of a lag window, is called a *spectral window*. A function such as Eq. (10.8), which is the Fourier transform of a data window, is called a *frequency window*. The forms of Eqs. (10.7) and (10.8) differ from the form of Eq. (10.2) due to the aliasing that occurs when the continuous-time window function is sampled to obtain a discrete-time window.

More software notes

The magnitude responses for continuous-time windows are generally computed using closed-form expressions such as Eq. (10.2). The base class ContinWindow-Response as well as the derived classes such as ContRectangularMagResp were

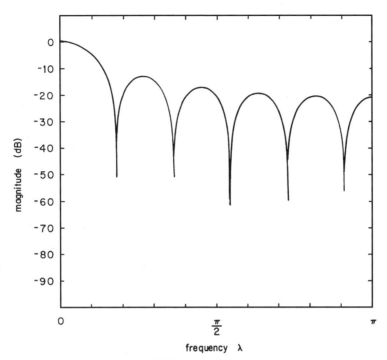

Figure 10.3 Magnitude of the DTFT for an 11-point rectangular window.

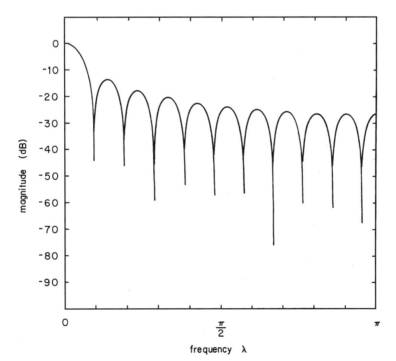

Figure 10.4 Magnitude of the DTFT for a 21-point rectangular window.

set up with this fact in mind. The approach taken for discrete-time windows and their spectra is very different.

A discrete-time window is also an FIR filter. Therefore, any method that can be used to compute the continuous-frequency response of an FIR filter can also be used to compute the continuous-frequency response of a discrete-time window. The approach used to generate the data for Fig. 10.4 consists of the following steps:

1. Generate the window coefficients. For the moment, let us assume that these coefficients have been placed in an array `window_coeff`. For the case of a rectangular window, this step is almost trivial, because all of the coefficients are 1.00. Some of the issues involved in the design of classes for generating window coefficients are easier to explore using more complicated windows. Therefore, these issues will be explored in Sec. 10.3 after the triangular window is introduced in Sec. 10.2.

2. Construct an FIR filter using these window coefficients. This construction can be accomplished using an alternate constructor for the `FirFilterDesign` class (introduced in Chap. 11) which is used as the base class for the derived class `FirIdealFilter` (discussed in Sec. 12.1)

```
fir_design = new FirFilterDesign(num_taps, window_coeff);
```

3. Compute the magnitude response of this FIR filter using the `FirFilter_Response` class (introduced in Chap. 11)

```
fir_response = new FirFilterResponse( fir_design, cin, cout );
fir_response->ComputeMagResp( );
fir_response->DumpMagResp( );
```

10.2 Triangular Window

A simple, but not particularly high-performance window is the *triangular window* shown in Fig. 10.5, and is defined by

$$w(t) = 1 - \frac{2|t|}{\tau} \quad |t| \leq \frac{\tau}{2} \tag{10.9}$$

Window functions are almost always even symmetric, and it is customary to show only the positive-time portion of the window, as in Fig. 10.6. The triangular window is sometimes called the *Bartlett window* after M. S. Bartlett, who described its use in a 1950 paper [5]. The Fourier transform of Eq. (10.9) is given by

$$W(f) = \frac{\tau}{2} \left[\frac{\sin(\pi f \tau/2)}{(\pi f \tau/2)} \right]^2 \tag{10.10}$$

The magnitude of Eq. (10.10) is plotted in Fig. 10.7. The peaks of the first through fourth sidelobes are attenuated by 26.5, 35.7, 41.6, and 46.0 dB, respectively. The data for Fig. 10.7 was generated using the method `ContTriangular_MagResponse` provided in file `con_tngl.cpp`.

Windows for Filtering and Spectral Analysis 187

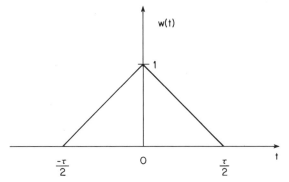

Figure 10.5 Triangular window.

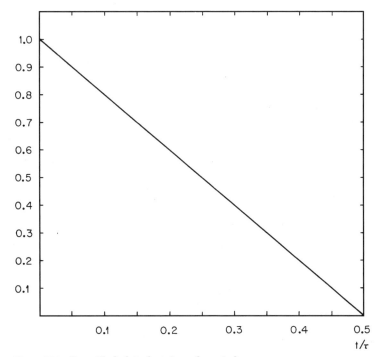

Figure 10.6 One-sided plot of a triangular window.

Discrete-time triangular window

If the function defined by Eq. (10.10) is sampled using $N = 2M + 1$ samples with $\tau = 2MT$, one sample at $t = 0$, and samples at nT for $n = \pm 1, \pm 2, \ldots, \pm M$, the sampled window function becomes the lag window defined by

$$w[n] = 1 - \frac{2|n|}{2M} \quad -M \leq n \leq M \quad (10.11)$$

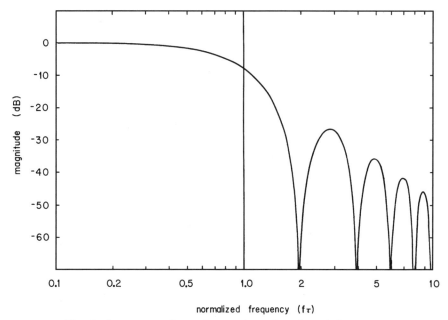

Figure 10.7 Magnitude response of a continuous-time triangular window.

for the normalized case of $T = 1$. This equation can be expressed in terms of the total number of samples N by substituting $(N-1)/2$ for M to obtain

$$w[n] = 1 - \frac{|2n|}{N-1} \quad \frac{-(N-1)}{2} \leq n \leq \frac{N-1}{2} \qquad (10.12)$$

In some texts (such as [6] and [7]), Eq. (10.12) is given as the definition of the discrete-time triangular window. However, evaluation of this equation reveals that $w[n] = 0$ for $n = \pm[(N-1)/2]$. This means that the two end-points do not contribute to the window contents and that the window length is effectively reduced to $N-2$ samples. To maintain a total of N *nonzero* samples, many authors substitute $(N+2)$ for N in Eq. (10.12) to obtain

Triangular Lag Window, N odd

$$w[n] = 1 - \frac{|2n|}{N+1} \quad \frac{-(N-1)}{2} \leq n \leq \frac{N-1}{2} \qquad (10.13)$$

For an even number of samples, the window values can be obtained by substituting $(n + 1/2)$ for n in Eq. (10.13) to obtain a window that is symmetrical about a line midway between $n = -1$ and $n = 0$.

Triangular Lag Window, N even, center at $-\frac{1}{2}$

$$w[n] = 1 - \frac{|2n+1|}{N+1} \quad \frac{-N}{2} \leq n \leq \frac{N}{2} - 1 \qquad (10.14)$$

Alternatively, we could substitute $(n-1/2)$ for n in Eq. (10.13) to obtain a window symmetric about a line midway between $n = 0$ and $n = 1$:

Triangular Lag Window, N even, center at $\frac{1}{2}$

$$w[n] = 1 - \frac{|2n-1|}{N+1} \quad \frac{-N}{2}+1 \leq n \leq \frac{N}{2} \tag{10.15}$$

An expression for the triangular *data* window can be obtained by substituting $[n-(N-1)/2]$ for n in Eq. (10.13) or by substituting $(n-N/2)$ for n in Eq. (10.14) to yield

Triangular Data Window

$$w[n] = 1 - \frac{|2n-N+1|}{N+1} \quad 0 \leq n \leq N-1 \tag{10.16}$$

Section 10.3 presents class designs for generating various forms of the discrete triangular window. These designs are generalized to support other types of windows to be introduced in subsequent sections.

Frequency and spectral windows

The spectral window obtained from the DTFT of the lag window Eq. (10.12) is given by

$$W(f) = \frac{1}{M}\left[\frac{\sin(M\pi f)}{\sin(\pi f)}\right]^2 \tag{10.17a}$$

$$W(\theta) = \frac{2}{N}\left\{\frac{\sin[(N/4)\theta]}{\sin[(1/2)\theta]}\right\}^2 \tag{10.17b}$$

where $M = \frac{N-1}{2}$
$\theta = \frac{2\pi f}{f_s}$

The form of Eq. (10.17) is closely related to the Fejer kernel $F_n(\cdot)$, which like the Dirichlet kernel presented in Sec. 10.1, has some variety in its definition:

$$F_n(x) \triangleq \frac{\sin^2(n\pi x)}{n\sin^2(\pi x)} \quad \text{per [2]}$$

$$F_n(x) \triangleq \frac{\sin^2(n\theta/2)}{2\pi n\sin^2(\theta/2)} \quad \text{per [3]}$$

The magnitude of Eq. (10.17) for $N = 11$ and $N = 21$ is plotted in Fig. 10.8.

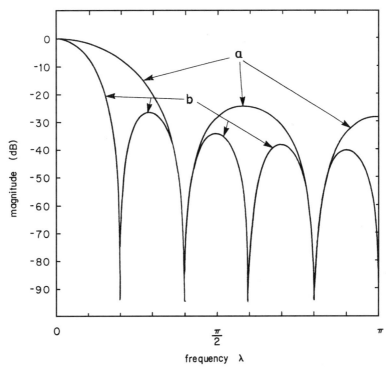

Figure 10.8 Magnitude of the DTFT for (a) an 11-point triangular window and (b) a 21-point triangular window.

10.3 Window Software

As shown in the previous section, a window function can come in a number of varieties—odd-length lag window, even-length lag window centered on $n = 1/2$, and so on. As was done for the triangular window, an explicit function for each variety can be derived. However, the task of designing and coding computer programs to generate window coefficients can be simplified somewhat if we view the different varieties from a slightly different perspective. Despite the apparent variety of specific formats, there really are only two basic forms that need to generated—one form for odd-length windows and one form for even-length windows. All of the specific varieties can be generated as horizontal translations of these two forms. Furthermore, since all of the windows considered in this book are symmetric, we need to generate the coefficients for only half of each window. An odd-length lag window is probably the most "natural" of the discrete-time windows. Consider the triangular window shown in Fig. 10.9, which has sample values indicated at $t = \pm nT$ for $n = 0, 1, 2, \ldots$. Because of symmetry, we will require our program to generate only the $(N+1)/2$ coefficients corresponding to $t = 0, 2T, 3T, \ldots, (N-1)T/2$ and place them in locations 0 through $(N-1)/2$ of an array called Half_Lag_Win. These coefficients can be obtained using Eq. (10.13). Next, we consider the

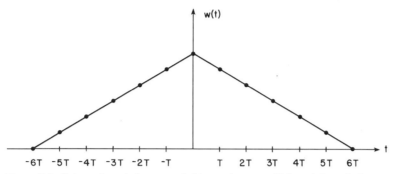

Figure 10.9 Triangular window sampled to produce an odd-length lag window.

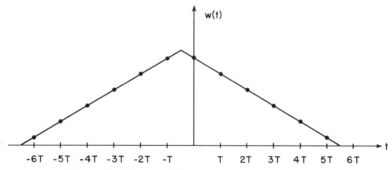

Figure 10.10 Triangular window shifted and sampled to produce an even-length lag window with axis of symmetry midway between $n = -1$ and $n = 0$.

triangular window shown in Fig. 10.10. This window has been shifted so that its axis of symmetry lies at $t = -T/2$. The sample values indicated in the figure can be obtained from Eq. (10.14). The sample values for either the even-length case of Fig. 10.10 or the odd-length case of 10.9 can be obtained from the combined formula

$$w[n] = 1 - \frac{|2x|}{N+1}$$

where

$$x = \begin{cases} n & \text{for } N \text{ odd} \\ n + \frac{1}{2} & \text{for } N \text{ even} \end{cases}$$

The class TriangularWindow, provided in file triangular.cpp, uses this formula to generate coefficients for both odd- and even-length discrete-time triangular windows. For N odd, the value placed in Half_Lag_Win[0] lies on the full window's axis of symmetry and is the value of the continuous-time window at $t = 0$. For N even, the value placed in Half_Lag_Win[0] lies one-half sample time to the right of the full window's axis of symmetry and is the value of the continuous-time window at $t = T/2$. The class TriangularWindow is derived from the base class GenericWindow, which is defined in file gen_win.cpp. This base class contains the attributes and methods which are common to all discrete-time windows. The class method GetDataWinCoeff can be used to obtain the data window coefficient for a specified value of sample index. This value actually is obtained by determining which sample in the half-lag window corresponds to the requested data window sample and then reading the appropriate location in Half_Lag_Win[-]. The method GetDataWindow allocates space for a complete data window and then fills this space by calling GetDataWindow for sample indices from 0 through Length-1.

10.4 von Hann Window

The continuous-time von Hann window function shown in Fig. 10.11 is defined by

$$w(t) = 0.5 + 0.5 \cos \frac{2\pi t}{\tau} \quad |t| \leq \frac{\tau}{2} \tag{10.18}$$

The corresponding frequency response, shown in Fig. 10.12, is given by

$$W(f) = 0.5\tau \operatorname{sinc}(\pi f \tau) + 0.25\tau \operatorname{sinc}[\pi \tau (f - \tau)] + 0.25\tau \operatorname{sinc}[\pi \tau (f + \tau)] \tag{10.19}$$

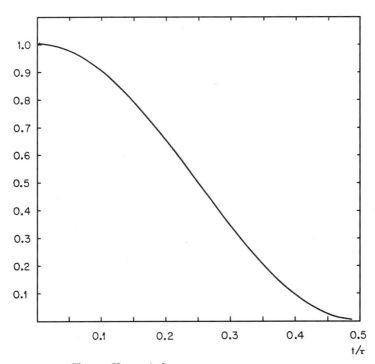

Figure 10.11 The von Hann window.

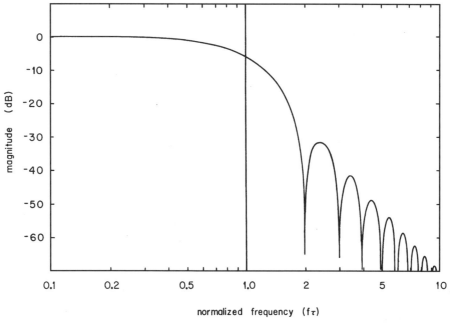

Figure 10.12 Magnitude response of the von Hann window.

The first sidelobe of this response is 31.4 dB below the mainlobe, and the mainlobe is twice as wide as the mainlobe of the rectangular window's response. The data for Fig. 10.12 was obtained using the class `ContHannMagResp` provided in file `con_hann.cpp`.

References to the von Hann window as the "hanning" window are widespread throughout the signal-processing literature. This is unfortunate for two reasons. First, the window gets its name from Julius von Hann, not someone named Hanning. Second, the term *hanning* is easily and often confused with *Hamming* (see Sec. 10.5).

This window is occasionally called a *raised-cosine window*.

Discrete-time von Hann window

If the function defined by Eq. (10.18) is sampled using $N = 2M+1$ samples with one sample at $t = 0$ and samples at nT for $n = \pm 1, \pm 2, \ldots, \pm M$, the sampled window function becomes

$$w[n] = 0.5 + 0.5 \cos \frac{\pi n}{M} \quad -M \leq n \leq M \tag{10.20}$$

for the normalized case of $T = 1$. Evaluation of Eq. (10.20) reveals that $w[n] = 0$ for $n = \pm M$. This means that two endpoints do not contribute to the window contents and that the window length is effectively reduced to $N - 2$ samples. In order to maintain a total of N nonzero samples, we must substitute $M + 1$ for

M in Eq. (10.20) to yield

$$w[n] = 0.5 + 0.5 \cos \frac{\pi n}{M+1} \quad -M \le n \le M \quad (10.21)$$

Equation (10.21) now can be recast in terms of N by substituting $(N-1)/2$ for M to obtain

von Hann Lag Window, N odd

$$w[n] = 0.5 + 0.5 \cos \frac{2\pi n}{N-1} \quad \frac{-(N-1)}{2} \le n \le \frac{N-1}{2} \quad (10.22)$$

For an even number of samples, the window values can be obtained by substituting either $(n+1/2)$ or $(n-1/2)$ for n in Eq. (10.22) to obtain

von Hann Lag Window, N even

$$w[n] = 0.5 + 0.5 \cos \frac{\pi(2n+1)}{N-1} \quad \frac{-N}{2} \le n \le \frac{N}{2}-1 \quad (10.23)$$

$$\text{centered at } \frac{-1}{2}$$

$$w[n] = 0.5 + 0.5 \cos \frac{\pi(2n-1)}{N-1} \quad \frac{-N}{2}+1 \le n \le \frac{N}{2} \quad (10.24)$$

$$\text{centered at } \frac{1}{2}$$

The class `HannWindow`, provided in file `hann_cpp`, generates coefficients for the von Hann window.

10.5 Hamming Window

The continuous-time Hamming window function shown in Fig. 10.13 is defined by

$$w(t) = 0.54 + 0.46 \cos \frac{2\pi t}{\tau} \quad |t| \le \frac{\tau}{2} \quad (10.25)$$

The Fourier transform of Eq. (10.25) is given by

$$W(f) = 0.54\tau \operatorname{sinc}(\pi f \tau) + 0.23\tau \operatorname{sinc}[\pi \tau(f-\tau)] + 0.23\tau \operatorname{sinc}[\pi \tau(f+\tau)] \quad (10.26)$$

The magnitude of Eq. (10.26) is plotted in Fig. 10.14. The highest sidelobe of this response is 42.6 dB below the mainlobe, and the mainlobe is twice as wide as the mainlobe of the rectangular window's response. The data for Fig. 10.14 was obtained using the class `ContHammingMagResp`, provided in file `con_hamm.cpp`. This window gets its name from R. W. Hamming, a pioneer in the areas of numerical analysis and signal processing, who opened his numerical analysis text [10]

Windows for Filtering and Spectral Analysis 195

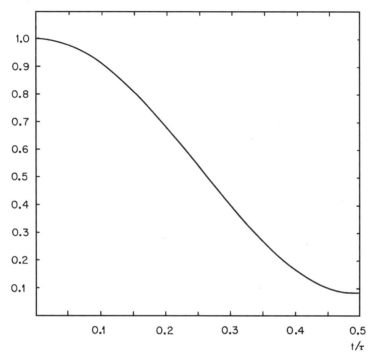

Figure 10.13 Hamming Window

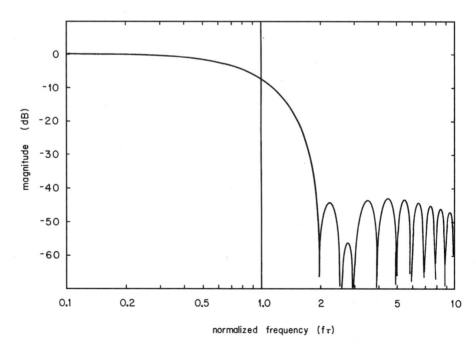

Figure 10.14 Magnitude response of the Hamming window.

with the now famous and oft-quoted pearl, "The purpose of computing is insight, not numbers."

Discrete-time Hamming windows

If the function defined by Eq. (10.25) is sampled using $N = 2M + 1$ samples with one sample at $t = 0$ and samples at nT for $n = \pm 1, \pm 2, \ldots, \pm M$, the sampled window function becomes the lag window defined by

$$w[n] = 0.54 + 0.46 \cos \frac{2\pi n}{2M} \quad -M \leq n \leq M \quad (10.27)$$

for the normalized case of $T = 1$. Equation (10.27) can be expressed in terms of the total number of samples N by substituting $(N-1)/2$ for M to obtain

Hamming Lag Window, N odd

$$w[n] = 0.54 + 0.46 \cos \frac{2\pi n}{N-1} \quad \frac{-(N-1)}{2} \leq n \leq \frac{N-1}{2} \quad (10.28)$$

For an even number of samples, the window values can be obtained by substituting $n + 1/2$ for n in Eq. (10.28) to obtain

Hamming Lag Window, N even

$$w[n] = 0.54 + 0.46 \cos \frac{\pi(2n+1)}{N-1} \quad \frac{-N}{2} \leq n \leq \frac{N}{2} - 1 \quad (10.29)$$

The data window form can be obtained by substituting $[n - (N-1)/2]$ for n in Eq. (10.28) or by substituting $(n - N/2)$ for n in Eq. (10.29) to yield

Hamming Data Window

$$w[n] = 0.54 - 0.46 \cos \frac{2\pi n}{N-1} \quad 0 \leq n \leq N-1 \quad (10.30)$$

(Note the change from + to − for the cosine term.)

Computer generation of window coefficients

The class HammingWindow, provided in file hamming.cpp, generates coefficients for the Hamming window. The output conventions for even and odd N are as described in Sec. 10.3.

10.6 Dolph-Chebyshev Window

The *Dolph-Chebyshev window* is somewhat different from the other windows in this chapter in that a closed-form expression for the time domain window is not known. Instead, this window is defined as the inverse Fourier transform of the sampled-frequency response, which is given by

$$W[k] = \begin{cases} \dfrac{\cosh(N\cosh^{-1}x)}{\cosh(N\cosh^{-1}\alpha)} & |x| > 1 \text{ (mainlobe)} \\ \dfrac{\cos(N\cos^{-1}x)}{\cosh(N\cosh^{-1}\alpha)} & |x| \leq 1 \text{ (sidelobes)} \end{cases} \quad (10.31)$$

$$-(N-1) \leq k \leq N-1$$

where $x = \alpha \cos(\pi k/N)$.

A sidelobe level of -80 dB is often claimed for this response, but, in fact, Eq. (10.31) defines a family of windows in which the minimum sidelobe attenuation is a function of α and N. An N-tap window with a sidelobe attenuation of 20β dB is obtained for a value of α given by

$$\alpha = \cosh\left[\frac{1}{N}\cosh^{-1}(10^\beta)\right] \quad (10.32)$$

Unlike other windows in which the sidelobe attenuation increases with increasing frequency, all of the sidelobes for a Dolph-Chebyshev window's response peak at the same level.

Coefficients for the time-domain window are obtained by evaluating Eq. (10.31) over the indicated range of k and then performing an inverse DFT on this sampled frequency response. The class DolphChebyWindow, provided in file dolph.cpp, generates coefficients for the Dolph-Chebyshev window.

Example 10.1 Determine the value of α needed for a 23-tap Dolph-Chebyshev window with a sidelobe attenuation of 70 dB, and use prog_10a to generate the window coefficients and frequency response.

solution The values for β and α are obtained as

$$\beta = \frac{70}{20} = 3.5$$

$$\alpha = \cosh\left[\frac{1}{23}\cosh^{-1}(10^{3.5})\right]$$

$$= 1.073279517$$

The sampled-frequency response given by Eq. (10.31) is plotted in Fig. 10.15. The time domain window coefficients are listed in Table 10.1. The continuous-frequency response of this window is plotted in Fig. 10.16.

The Dolph-Chebyshev window often is touted as being the ultimate window function because it has the narrowest mainlobe for a given sidelobe level, and it theoretically can achieve extremely large values of sidelobe attenuation. However, several authors (e.g. [16, 17]) claim that the Dolph-Chebyshev window is

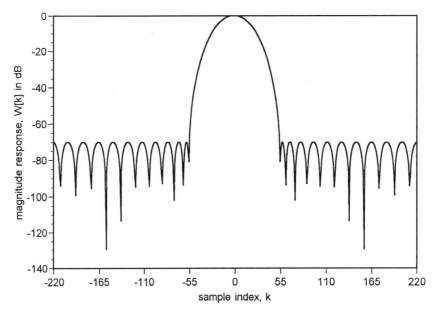

Figure 10.15 Sampled-frequency response of Dolph-Chebyshev window for Example 10.1.

extremely sensitive to coefficient quantization and round-off error. The seriousness of these claims depends on one's sense of what is extremely sensitive. When the coefficients of Example 10.1 are rounded to 15 bits, the resulting frequency response is nearly indistinguishable from the original response. Figure 10.17 shows the frequency response resulting from rounding the coefficients to 10 bits. The degradation is noticeable, but the sidelobe attenuation still outperforms many simpler windows.

TABLE 10.1 Time-domain Coefficients for the 23-tap Dolph-Chebyshev Window of Example 10.1

$$h[0] = 1.000000000$$
$$h[1] = h[-1] = 0.971701394$$
$$h[2] = h[-2] = 0.891044602$$
$$h[3] = h[-3] = 0.769734078$$
$$h[4] = h[-4] = 0.624495599$$
$$h[5] = h[-5] = 0.473542567$$
$$h[6] = h[-6] = 0.333205337$$
$$h[7] = h[-7] = 0.215193225$$
$$h[8] = h[-8] = 0.125422872$$
$$h[9] = h[-9] = 0.064096225$$
$$h[10] = h[-10] = 0.027206193$$
$$h[11] = h[-11] = 0.008655853$$

Windows for Filtering and Spectral Analysis   199

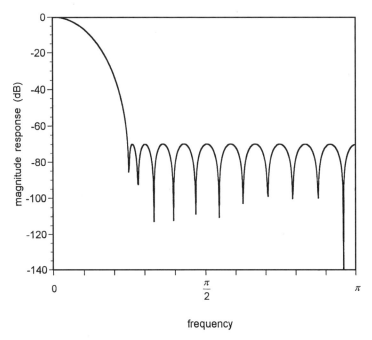

Figure 10.16 Continuous-frequency response of Dolph-Chebyshev window for Example 10.1.

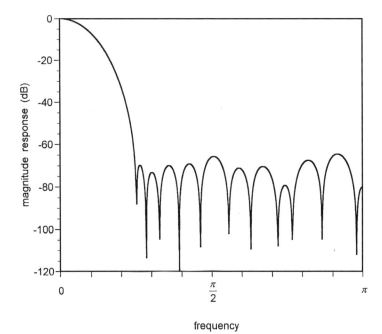

Figure 10.17 Frequency response of 23-tap Dolph-Chebyshev window with coefficients rounded to 10 bits.

The Dolph-Chebyshev window takes its name from C. L. Dolph and Pafnuti Chebyshev. The function $W[k]$ is a normalized form of the function developed by Dolph [11] for specifying an antenna pattern optimized to achieve a narrow mainlobe while simultaneously restricting the sidelobe response. Chebyshev polynomials play a central role in Dolph's development. Helms [12] applied Dolph's result to the analogous problem of optimizing a filter response for a narrow transition band while simultaneously restricting sidelobe response.

10.7 Kaiser Window

One approach for defining an optimal window format would be to find the particular time-limited function having the minimum energy outside the mainlobe of its frequency response. The exact solution [13, 14] to this optimization problem involves prolate spheroidal wave functions, and is not well suited for practical applications. However, Kaiser [15] showed that I_0-sinh functions can be used to approximate the zero-order prolates and obtain for N odd a window of the form

Kaiser Lag Window, N odd

$$w[n] = \frac{I_0\left[\pi\alpha\sqrt{1-(2n/N)^2}\right]}{I_0(\pi\alpha)} \qquad \frac{-(N-1)}{2} \leq n \leq \frac{N-1}{2} \qquad (10.33)$$

where I_0 is the zero-order modified Bessel function of the first kind. This window is referred to as the *Kaiser window* or sometimes as the *Kaiser-Bessel window*. Based on Eq. (10.34), $I_0(x)$ can be expanded in an infinite series as

$$I_0(x) = \sum_{m=0}^{\infty}\left[\frac{(x/2)^m}{m!}\right]^2 \qquad (10.34)$$

For most practical applications, adequate precision usually can be obtained by summing only the terms for $0 \leq m \leq 32$. Increasing the value of α will increase both the sidelobe attenuation and the width of the mainlobe.

For an even number of samples, the window values can be obtained by substituting $n + 1/2$ for n in Eq. (10.33) to obtain

Kaiser Lag Window, N even

$$w[n] = \frac{I_0\left\{\pi\alpha\sqrt{1-[(2n+1)/N]^2}\right\}}{I_0(\pi\alpha)} \qquad \frac{-N}{2} \leq n \leq \frac{N}{2} - 1 \qquad (10.35)$$

The data window form can be obtained by substituting $[n-(N-1)/2]$ for n in Eq. (10.33) or by substituting $(n - N/2)$ for n in Eq. (10.35) to yield

Kaiser Data Window

$$w[n] = \frac{I_0\left\{\pi\alpha\sqrt{1-[(2n-N+1)/N]^2}\right\}}{I_0(\pi\alpha)} \qquad 0 \leq n \leq N-1 \qquad (10.36)$$

Computer generation of window coefficients

The class KaiserWindow, provided in file kaiser.cpp, generates coefficients for the Kaiser window. The output conventions for even and odd N are as described in Sec. 10.3.

10.8 Impacts of Quantization

The methods ScaleCoefficients and QuantizeCoefficients have been added to class FirFilterDesign to provide a convenient way to scale and quantize window coefficients.

Example 10.2 Obtain the coefficients for a 21-tap triangular window. Compare the DTFTs obtained for the quantized window obtained by *truncating* the window coefficients to 7 bits and the quantized window obtained by *rounding* the window coefficients to 7 bits. Do not include the zero-valued endpoints of the window as part of the total tap count.

solution Since window coefficients are always positive, all of the available bits can be used to represent magnitude. Therefore a value of $2^7 = 128$ is used for quant_factor in the call to QuantizeCoefficients. Because the peak window coefficient is 1, all of the window coefficients were scaled by the factor 127/128 prior to quantization so that the peak value will fit into 7 bits. The unquantized coefficients and the corresponding quantized values are listed in Table 10.2, and the DTFT magnitudes for the truncated and rounded coefficients are plotted in Figs. 10.18 and 10.19. The DTFT spectrum for rounded coefficients has deeper nulls than the spectrum for truncated coefficients, but otherwise the two spectra are very similar. Comparison of these figures and trace (*b*) of Fig. 10.8 reveals no significant degradation due to quantization.

Unless explicitly stated to the contrary, all of the quantization performed in the remainder of this chapter will employ rounding rather than simple truncation.

Example 10.3 Obtain the coefficients for a 21-tap von Hann window, quantize each coefficient to 7 bits, and compute the DTFT for the resulting quantized window.

solution The unquantized coefficients and the corresponding quantized values are listed in Table 10.3, and the DTFT magnitudes are plotted in Figs. 10.20 and 10.21. There is no significant difference in the spectra for normalized frequencies below $\pi/4$. The third and fourth lobes of the quantized response are actually a few dB lower than the corresponding lobes of the unquantized response. The null separating the 6th and 7th lobes of the unquantized response does not appear at all in the quantized response. The most noticeable differences occur at normalized frequencies between $3\pi/4$ and π where the attenuation of the quantized response is degraded by more than 10 dB.

TABLE 10.2 Coefficients for the 21-tap Triangular Window of Example 10.2

n	Original $h[n]$	Truncated $h[n]$	Quantized $h[n]$
0, 20	0.090909	0.085938	0.093750
1, 19	0.181818	0.179688	0.179688
2, 18	0.272727	0.265625	0.273438
3, 17	0.363636	0.359375	0.359375
4, 16	0.454545	0.445313	0.453125
5, 15	0.545454	0.539063	0.539063
6, 14	0.636364	0.625000	0.632813
7, 13	0.727273	0.718750	0.718750
8, 12	0.818182	0.804688	0.812500
9, 11	0.909091	0.898438	0.898438
10	1.00	0.992188	0.992188

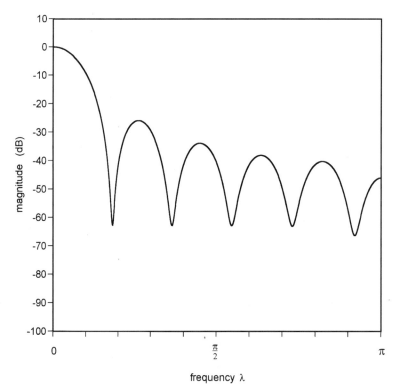

Figure 10.18 Magnitude of the DTFT for a 21-point triangular window with quantized coefficients truncated to 7 bits.

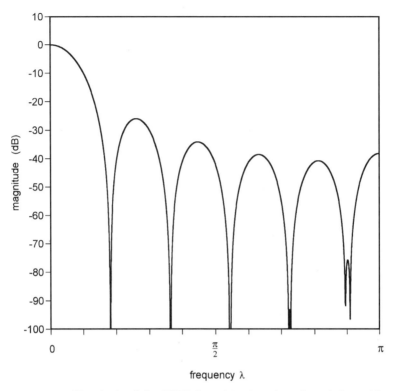

Figure 10.19 Magnitude of the DTFT for a 21-point triangular window with quantized coefficients rounded to 7 bits.

TABLE 10.3 Coefficients for the 21-tap von Hann Window of Example 10.3

n	Original $h[n]$	Quantized $h[n]$
0, 20	0.020254	0.023438
1, 19	0.079373	0.078125
2, 18	0.172570	0.171875
3, 17	0.292292	0.289063
4, 16	0.428843	0.421875
5, 15	0.571157	0.570313
6, 14	0.707708	0.703125
7, 13	0.827430	0.820313
8, 12	0.920627	0.914063
9, 11	0.979746	0.968750
10	1.00	0.992188

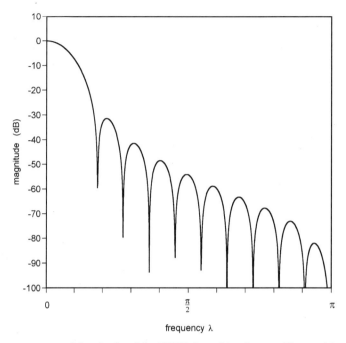

Figure 10.20 Magnitude of the DTFT for a 21-point von Hann with window.

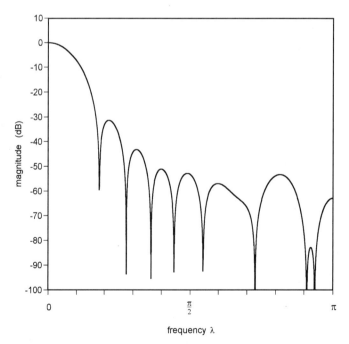

Figure 10.21 Magnitude of the DTFT for a 21-point von Hann with window with coefficients quantized to 7 bits.

References

1. Stanley, W. D. *Digital Signal Processing*, Reston, Reston, VA, 1975.
2. Priestley, M. B. *Spectral Analysis and Time Series*, vol. 1: *Univariate Series*, Academic, London, 1981.
3. Dym, H. and H. P. McKean. *Fourier Series and Integrals*, Academic, New York, 1972.
4. Weaver, H. J. *Theory of Discrete and Continuous Fourier Analysis*, Wiley, New York, 1989.
5. Bartlett, M. S. "Periodogram Analysis and Continuous Spectra," *Biometrika*, vol. 37, pp. 1–16, 1950.
6. Marple, S. L. *Digital Spectral Analysis with Applications*, Prentice-Hall, Englewood Cliffs, NJ, 1987.
7. Kay, S. M. *Modern Spectral Estimation: Theory & Application*, Prentice-Hall, Englewood Cliffs, NJ, 1985.
8. Oppenheim, A. V. and R. W. Schafer. *Digital Signal Processing*, Prentice-Hall, Englewood Cliffs, NJ, 1975.
9. Blackman, R. B. and J. W. Tukey. *The Measurement of Power Spectra*, Dover, New York, 1958.
10. Hamming, R. W. *Numerical Methods for Engineers and Scientists*, McGraw-Hill, New York, 1962.
11. Dolph, C. L. "A Current Distribution for Broadside Arrays Which Optimizes the Relationship Between Beam Width and SideLobe Level," *Proc. IRE*, vol. 35, pp. 335–348, June 1946.
12. Helms, H. D. "Nonrecursive Digital Filters: Design Methods for Achieving Specifications on Frequency Response," *IEEE Trans. Audio and Electroacoust.*, vol. AU-16, pp. 336–342, September 1968.
13. Slepian, D. and H. Pollak. "Prolate-Speroidal Wave Functions, Fourier Analysis, and Uncertainty—I," *Bell System Technical Journal*, vol. 40, pp. 43–64, Jan. 1961.
14. Landau, H. and H. Pollak. "Prolate-Speroidal Wave Functions, Fourier Analysis, and Uncertainty—II," *Bell System Technical Journal*, vol. 40, pp. 65–84, Jan. 1961.
15. Kaiser, J. F. "Nonrecursive Digital Filter Design Using the I_0-sinh Window Function," *Proc. 1974 IEEE Int. Symp. on Circuits and Syst.*, pp. 20–23, April 22–25, 1974.
16. Loy, N. J. *An Engineer's Guide to FIR Digital Filters*, Prentice-Hall, Englewood Cliffs, NJ, 1988.
17. Ifeachor, E. C. and B. W. Jervis. *Digital Signal Processing: A Practical Approach*, Addison-Wesley, Reading, Mass., 1993.

Chapter 11

FIR Filter Fundamentals

11.1 Introduction to FIR Filters

The general form for a linear time-invariant FIR system's output $y[k]$ at time k is given by

$$y[k] = \sum_{n=0}^{N-1} h[n]x[k-n] \tag{11.1}$$

where $h[n]$ is the system's impulse response. As Eq. (11.1) indicates, the output is a linear combination of the present input and the N previous inputs. This chapter is devoted to the basic properties and realization issues for FIR filters. Specific design approaches for selecting the coefficients b_n are covered in Chaps. 12–14.

Software design

The essential data components for representing an FIR filter consist only of an integer for the number of taps and a real-valued array for the coefficient values. Member functions for the class `FirFilterDesign` are provided in file `fir_dsgn.cpp`. The data portion of the class definition for `FirFilterDesign` consists of only two items:

```
int Num_Taps;
double *Imp_Resp_Coeff;
```

The default constructor contains no executable statements, and invoking this constructor simply causes storage to be allocated for these two data items. The default constructor will be used in later chapters when `FirFilterDesign` is used as a base class for more specialized derived classes that will be used to represent specific types of FIR filters. For many of these applications, `FirFilterDesign` must be allocated before the number of filter taps is know—hence the need for a default constructor. Whenever an `FirFilterDesign` object is instantiated via the default constructor, the Initialize method must subsequently be used to set the value of Num_Taps and allocate space on the heap for holding the array of filter coefficients.

When the number of taps is known prior to instantiation of FirFilterDesign, the constructor form FirFilterDesign(int num_taps) can be used to set Num_Taps and allocate the coefficient array from within the constructor. A third form of the constructor, FirFilterDesign(int num_taps, double *coeff) can be used to both allocate and initialize the coefficient array. The method CopyCoefficients(double *coeff) can be used to initialize the coefficients in instances of FirFilterDesign that were created via either of the first two constructors. This method could also be used to re-initialize the coefficients in any instance of FirFilterDesign provided that the number of coefficients is held constant. Additional "housekeeping" functions provided in FirFilterDesign include GetNumTaps(), GetCoefficients() and DumpCoefficients(ofstream*).

11.2 Evaluating the Frequency Response of FIR Filters

A digital filter's impulse response $h[n]$ is related to the frequency response $H(e^{j\lambda})$ via the DTFT:

$$H(e^{j\lambda}) = \sum_{n=-\infty}^{\infty} h[n] e^{-jn\lambda} \qquad (11.2)$$

For an FIR filter, $h[n]$ is nonzero only for $0 \leq n < N$. Therefore, the limits of the summation can be changed to yield

$$H(e^{j\lambda}) = \sum_{n=0}^{N-1} h[n] e^{-jn\lambda} \qquad (11.3)$$

This equation can be evaluated directly at any desired value of λ.

We now take note of the fact that $\lambda = \omega T$ and that the value of continuous radian frequency ω_n corresponding to the discrete-frequency index m is given by

$$\omega_m = 2\pi m F \qquad (11.4)$$

Substituting $2\pi m F T$ for λ, and $H[m]$ for $H(e^{j\lambda})$ in Eq. (11.3) yields the discrete Fourier transform:

$$H[m] = \sum_{n=0}^{N-1} h[n] \exp(-2\pi j n m F T) \qquad (11.5)$$

Thus, the DTFT can be evaluated at a set of discrete frequencies $\omega = \omega_m$, $0 \leq m < N$, by using the DFT, which in turn may be evaluated in a computationally efficient fashion using one of the various FFT algorithms.

Software design

From a theoretical perspective, we could easily view the continuous-frequency response of an FIR filter as being an intrinsic property or *attribute* of the filter itself, and therefore we might conclude that the software for producing a filter's frequency response should be made part of the class FirFilterDesign that implements the filter. However, once we begin to think about designing software to calculate a filter's frequency response, we quickly discover that, in practice, this response possesses a number of attributes that are most definitely not attributes of the filter itself. In theory, the frequency response is a function of continuous frequency defined over all frequencies. In practice we can evaluate the response at only a finite number of discrete frequencies over a finite range. Therefore, to define a practical response plot, we will need to know the answers to a number of questions:

- At how many frequencies is the response to be evaluated?
- What is the minimum frequency at which the response is to be evaluated?
- What is the maximum frequency at which the response is to be evaluated?
- How are the evaluation frequencies to be spaced between the minimum and the maximum? (The obvious choice of uniform spacing may not always be the correct answer—for responses to be plotted against a logarithmic frequency axis, we may want to space the frequency values so that they are uniformly spaced in the plot.)
- Should values of phase response be produced in degrees or in radians?
- Should values of magnitude response be produced in linear numeric units or in decibels?
- Should the "raw" magnitude response be left as-is or should the values be normalized so that the peak passband value is either 1.0 (linear) or 0.0 (dB)?

Because the software that implements the frequency response must "know" or obtain the answers to all of these questions, this software has been implemented in the separate class FirFilterResponse which is provided in file fir_resp.cpp. The data member definitions for this class are listed in Table 11.1.

TABLE 11.1 Data Members for Class FirFilterResponse

```
FirFilterDesign *Filter_Design;
int Num_Resp_Pts;
logical Db_Scale_Enabled
logical Normalize_Enabled
ofstream *Response_File
int Num_Taps
double *Mag_Resp
```

The FirFilterResponse class has been designed with two different constructors. The first constructor takes all the necessary configuration parameters plus a pointer to an FirFilterDesign object as input arguments. The second constructor's input arguments are limited to a pointer to an FirFilterDesign object plus references to an input stream (&uin) and output stream (&uout). These streams are used to conduct a short interactive exchange with the user to obtain the necessary configuration parameters. (Questions are inserted into uout and user responses are extracted from uin.) Typically this constructor would be called with cout as the output stream and cin as the input stream.

The method ComputeMagResp will compute the desired magnitude response, and if Normalize_Enabled is set TRUE, call the method NormalizeResponse. Both ComputeMagResp and NormalizeResponse are provided as separate public methods so that the filter's magnitude response can be recomputed if the coefficients in FirFilterDesign are changed. The "pure" object-oriented approach for dealing with changes to FirFilterDesign would be to delete the obsolete FirFilterResponse object and construct a new one using the updated coefficients from FirFilterDesign. This approach was not taken here because FirFilterResponse will be used in Chap. 15 as part of an iterative design procedure that will make *many* small changes to FirFilterDesign's coefficients and then recompute the magnitude spectrum after each change. Continued deletion and allocation of new FirFilterResponse objects would excessively fragment heap memory and possibly cause an out-of-memory condition on many smaller computers. As long as none of the basic configuration parameters is changed, it is perfectly safe to change coefficients in FirFilterDesign and then recompute the magnitude response without constructing a new FirFilterResponse object each time. However, attempting to change Num_Taps will most likely lead to disaster.

Another advantage in making ComputeMagResp a separate function is that when FirFilterResponse is used as a base class, it will be easier for a derived class to redefine the method that used to compute the magnitude response. As we will see in Sec. 13.3 there are ways to compute the magnitude response of a linear phase FIR filter that require less computation than the general method implemented in FirFilterResponse. These methods will be implemented in a derived class that inherits from FirFilterResponse.

For the class design provided in fir_resp.cpp, the questions posed earlier are answered in the following ways:

- The number of evaluation frequencies is a user-supplied input—either passed in as an argument to the first constructor or read in from stream uin by the interactive constructor.
- The minimum evaluation frequency is assumed to be zero. This assumption is hard-coded into the method ComputeMagResp.
- The maximum evaluation frequency is assumed to be π radians per second. This assumption is hard-coded into the method ComputeMagResp.
- The spacing of the evaluation frequencies is assumed to be uniform between the minimum and the maximum.

- A user-supplied input db_scale_enabled is used to select between linear numeric units or decibels for magnitude response values. This flag is either passed in as an argument to the first constructor or read in from stream uin by the interactive constructor.
- A user-supplied input normalize_enabled is used to select between raw magnitude response values or magnitude responses that are normalized to have a peak passband value of 1.0 (for db_scale_enabled==FALSE) or 0.0 (for db_scale_enabled==TRUE).

11.3 Linear Phase FIR Filters

As discussed in Sec. 4.7, constant group delay is a desireable property for filters to have since nonconstant group delay will cause envelope distortion in modulated-carrier signals and pulse shape distortion in baseband digital signals. A filter's frequency response $H(e^{j\omega})$ can be expressed in terms of amplitude response $A(\omega)$ and phase response $\theta(\omega)$ as

$$H(e^{j\omega}) = A(\omega) e^{j\theta(\omega)}$$

If a filter has a linear phase response of the form

$$\theta(\omega) = -\alpha\omega \quad -\pi \leq \omega \leq \pi \tag{11.6}$$

it will have both constant phase delay τ_p and constant group delay τ_g. In fact, in this case $\tau_p = \tau_g = \alpha$. It can be shown (e.g., [1]) that for $\alpha = 0$, the impulse response is an impulse of arbitrary strength:

$$h[n] = \begin{cases} c & n = 0 \\ 0 & n \neq 0 \end{cases}$$

For nonzero α, it can be shown that Eq. (11.6) is satisfied if and only if

$$\alpha = \frac{N-1}{2} \tag{11.7a}$$

$$h[n] = h[N-1-n] \tag{11.7b}$$

Within the constraints imposed by Eq. (11.7), the possible filters are usually separated into two types. Type 1 filters satisfy Eq. (11.7) with N odd, and type 2 filters satisfy Eq. (11.7) with N even. For type 1 filters, the axis of symmetry for $h[n]$ lies at $n = (N-1)/2$ as shown in Fig. 11.1. For type 2 filters, the axis of symmetry lies midway between $n = N/2$ and $n = (N-2)/2$ as shown in Fig. 11.2.

Filters can have constant group delay without having constant phase delay if the phase response is a straight line that does not pass through the origin. Such a phase response is given by

$$\theta(\omega) = \beta + \alpha\omega \tag{11.8}$$

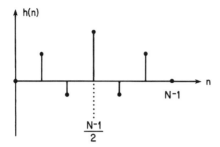

Figure 11.1 Impulse response for a type 1 linear phase FIR filter showing even symmetry about $n = (N-1)/2$.

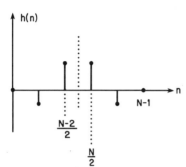

Figure 11.2 Impulse response for a type 2 linear phase FIR filter showing even symmetry about the abscissa midway between $n = (N-2)/2$ and $n = N/2$.

The phase of a filter will satisfy Eq. (11.8) if

$$\alpha = \frac{N-1}{2} \tag{11.9a}$$

$$\beta = \pm\frac{\pi}{2} \tag{11.9b}$$

$$h[n] = -h[N-1-n] \quad 0 \le n \le N-1 \tag{11.9c}$$

An impulse response satisfying Eq. (11.9c) is said to be *odd symmetric*, or *antisymmetric*. Within the constraints imposed by Eq. (11.9), the possible filters can be separated into two types that are commonly referred to as type 3 and type 4 *linear phase* filters despite the fact that the phase response is *not truly linear*. [The phase response is a straight line, but it does not pass through the origin, and consequently $\theta(\omega_1 + \omega_2)$ does not equal $\theta(\omega_1) + \theta(\omega_2)$.] Type 3 filters satisfy Eq. (11.9) with N odd, and type 4 filters satisfy Eq. (11.9) with N even. For type 3 filters, the axis of antisymmetry for $h[n]$ lies at $n = (N-1)/2$ as shown in Fig. 11.3. When $n = (N-1)/2$ with N even, Eq. (11.9c) yields

$$h\left[\frac{N-1}{2}\right] = -h\left[\frac{N-1}{2}\right]$$

Therefore, $h[(N-1)/2]$ must always equal zero in type 3 filters. For type 4 filters, the axis of antisymmetry lies midway between $n = N/2$ and $n = (N-2)/2$ as shown in Fig. 11.4.

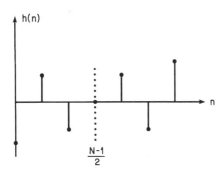

Figure 11.3 Impulse response for a type 3 linear phase FIR filter showing odd symmetry about $n = (N-2)/2$.

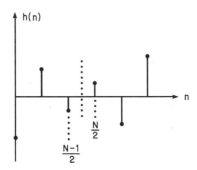

Figure 11.4 Impulse response for a type 4 linear phase FIR filter showing odd symmetry about the abscissa midway between $n = (N-2)/2$ and $n = N/2$.

Software design

The definition for derived class `LinearPhaseFirDesign` is provided in file `lin_dsgn.cpp`. This class does not add much functionality to that already provided in the base class `FirFilterDesign`. In fact, this class is not meant to be instantiated as an object on its own. It was created simply to provide a common base from which several specific types of linear phase FIR filter classes can be derived. (In Chap. 12, we will derive a class `FirIdealFilter` for representing linear phase FIR approximations to *ideal* filters, and in Chap. 13 we will derive the class `FreqSampFirDesign` for representing a type of linear phase FIR that is designed by optimizing the value of samples in the transition band of the filter's magnitude response.) The important additions provided by `LinearPhaseFirFilter` are the two new data members, `Band_Config` and `Fir_Type`. These are two attributes of linear phase FIR filters that are not included in the base class `FirFilterDesign`. (Note: The type `BAND_CONFIG_TYPE` is an enumeration that is defined in file `typedefs.h`.)

Frequency response of linear phase FIR filters

The discrete-time Fourier transform (DTFT) can be used directly, as it is in `FirFilterResponse`, to obtain the frequency response of any FIR filter. However, for the special case of linear phase FIR filters, the symmetry and *realness* properties of the impulse response can be used to modify the general DTFT to obtain dedicated formulas having reduced computational burdens.

TABLE 11.2 Frequency Response Formulas for Linear Phase FIR Filters

$h(nT)$ symmetry	N	$H(e^{j\omega T})$	$A(e^{j\omega T})$
Even	Odd	$e^{-j\omega(N-1)T/2}A(e^{j\omega T})$	$\sum_{k=0}^{(N-1)/2} a_k \cos \omega kT$
Even	Even	$e^{-j\omega(N-1)T/2}A(e^{j\omega T})$	$\sum_{k=1}^{N/2} b_k \cos\left[\omega\left(k-\frac{1}{2}\right)T\right]$
Odd	Odd	$e^{-j[\omega(N-1)T/2-\pi/2]}A(e^{j\omega T})$	$\sum_{k=1}^{(N-1)/2} a_k \sin \omega kT$
Odd	Even	$e^{-j[\omega(N-1)T/2-\pi/2]}A(e^{j\omega T})$	$\sum_{k=1}^{N/2} b_k \sin\left[\omega\left(k-\frac{1}{2}\right)T\right]$

NOTE: $a_0 = h\left[\frac{(N-1)/T}{2}\right]$ $a_k = 2h\left[\left(\frac{N-1}{2}-k\right)T\right]$ $b_k = 2h\left[\left(\frac{N}{2}-k\right)T\right]$

The frequency response $H(e^{j\omega T})$ and amplitude response $A(e^{j\omega T})$ are listed in Table 11.2 for the four types of linear phase FIR filters. The properties of these four types are summarized in Table 11.3.

At first glance, the fact that $A(\omega)$ is periodic with a period of 4π for type 2 and type 4 filters seems to contradict the fundamental relationship between sampling rate and folding frequency that was established in Chap. 6. The difficulty lies in how we have defined $A(\omega)$. The frequency response $H(\omega)$ is, in fact, periodic in 2π for all four types, as we would expect. Both $\text{Re}[H(\omega)]$ and $\text{Im}[H(\omega)]$ are periodic in 2π, but factors of -1 are allocated between $A(\omega)$ and $\theta(\omega)$ differently over the intervals $(0, 2\pi)$ and $(2\pi, 4\pi)$ so that $\theta(\omega)$ can be made linear [and $A(\omega)$ can be made analytic].

Some of the properties listed in Table 11.3 have an impact on which types can be used in particular applications. As a consequence of odd symmetry about $\omega = 0$, types 3 and 4 always have $A(0) = 0$ and should therefore not be used for lowpass or bandstop filters. As a consequence of their odd symmetry about $\omega = \pi$, types 2 and 3 always have $A(\pi) = 0$ and should therefore not be used for highpass or bandstop filters. Within the bounds of these

TABLE 11.3 Properties of FIR Filters Having Constant Group Delay

	Type			
	1	2	3	4
Length, N	Odd	Even	Odd	Even
Symmetry about $\omega = 0$	Even	Even	Odd	Odd
Symmetry about $\omega = \pi$	Even	Odd	Odd	Even
Periodicity	2π	4π	2π	4π

restrictions, the choice between an odd-length or even-length filter is often made so that the desired transition frequency falls as close as possible to the midpoint between two sampled frequencies. The phase response of types 3 and 4 includes a constant component of 90° in addition to the linear component. Therefore, these types are suited for use as differentiators and Hilbert transformers (see [1]).

Software design

The equations of Table 11.2 are implemented in the class `LinearPhaseFirResponse`. The two member functions for this derived class are provided in file `lin_resp.cpp`. The constructor simply invokes the constructor for the base class `FirFilterResponse` using initializer syntax. The member function `ComputeMagResp` redefines the like-named function that is declared virtual in the base class. This redefined function will compute the magnitude response using the appropriate equation from Table 11.2, and if `Normalize_Enabled` is set TRUE, call the base class method `NormalizeResponse`.

11.4 Structures for FIR Realizations

A number of different structures can be used to realize FIR filters. For any given set of filter coefficients, all of these different structures are equivalent, assuming that they are implemented with infinite-precision arithmetic. However, when implemented using finite-precision arithmetic with quantized coefficients and quantized signals, the different structures can exhibit vastly different behavior. Therefore, some care needs to be exercised in the selection of a structure for implementing a particular filter. This section will explore several of the most common implementation structures used for FIR filters.

Direct form

As revealed in prior sections, an FIR filter can be represented by a difference equation of the form

$$y[n] = \sum_{k=0}^{N-1} h[k]\, x[n-k] \tag{11.10}$$

where $h[k]$ is the impulse response of the filter. Equation (11.10) maps directly into the *direct form* FIR structure shown in Fig. 11.5. Recall that the interpretation of a signal flow graph such as Fig. 11.5 is subject to the following rules:

- The gain of each branch is indicated near an arrowhead in the center of the branch.
- An unlabeled branch is assumed to have unity gain.

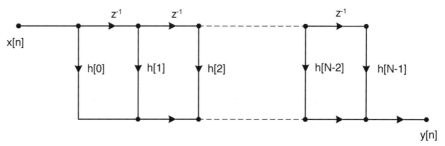

Figure 11.5 Direct form realization of an FIR filter.

- An indicated gain of z^{-M} is used to denote a delay of M sample times, hence an indicated gain of z^{-1} denotes a unit delay.
- All of the branch signals going into a node are added together to obtain the signal out of the node.

In some of the literature that discusses applications such as adative equalization, the structure of Fig. 11.5 is referred to as a *tapped delay line* or *transversal filter* and is drawn in the form shown in Fig. 11.6.

Transposed direct form

The transposition theorem can be applied to the direct form structure of Fig. 11.5 to obtain the *transposed direct form* structure shown in Fig. 11.7.

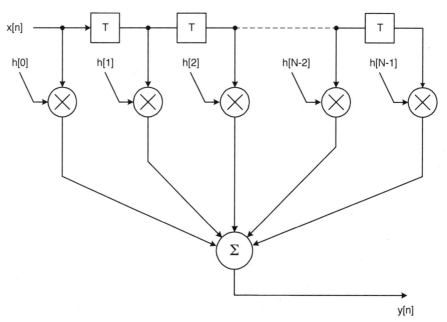

Figure 11.6 Tapped delay line representation of an FIR filter.

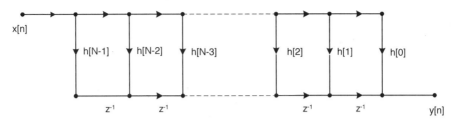

Figure 11.7 Transposed direct form realization of an FIR filter.

Cascade form

The polynomial system function $H(z)$ for an FIR filter can be factored into a product of first- and second-order polynomials to yield:

$$H(z) = \sum_{n=0}^{N} h[n]z^{-n} = \prod_{k=1}^{N_1} \left(\beta_{0k} + \beta_{1k}z^{-1}\right) \prod_{k=1}^{N_2} \left(b_{0k} + b_{1k}z^{-1} + b_{2k}z^{-2}\right) \quad (11.11)$$

where each of the first-order polynomials has as its root a *real* root of $H(z)$, each of the second-order polynomials has as its roots a complex-conjugate pairs of roots from $H(z)$, and N_1, N_2 satisfy $N = N_1 + 2N_2$. For all of the even-order FIR filters examined in this book, the roots of $H(z)$ will always occur in complex-conjugate pairs. For all of the odd-order FIR filters examined in this book, $H(z)$ will have a single real root (usually at $z = -1$), and all of the remaining roots will occur in complex-conjugate pairs. Therefore we can rewrite Eq. (11.11) as

$$H(z) = \begin{cases} \prod_{n=1}^{N/2} \left(b_{0k} + b_{1k}z^{-1} + b_{2k}z^{-2}\right) & N \text{ even} \\ \left(b_{00} + b_{10}z^{-1}\right) \prod_{n=1}^{(N-1)/2} \left(b_{0k} + b_{1k}z^{-1} + b_{2k}z^{-2}\right) & N \text{ odd} \end{cases} \quad (11.12)$$

The form of Eq. (11.12) suggests that the corresponding filter can be implemented as a cascade of second-order sections plus, for odd-order filters, one first-order section. Cascade implementations that use direct form sections are shown in Figs. 11.8 and 11.9. Notice that the single first-order section in Fig. 11.9 is

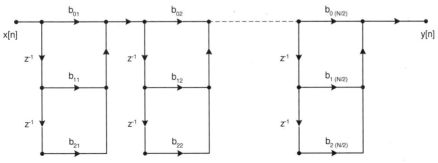

Figure 11.8 Cascade form realization of an even-length FIR filter.

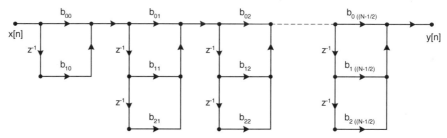

Figure 11.9 Cascade form realization of an odd-length FIR filter.

equivalent to one of the section-order sections with the coefficient b_{2k} set to zero. Often this fact is exploited to collapse the two cases in Eq. (11.12) into the following single case

$$H(z) = \prod_{n=1}^{N_S} \left(b_{0k} + b_{1k}z^{-1} + b_{2k}z^{-2}\right)$$

where $N_S = \lfloor (N+1)/2 \rfloor$, and it is understood that for odd-order filters one of the b_{2k} is zero. (The notation $\lfloor x \rfloor$ represent the *floor* of x which is largest integer that does not exceed x.)

It is possible to obtain additional cascade forms by using the transpose direct form for the second-order sections as in Fig. 11.10, or by applying the transpose theorem to the entire structure of Fig. 11.8. If this second approach is used, each of the sections will be like the sections in Fig. 11.10, but the sequence of the sections will be reversed with b_{0N_S} coming right after the input and b_{01} coming right before the output.

Structures for linear phase FIR filters

The various types of linear phase filters discussed in Sec. 13.2 have impulse responses which exhibit symmetry. This symmetry can be exploited to devise structures that implement linear phase filters using only half the number of multiplications that are required for non-symmetric filters of the same length.

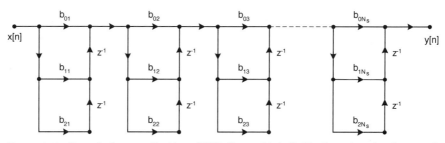

Figure 11.10 Cascade form realization of FIR filter with individual sections implemented in transposed direct form.

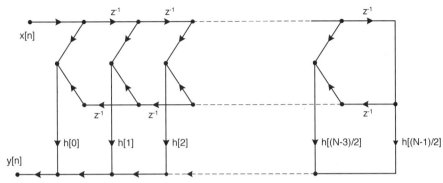

Figure 11.11 Direct form structure for type 1 linear phase FIR filter.

For type 1 filters (which have odd length and even symmetry), the difference equation relating output to input can be written as

$$y[n] = h\left[\frac{N-1}{2}\right] x\left[n - \frac{N-1}{2}\right] + \sum_{k=0}^{(N-3)/2} h[k]\left(x[n-k] + x[n+k+1-N]\right)$$

The corresponding implementation structure is shown in Fig. 11.11.

For type 2 filters (which have even length and even symmetry), the difference equation relating output to input can be written as

$$y[n] = \sum_{k=0}^{N/2-1} h[k]\left(x[n-k] + x[n+k+1-N]\right)$$

The corresponding implementation structure is shown in Fig. 11.12.

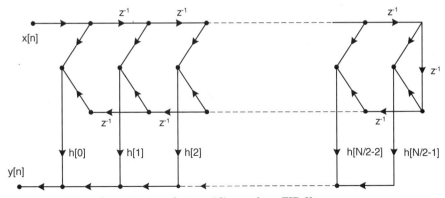

Figure 11.12 Direct form structure for type 3 linear phase FIR filter.

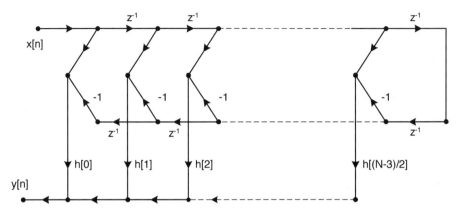

Figure 11.13 Direct form structure for type 3 linear phase FIR filter.

For type 3 filters (which have odd length and odd symmetry), the difference equation relating output to input can be written as

$$y[n] = \sum_{k=0}^{(N-3)/2} h[k]\,(x[n-k] - x[n+k+1-N])$$

The corresponding implementation structure is shown in Fig. 11.13.

Finally, for type 4 filter (which have even length and odd symmetry), the difference equation relating output to input can be written as

$$y[n] = \sum_{k=0}^{N/2-1} h[k]\,(x[n-k] - x[n+k+1-N])$$

The corresponding implementation structure is shown in Fig. 11.14.

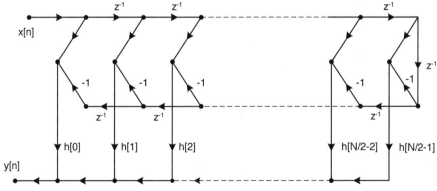

Figure 11.14 Direct form structure for type 4 linear phase FIR filter.

11.5 Assessing the Impacts of Quantization and Finite Precision Arithmetic

Using digital words of finite length to implement FIR filters introduces performance degradations in several different ways. The first source of degradation, namely *coefficient quantization*, is the inability to implement the precise coefficient values that have been determined by the filter design algorithm. In other words, coefficient quantization essentially forces us to implement a filter that is not exactly the filter that we want to implement. However, the resulting filter is still a *linear* system, and all of the linear system analysis techniques from previous chapters still apply. As we will discover in subsequent chapters, some of the design algorithms can take quantization into account and partially mitigate its impact, while other *open-loop* algorithms cannot.

A second source of degradation, *signal quantization*, forces our filter to operate on signal values that may be only approximations of the *true* unquantized signal values. Signal quantization is not a linear process—the quantized sum of two unquantized signals generally is not equal to the sum of the two signals after they have been individually quantized.

A third source of degradation, *finite precision arithmetic*, appears closely related to signal quantization, but there are subtle differences. FIR filter implementations tend to multiply fractional signal values by fractional coefficient values, producing product terms that are smaller than either of the two constituent factors. Then a large number of these small terms are added together to produce output values that are in roughly the same size range as the input values. A quantization scheme that preserves adequate precision in the inputs, outputs and coefficients still can yield terrible performance if the precision of the small intermediate product terms is not preserved.

Direct form realizations

Consider the direct form realization of an FIR filter as depicted in Figs 11.5 and 11.6. Each of the input samples $x[k]$ and each of the coefficients $h[n]$ will be quantized to some finite number of bits. If each $x[k]$ is quantized to B_x bits and each $h[n]$ is quantized to B_h bits, a total of $(B_x + B_h)$ bits will be needed to represent each product term $x[k] \cdot h[n]$ without further loss of precision. Still more bits will be needed to precisely represent the sum of all of these product terms. Most digital filter designs represent a compromise—some performance is sacrificed to allow for easier implementations. How severely will performance be degraded if the filter is implemented with fewer than the required number of bits used to represent each of the arithmetic results? Before we can begin to answer this question, we need to take a closer look at how the filter's performance should be gauged.

In a truely linear system we can take several different paths to arrive at the same answer. For example, we can compute a linear FIR filter's frequency response using two different methods:

1. In any shift-invariant linear discrete-time system, the frequency response can be obtained as the DTFT of the system's impulse response sequence. In the case of a discrete-time linear FIR filter, this means that the frequency response can be obtained as the DTFT of the filter's coefficients.

2. The response of the filter at any one specific frequency can be measured directly by using a sinusoid of the specific frequency as input to the filter and then comparing the amplitude and phase of the output sinusoid to the amplitude and phase of the input sinusoid. If this process is repeated over many closely spaced frequencies, the result will be a sampled version of the frequency response.

For a linear system, the results obtained via these two methods should agree. Unfortunately, signal quantization is not a linear operation. If we use the two methods given above to compute the response for an FIR filter that uses quantized signal values and finite-precision arithmetic, we have no guarantee that the two different methods will yield the same result. In fact, if just the second method is repeated using several different amplitudes for the input sinusoid, the results will generally be slightly different for each amplitude. Coarser quantization will result in larger differences for different input amplitudes; finer quantization will result in smaller differences. Just how different will these results be for a given filter design? Which is the *correct* amplitude to use for evaluating the effects of quantization upon a particular design? Software-based experiments in Secs. 12.3 and 13.7 will help answer these questions.

Software design

Let's assume that an FIR filter is to be implemented in the direct-form tapped delay line structure depicted in Fig. 11.6. We need to design some software to emulate the behavior of this structure including quantization and finite precision effects. For ease of implementation we will perform the arithmetic using the second method's complement arithmetic. In the analysis of quantization effects, it is customary to assume that both the filter coefficients and the signal values have magnitudes that are *strictly less than* 1. For virtually all practical filter designs, this assumption about the coefficients is satisfied. However, all of the windows presented in Chap. 10 have a peak value that *equals* 1. If we multiply a value of 1 by 2^k, the result in binary form is a single *1* followed by k zeros. Such a result is actually a $(k+1)$-bit value. So if we actually were to implement a window using k-bit representations of the coefficients, the coefficients would have to be scaled to reduce the peak to something less than 1. This only becomes an issue if the windows are applied directly to data, as they sometimes are in DFT-based spectral analysis schemes.

A coefficient value can be quantized by multiplying the coefficient h by some quantization factor Q and then truncating the fractional part of the result:

$$h_T = \lfloor hQ \rfloor \tag{11.13}$$

If $Q = 2^k$, the result h_T will be a k-bit integer value. For the quantized coefficient to be a *rounded* approximation of the original coefficient, we need to add 0.5 to hQ before truncating

$$h_R = \left\lfloor hQ + \frac{1}{2} \right\rfloor \tag{11.14}$$

When the coefficient values are signed (i.e., when the coefficients can take on positive or negative values) one bit of the coefficients' binary representation must be reserved for the sign. This means that for k-bit quantization of signed values, a quantization factor of $Q = 2^{k-1}$ should be used instead of $Q = 2^k$. The quantized value of the coefficient often will be more convenient to work with in filter *design* software if it is returned to a floating point representation by dividing by Q.

The class DirectFormFir is provided in file dirform1.cpp. As its name implies, this class is a software implementation of a direct form 1 FIR filter. The constructor accepts an array of filter coefficients, plus two quantizing factors. One of these factors is used by the constructor to quantize the coefficient values, and the other is stored in a class attribute so that it can be used later by the ProcessSample function for quantizing the input signal samples. DirectFormFir is used by a second class, SweptResponse, which is provided in file swept.cpp. SweptResponse generates sinusoids with frequencies that range from nearly zero up to a maximum that equals half the filter's sampling rate. Samples of these sinusoids are provided as input to an instance of DirectFormFir. After each step in frequency, the filter output is allowed to settle for a number of sample times, and then the strength of the output is measured to determine the filter's attenuation at that particular frequency.

Reference

1. Rabiner, L. R. and B. Gold. *Theory and Application of Digital Signal Processing*, Prentice-Hall, Englewood Cliffs, NJ, 1975.

Chapter 12

FIR Filter Design: Window Method

The window method of FIR filter design is built on a more fundamental approach usually called the *Fourier series method*. This chapter begins with an exploration of the Fourier series method before moving on to discuss windowing issues in subsequent sections.

12.1 Fourier Series Method

The Fourier series method of FIR design is based on the fact that the frequency response of a digital filter is periodic, and is therefore representable as a Fourier series. A desired target frequency response is selected and expanded as a Fourier series. This expansion is truncated to a finite number of terms that are then used as the filter coefficients or tap weights. The resulting filter has a frequency response that approximates the original desired target response.

Algorithm 12.1 Designing FIR filters via the Fourier series method.

1. Specify a desired frequency response $H_d(\lambda)$.
2. Specify the desired number of filter taps N.
3. Compute the filter coefficients $h[n]$ for $n = 0, 1, 2, \ldots, N-1$ using

$$h[n] = \frac{1}{2\pi} \int_{-\pi}^{\pi} H_d(\lambda)[\cos(m\lambda) + j\sin(m\lambda)]\, d\lambda \qquad (12.1)$$

where $m = n - (N-1)/2$. When $H_d(\lambda)$ is even symmetric, Eq. (12.1) can be simplified to

$$h[n] = \frac{1}{\pi} \int_{0}^{\pi} H_d(\lambda) \cos(m\lambda)\, d\lambda \qquad (12.2)$$

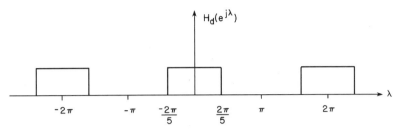

Figure 12.1 Desired frequency response for Example 12.1.

4. Using the techniques presented in Sec. 11.2, compute the actual frequency response of the resulting filter. If the performance is not adequate, change N or $H_d(\lambda)$ and go back to step 3.

Example 12.1 Use the Fourier series method to design a 19-tap FIR filter that approximates the amplitude response of an ideal lowpass filter with a cutoff frequency of 2 kHz assuming a sampling frequency of 5 kHz.

solution The normalized cutoff is $\lambda = 2\pi/5$. The desired frequency response is depicted in Fig. 12.1. Using Eq. (12.2), we can immediately write

$$h[n] = \frac{1}{\pi} \int_0^{2\pi/5} \cos(m\lambda)\, d\lambda$$

Therefore,
$$h[n] = \left. \frac{\sin(m\lambda)}{m\pi} \right|_{\lambda=0}^{2\pi/5}$$

$$= \frac{\sin(2m\pi/5)}{m\pi} \tag{12.3}$$

where $m = n - 9$.

L'Hospital's rule can be used to evaluate Eq. (12.3) for the case of $m = 0$ (that is, $n = 9$):

$$h[9] = \left. \frac{\frac{d}{dm}\sin(2m\pi/5)}{\frac{d}{dm} m\pi} \right|_{m=0}$$

$$= \left. \frac{(2\pi/5)\cos(2m\pi/5)}{\pi} \right|_{m=0}$$

$$= \frac{2}{5} = 0.4$$

Evaluation of Eq. (12.3) is straightforward for $n \neq 9$. The values of $h[n]$ are listed in Table 12.1, and the corresponding magnitude response is shown in Figs. 12.2 and 12.3. Usually, the passband ripples are more pronounced when the vertical axis is in linear units such as numeric magnitude or percentage of peak magnitude, as in

TABLE 12.1 Impulse Response Coefficients for the 19-tap Lowpass Filter of Example 12.1

$$h[0] = h[18] = -0.033637$$
$$h[1] = h[17] = -0.023387$$
$$h[2] = h[16] = 0.026728$$
$$h[3] = h[15] = 0.050455$$
$$h[4] = h[14] = 0.000000$$
$$h[5] = h[13] = 0.075683$$
$$h[6] = h[12] = -0.062366$$
$$h[7] = h[11] = 0.093549$$
$$h[8] = h[10] = 0.302731$$
$$h[9] = 0.400000$$

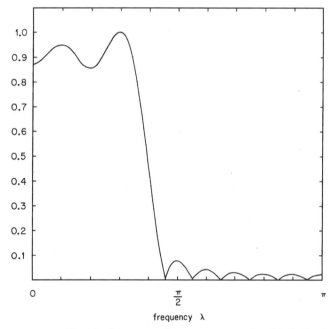

Figure 12.2 Magnitude response (as a percentage of peak) obtained from the 19-tap lowpass filter of Example 12.1.

Fig. 12.2. On the other hand, details of the stopband response are usually more clearly displayed when the vertical axis is in decibels, as in Fig. 12.3.

Properties of the Fourier series method

1. Filters designed using Algorithm 12.1 will exhibit the linear phase property discussed in Sec. 11.3, provided that the target frequency response $H_d(\lambda)$ is either symmetric or antisymmetric.

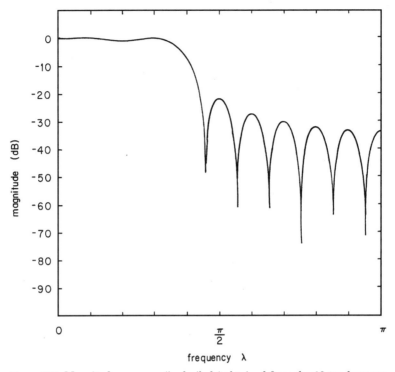

Figure 12.3 Magnitude response (in decibels) obtained from the 19-tap lowpass filter of Example 12.1.

2. As a consequence of the Gibbs phenomenon, the frequency response of filters designed with Algorithm 12.1 will contain undershoots and overshoots at the band edges as exhibited by the responses shown in Figs. 12.2 and 12.3. As long as the number of filter taps remains finite, these disturbances cannot be eliminated by increasing the number of taps. Windowing techniques to reduce the effects of the Gibbs phenomena will be presented later in this chapter.

Algorithm 12.2 can be used to generate impulse response coefficients for an FIR approximation of the ideal lowpass amplitude response shown in Fig. 12.4.

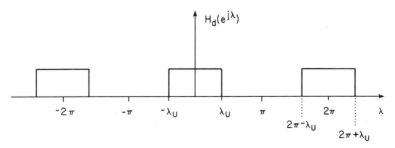

Figure 12.4 Frequency response of ideal lowpass digital filter.

Algorithm 12.2 FIR approximation for ideal lowpass filter.

1. Specify the desired number of filter taps N.
2. For $n = 0, 1, \ldots, \lceil \frac{N-1}{2} \rceil - 1$, compute $h[n]$ and $h[N-1-n]$ as

$$h[n] = h[N-1-n] = \frac{\sin(m\lambda_U)}{m\pi} \tag{12.4}$$

where $m = n - (N-1)/2$.

3. For N odd, compute $h[(N-1)/2]$ as

$$h\left[\frac{N-1}{2}\right] = \frac{\lambda_U}{\pi} \tag{12.5}$$

Software for FIR approximations of ideal filters. An FIR approximation of an ideal filter can be represented in software by the class `FirIdealFilter` which is provided in file `firideal.cpp`. This class is derived from the class `LinearPhaseFirDesign` which in turn is derived from base class `FirFilterDesign` that was described in Chap. 11. The data portion of the derived class is simple, consisting only of two `double` values which hold the ideal filter's cutoff frequencies. (For lowpass and highpass configurations, only one of the frequency values is actually used; both are used for bandpass and bandstop). The derived class also uses two data items that are members of the base class: an `int` value for holding the number of filter taps, and a pointer to a `double` array that will hold the filter coefficients. The base class constructor allocates the array of `double` for the filter coefficients, then the derived class constructor calls the appropriate method for computing the coefficients according to the input argument `band_config`. Each of these four different methods—lowpass, highpass, bandpass, and bandstop—will assume that the specified cutoff frequencies are normalized such that the folding frequency will be π radians/sec (in other words the frequencies are normalized for a sampling interval of $1/(2\pi)$). The method `FirIdealFilter::IdealLowpass` implements Algorithm 12.2.

Highpass filters

The amplitude response for an ideal highpass digital filter is shown in Fig. 12.5.

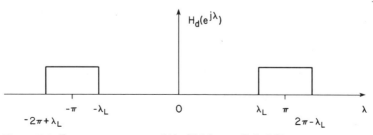

Figure 12.5 Frequency response of ideal highpass digital filter.

TABLE 12.2 Impulse Response Coefficients for the 19-tap Highpass Filter of Example 12.2

$h[0] = h[18] =$	0.033637
$h[1] = h[17] =$	-0.023387
$h[2] = h[16] =$	-0.026728
$h[3] = h[15] =$	0.050455
$h[4] = h[14] =$	0.000000
$h[5] = h[13] =$	-0.075683
$h[6] = h[12] =$	0.062366
$h[7] = h[11] =$	0.093549
$h[8] = h[10] =$	-0.302731
$h[9] =$	0.400000

Algorithm 12.3 FIR approximation for ideal highpass filter.

1. Specify the desired number of filter taps N.
2. For $n = 0, 1, \ldots, \lceil \frac{N-1}{2} \rceil - 1$, compute $h[n]$ and $h[N-1-n]$ as

$$h[n] = h[N-1-n] = \frac{-\sin(m\lambda_L)}{m\pi}$$

   where $m = n - (N-1)/2$.
3. For N odd, compute $h[(N-1)/2]$ as

$$h\left[\frac{N-1}{2}\right] = 1 - \frac{\lambda_L}{\pi}$$

The method `FirIdealFilter::IdealHighpass` implements Algorithm 12.3.

Example 12.2 Use Algorithm 12.3 to design a 19-tap FIR filter that approximates the amplitude response of an ideal highpass filter with a normalized cutoff frequency of $\lambda_U = 3\pi/5$.

solution The coefficients $h[n]$ are listed in Table 12.2, and the resulting frequency response is shown in Figs. 12.6 and 12.7.

Bandpass filters

The amplitude response for an ideal bandpass digital filter is shown in Fig. 12.8.

Algorithm 12.4 FIR approximation for ideal bandpass filter.

1. Specify the desired number of filter taps N.
2. For $n = 0, 1, \ldots, \lceil \frac{N-1}{2} \rceil - 1$, compute $h[n]$ and $h[N-1-n]$ as

$$h[n] = h[N-1-n] = \frac{1}{m\pi}[\sin(m\lambda_U) - \sin(m\lambda_L)] \tag{12.6}$$

   where $m = n - (N-1)/2$.

FIR Filter Design: Window Method   231

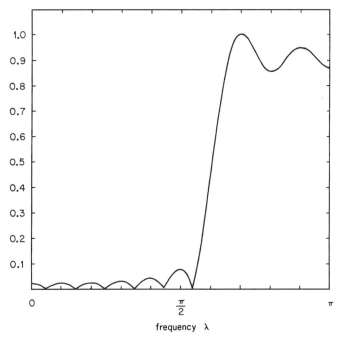

Figure 12.6 Magnitude response (as a percentage of peak) obtained from the 19-tap highpass filter of Example 12.2.

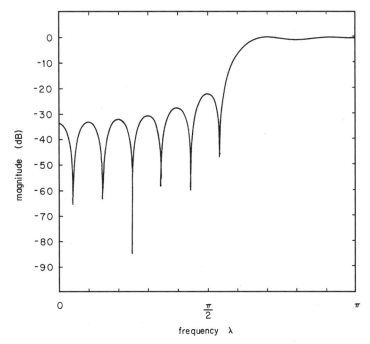

Figure 12.7 Magnitude response (in decibels) obtained from 19-tap highpass filter of Example 12.2.

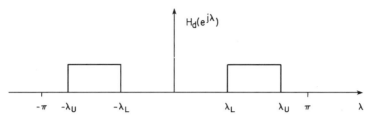

Figure 12.8 Frequency response of ideal bandpass digital filter.

TABLE 12.3 Impulse Response Coefficients for the 17-tap Bandpass Filter of Example 12.3

$h[0] = h[16] =$	0.046774
$h[1] = h[15] =$	0.000000
$h[2] = h[14] =$	−0.100910
$h[3] = h[13] =$	0.000000
$h[4] = h[12] =$	0.151365
$h[5] = h[11] =$	0.000000
$h[6] = h[10] =$	−0.187098
$h[7] = h[9] =$	0.000000
$h[8] =$	0.200000

3. For N odd, compute $h[(N-1)/2]$ as

$$h[(N-1)/2] = \frac{\lambda_U - \lambda_L}{\pi} \qquad (12.7)$$

The method `FirIdealFilter::IdealBandpass` implements Algorithm 12.4.

Example 12.3 Use Algorithm 12.4 to design a 17-tap FIR filter that approximates the amplitude response of an ideal bandpass filter with a passband that extends from $\lambda_L = 2\pi/5$ to $\lambda_U = 3\pi/5$.

solution The coefficients $h[n]$ are listed in Table 12.3, and the resulting frequency response is shown in Fig. 12.9.

Bandstop filters

The amplitude response for an ideal bandstop digital filter is shown in Fig. 12.10.

Algorithm 12.5 FIR approximation for ideal bandstop filter.

1. Specify the desired number of filter taps N.
2. For $n = 0, 1, \ldots, \lceil \frac{N-1}{2} \rceil - 1$, compute $h[n]$ and $h[N-1-n]$ as

$$h[n] = h[N-1-n] = \frac{1}{n\pi}[\sin(m\lambda_L) - \sin(m\lambda_U)] \qquad (12.8)$$

where $m = n - (N-1)/2$.

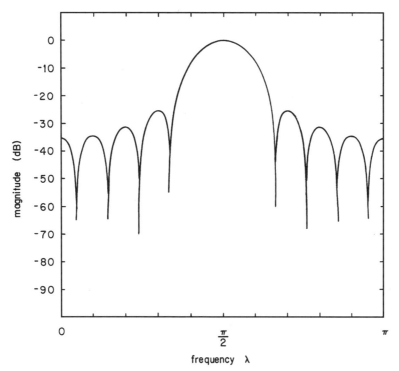

Figure 12.9 Magnitude response (in decibels) obtained from the 17-tap bandpass filter of Example 12.3.

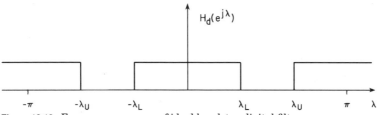

Figure 12.10 Frequency response of ideal bandstop digital filter.

3. For N odd, compute $h[(N-1)/2]$ as

$$h[(N-1)/2] = 1 + \frac{\lambda_L - \lambda_U}{\pi} \qquad (12.9)$$

The method `FirIdealFilter::IdealBandstop` implements Algorithm 12.5.

Example 12.4 Use Algorithm 12.5 to design a 29-tap FIR filter that approximates the amplitude response of an ideal bandstop filter with a stopband that extends from $\lambda_L = 2\pi/5$ to $\lambda_U = 3\pi/5$.

solution The coefficients $h[n]$ are listed in Table 12.4, and the resulting frequency response is shown in Figs. 12.11 and 12.12.

TABLE 12.4 Impulse Response Coefficients for the 29-tap Bandstop Filter of Example 12.4

$h[0] = h[28] =$	-0.043247
$h[1] = h[27] =$	0.000000
$h[2] = h[26] =$	0.031183
$h[3] = h[25] =$	0.000000
$h[4] = h[24] =$	0.000000
$h[5] = h[23] =$	0.000000
$h[6] = h[22] =$	-0.046774
$h[7] = h[21] =$	0.000000
$h[8] = h[20] =$	0.100910
$h[9] = h[19] =$	0.000000
$h[10] = h[18] =$	-0.151365
$h[11] = h[17] =$	0.000000
$h[12] = h[16] =$	0.187098
$h[13] = h[15] =$	0.000000
$h[14] =$	0.800000

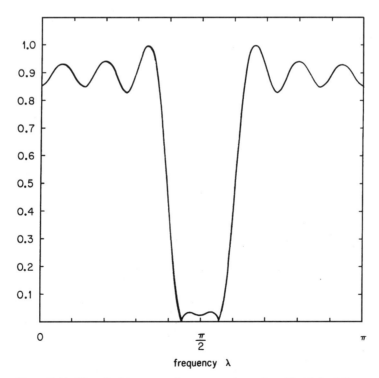

Figure 12.11 Magnitude response (as a percentage of peak) obtained from 29-tap bandstop filter of Example 12.4.

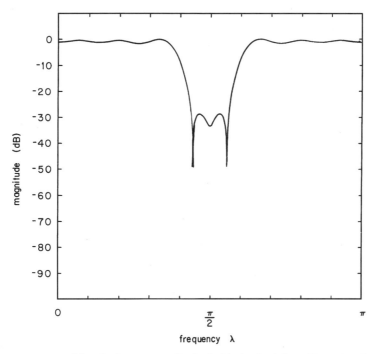

Figure 12.12 Magnitude response (in decibels) obtained from 29-tap bandstop filter of Example 12.4.

12.2 Applying Windows to Fourier Series Filters

Conceptually, a tapering window such as the triangular window is applied to the input of an FIR approximation to an ideal filter. However, since multiplication is associative, a much more computationally efficient implementation can be had by multiplying the window coefficients and the original filter coefficients to arrive at a modified set of filter coefficients. We can accomplish this modification quite simply by adding the following method to the class FirFilterDesign that was introduced in Chap. 11:

```
void FirFilterDesign::ApplyWindow( GenericWindow *window)
   {
   for(int n=0; n<Num_Taps; n++)
        { Imp_Resp_Coeff[n] *= window->GetDataWinCoeff(n); }
   }
```

The following code fragment illustrates how this method can be used in conjunction with a window class and an ideal filter class to produce coefficients for a windowed filter design, then compute the frequency response of the resulting design.

```
disc_window = new TriangularWindow( cin, cout );
num_taps = disc_window->GetNumTaps( );
filter_design = new FirIdealFilter( num_taps, cin, cout );
filter_design->ApplyWindow( disc_window );
```

TABLE 12.5 Coefficients for a 19-tap Triangular-Windowed Lowpass Filter

n	$h[n]$	$w[n]$	$w[n] \cdot h[n]$
0, 18	−0.033673	0.1	−0.0033673
1, 17	−0.023387	0.2	−0.0046774
2, 16	0.026728	0.3	0.0080184
3, 15	0.050455	0.4	0.0201820
4, 14	0.000000	0.5	0.000000
5, 13	−0.075683	0.6	−0.0454098
6, 12	−0.062366	0.7	−0.0436562
7, 11	0.093549	0.8	0.0748392
8, 10	0.302731	0.9	0.2724579
9	0.400000	1.00	0.400000

```
filter_design->DumpCoefficients( &CoeffFile );
filter_response = new FirFilterResponse( filter_design, cin, cout );
filter_response->ComputeMagResp( );
filter_response->DumpMagResp( );
```

This fragment uses the generic FIR magnitude response computation implemented by the class FirFilterResponse. To make use of the more efficient response computations that were derived in Chap. 11 for linear phase FIR filters, the sixth line of the fragment given above must be replaced by the following:

```
filter_response = new LinearPhaseFirResponse(
                (LinearPhaseFirDesign*)filter_design,
                cin, cout );
```

The file prog_12a.cpp contains a main program that provides some user interface support and that allows any of the windowing functions discussed in Chap. 10 to be applied to a Fourier series filter. The filters for Example 12.1 through 12.4 were obtained from this program using a rectangular window.

Example 12.5 Apply a triangular window to the 19-tap lowpass filter of Example 12.1.

solution Table 12.5 lists the original values of the filter coefficients, the corresponding discrete-time window coefficients, and the final value of the filter coefficients after the windowing has been applied. The frequency response of the windowed filter is shown in Figs. 12.13 and 12.14. The response looks pretty good when plotted against a linear axis as in Fig. 12.13, but the poor stopband performance is readily apparent when the response is plotted on a decibel scale as in Fig. 12.14.

Example 12.6 Apply a von Hann window to the 19-tap lowpass filter of Example 12.1.

solution Table 12.6 lists the original values of the filter coefficients, the corresponding discrete-time window coefficients, and the final value of the filter coefficients after the windowing has been applied. The frequency response of the windowed filter is shown in Fig. 12.15.

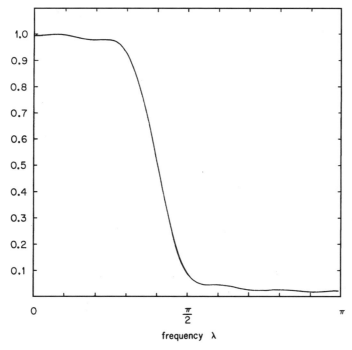

Figure 12.13 Magnitude response (as a percentage of peak) for a triangular-windowed 19-tap lowpass filter.

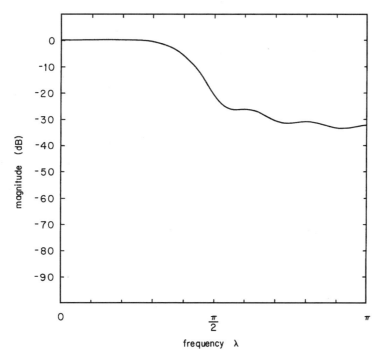

Figure 12.14 Magnitude response (in decibels) for a triangular-windowed 19-tap lowpass filter.

TABLE 12.6 Coefficients for a 19-tap von Hann-Windowed Lowpass Filter

n	$h[n]$	$w[n]$	$w[n] \cdot h[n]$
0, 18	−0.033673	0.0244717	−0.000824036
1, 17	−0.023387	0.0954915	−0.002233260
2, 16	0.026728	0.206107	0.005508828
3, 15	0.050455	0.345492	0.017431799
4, 14	0.000000	0.500000	0.000000000
5, 13	−0.075683	0.654508	−0.049535129
6, 12	−0.062366	0.793893	−0.049511931
7, 11	0.093549	0.904508	0.084615819
8, 10	0.302731	0.975528	0.295322567
9	0.400000	1.00	0.400000000

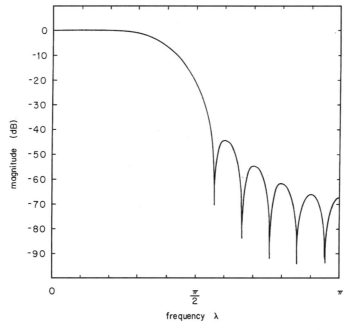

Figure 12.15 Magnitude response for a von Hann-windowed 19-tap lowpass filter.

Example 12.7 Apply a Hamming window to the 19-tap lowpass filter of Example 12.1.

solution The windowed values of $h[k]$ are listed in Table 12.7, and the corresponding frequency response is shown in Fig. 12.16.

12.3 Impacts of Quantization

Designing an FIR filter with quantized coefficients using the window method is straightforward, because the quantization step can be tacked onto the end of

TABLE 12.7 Coefficients for a 19-tap Hamming-Windowed Lowpass Filter

n	$h[n]$	$w[n]$	$w[n] \cdot h[n]$
0, 18	−0.033637	0.080000000	−0.002693840
1, 17	−0.023387	0.107741394	−0.002519748
2, 16	0.026728	0.187619556	0.005014695
3, 15	0.050455	0.310000000	0.015641050
4, 14	0.000000	0.460121838	0.000000000
5, 13	−0.075683	0.619878162	−0.046914239
6, 12	−0.062366	0.770000000	−0.048021820
7, 11	0.093549	0.892380444	0.083481298
8, 10	0.302731	0.972258606	0.294332820
9	0.400000	1.00	0.400000000

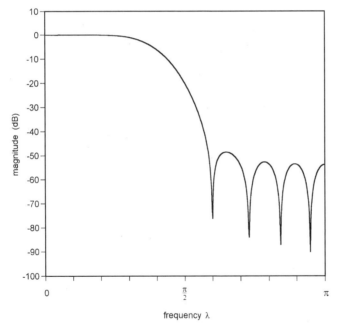

Figure 12.16 Magnitude response for a Hamming-windowed 19-tap lowpass filter.

the usual procedure. Simply obtain the coefficients via the methods of Sec. 12.2 using double precision floating-point arithmetic and then quantize these coefficients to the desired number of bits.

Coefficient quantization

As shown in Chap. 10, the impact of quantization on window responses does not seem to be much cause for concern. Does this also hold for windowed filter

TABLE 12.8 Coefficients for the 19-tap Hamming Windowed Filter of Example 12.8

n	Original $h[n]$	Quantized $h[n]$
0, 18	−0.0026909395	0.0000000
1, 17	−0.0025197730	0.0000000
2, 16	−0.0050147453	0.0078125
3, 15	0.0156410857	0.0156250
4, 14	0.0000000000	0.0000000
5, 13	−0.0469140361	−0.0390625
6, 12	−0.0480217832	−0.0390625
7, 11	−0.0834812342	0.0859375
8, 10	0.2943325199	0.2968750
9	0.4000000000	0.3984375

designs? The coefficients for windowed filters are obtained by multiplying window coefficients and coefficients for an FIR approximation of an ideal filter. Since it is these modified coefficients that will be used to implement the filter, the quantization should be applied to them rather than to the coefficients for either the window or the original filter. In our design software this can be accomplished by adding the method ApplyWindow to the FirFilterDesign class. The filter in question is initialized as the appropriate *ideal* filter and then the ApplyWindow method is called to multiply the filter coefficients by the specified window coefficients. Finally, the method QuantizeCoefficients is called to quantize the windowed coefficients to the desired number of bits.

Example 12.8 Apply a Hamming window to a 19-tap FIR filter that approximates the amplitude response of an ideal lowpass filter with a normalized cutoff frequency of $2\pi/5$. Assess the performance degradation caused by quantizing the coefficients to 8 bits.

solution Since the filter coefficients are signed values, one of the 8 bits must be reserved for the sign, leaving only 7 bits to represent magnitude. Therefore a value of $2^7 = 128$ is used for quant_factor in the call to QuantizeCoefficients. Because the peak window coefficient is 1, and all the *ideal* coefficients will always have magnitudes strictly less than 1, we know that the windowed coefficients will always have magnitudes strictly less than 1. Therefore, it will not be necessary to scale the coefficients prior to quantization. The unquantized coefficients and the corresponding quantized values are listed in Table 12.8, and the DTFT magnitudes for the unquantized and quantized filters are plotted in Fig. 12.17. The spectrum for the quantized filter shows a significant ($\approx$10 dB) degradation in stopband attenuation.

After quantization, the values for coefficients $h[0]$, $h[1]$, $h[17]$, and $h[18]$ have all become zero, thus turning the original 19-tap filter into a 15-tap filter. Since FIR filter coefficients generally tend to have smaller magnitudes close to the

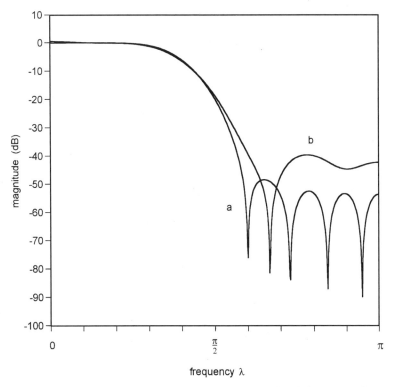

Figure 12.17 Magnitude response for the (a) unquantized and (b) quantized 21-tap Hamming windowed lowpass filter of Example 12.8.

ends or tails of the impulse response, it is common for quantization to reduce the length of the filter by changing one or more of the *tail* coefficients to zero values, as it has in this example. Assuming we had the capability to implement a 19-tap filter, this design would not fully utilize this capability. What we need is a filter design with more than 19 taps that becomes a 19-tap filter after the coefficients are quantized. If we insist on 8-bit quantization of Hamming-windowed coefficients, such a filter cannot be found. A 27-tap filter becomes a 15-tap filter after quantization, and a 29-tap filter becomes a 23-tap filter after quantization. In fact, filters with 29, 31, or 33 taps all become 23-tap filters after quantization.

Example 12.9 An 8-bit quantized 23-tap Hamming-windowed lowpass filter with a normalized cutoff frequency of $2\pi/5$ can be obtained from the corresponding unquantized filter with 29, 31, or 33 taps. Determine which of the three unquantized filters results in the 23-tap quantized design with the best stopband attenuation performance.

solution The coefficients for a 23-tap unquantized filter the 23 *active* coefficients for each of the three quantized filters are listed in Table 12.9, and the DTFT magnitudes for each of the filters are plotted in Figs. 12.18 through 12.21. Each of the three

TABLE 12.9 Coefficients for the 23-tap Hamming Windowed Filters of Example 12.9

n	Unquantized 23-tap	Quantized 29-tap	Quantized 31-tap	Quantized 33-tap
0, 22	0.00220167727	0.0078125	0.0078125	0.0078125
1, 21	−0.00000000193	0.0000000	0.0000000	0.0000000
2, 20	−0.00514720893	0.0000000	−0.0078125	−0.0078125
3, 19	−0.00558402677	0.0000000	0.0000000	−0.0078125
4, 18	0.00932574033	0.0156250	0.0156250	0.0156250
5, 17	0.02394272404	0.0312500	0.0312500	0.0390625
6, 16	−0.00000001184	0.0000000	0.0000000	0.0000000
7, 15	−0.05533091829	−0.0546875	−0.0546875	−0.0546875
8, 14	−0.05246446701	−0.0468750	−0.0468750	−0.0468750
9, 13	0.08671768458	0.0859375	0.0859375	0.0937500
10, 12	0.29708983428	0.2968750	0.2968750	0.2968750
11	0.39999998044	0.3984375	0.3984375	0.3984375

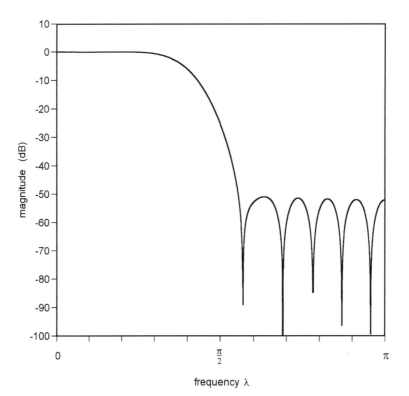

Figure 12.18 Magnitude response for the unquantized 23-tap Hamming-windowed lowpass filter of Example 12.9.

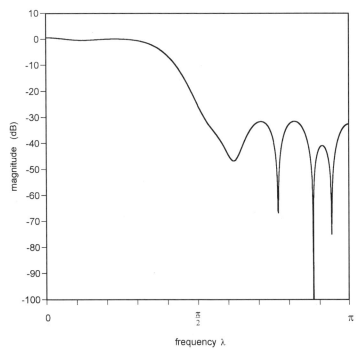

Figure 12.19 Magnitude response for the 23-tap filter that is obtained in Example 12.9 by quantizing a 29-tap Hamming-windowed lowpass filter.

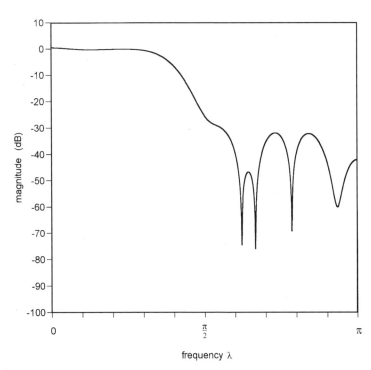

Figure 12.20 Magnitude response for the 23-tap filter that is obtained in Example 12.9 by quantizing a 31-tap Hamming-windowed lowpass filter.

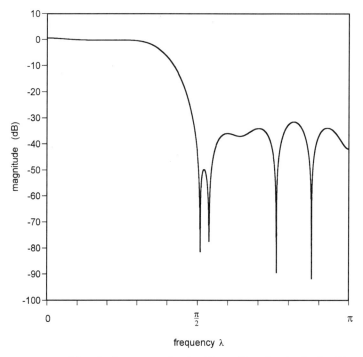

Figure 12.21 Magnitude response for the 23-tap filter that is obtained in Example 12.9 by quantizing a 33-tap Hamming-windowed lowpass filter.

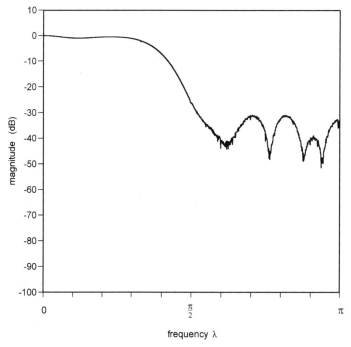

Figure 12.22 Swept-tone magnitude response for the 23-tap filter that is break obtained in Example 12.10 by quantizing a 29-tap Hamming-windowed lowpass filter.

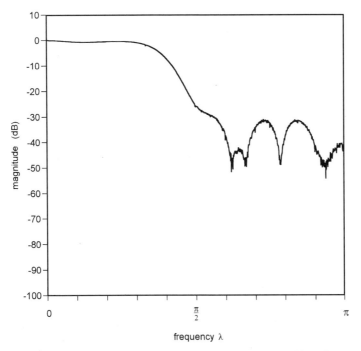

Figure 12.23 Swept-tone magnitude response for the 23-tap filter that is obtained in Example 12.10 by quantizing a 31-tap Hamming-windowed lowpass filter.

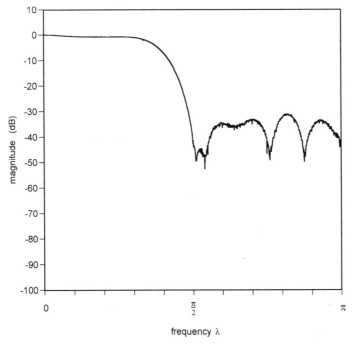

Figure 12.24 Swept-tone magnitude response for the 23-tap filter that is obtained in Example 12.10 by quantizing a 33-tap Hamming-windowed lowpass filter.

quantized filters has approximately the same level of peak ripple (≈ -31.5 dB) in the stopband. The 23-tap filter obtained from quantization of a 33-tap filter clearly shows the best transition performance dropping into a deep (≈ -70 dB) null around a normalized frequency of 0.51π. The other two quantized filters have attenuations that are less than 28 dB at this same frequency.

Signal quantization

So far we have looked at how quantized coefficient values can degrade the performance of windowed filter designs. Now we will use the swept-tone method that was described in Sec. 11.5 to see how the combination of coefficient quantization and signal quantization might further degrade performance.

Example 12.10 Repeat Example 12.9 using the swept-tone method to determine how each of the three quantized filters will perform when the input signal is quantized to 8 bits.

solution The swept responses for the three quantized filters are plotted in Figs. 12.22 through 12.24. As shown in the figures, quantization of the input signals degrades performance by only a small amount beyond the degradation caused by quantization of the coefficients alone.

Chapter 13

FIR Filter Design: Frequency Sampling Method

13.1 Introduction

In Chap. 12, the desired frequency response for an FIR filter is specified in the continuous-frequency domain, and the discrete-time impulse response coefficients are obtained via the Fourier series. We can modify this procedure so that the desired frequency response is specified in the discrete-frequency domain and then use the *inverse discrete Fourier transform* (IDFT) to obtain the corresponding discrete-time impulse response.

Example 13.1 Consider the case of a 21-tap lowpass filter with a normalized cutoff frequency of $\lambda_U = 3\pi/7$. The sampled magnitude response for positive frequencies is shown in Fig. 13.1. The normalized cutoff frequency λ_U falls midway between $n = 4$ and $n = 5$, and the normalized folding frequency of $\lambda = \pi$ falls midway between $n = 10$ and $n = 11$. (Note that $4.5/10.5 = 3/7$.) We assume that $H_d(-n) = H_d(n)$ and use the inverse DFT to obtain the filter coefficients listed in Table 13.1. The actual continuous-frequency response of an FIR filter having these coefficients is shown in Figs. 13.2 and 13.3. Figure 13.2 is plotted against a linear ordinate, and dots are placed at points corresponding to the discrete frequencies specified in Fig. 13.1. Figure 13.3 is included to provide a convenient baseline for comparison of subsequent plots that will have to be plotted against decibel ordinates in order to show low stopband levels.

The ripple performance in both the passband and stopband responses can be improved by specifying one or more transition-band samples at values somewhere between the passband value of $H_d(m) = 1$ and the stopband value of $H_d(m) = 0$. Consider the case depicted in Fig. 13.4, where we have modified the response of Fig. 13.1 by introducing a one-sample transition band by setting $H_d(5) = 0.5$. The continuous-frequency response of this modified filter is shown in Fig. 13.5, and the coefficients are listed in Table 13.2.

The peak stopband ripple has been reduced by 13.3 dB. An even greater reduction can be obtained if the transition-band value is optimized rather than

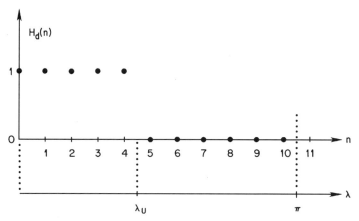

Figure 13.1 Desired discrete-frequency magnitude response for a low-pass filter with $\lambda_U = 3\pi/7$.

TABLE 13.1 Coefficients for the 21-tap Filter of Example 13.1

$h[0] = h[20] =$	0.037334
$h[1] = h[19] =$	-0.021192
$h[2] = h[18] =$	-0.049873
$h[3] = h[17] =$	0.000000
$h[4] = h[16] =$	0.059380
$h[5] = h[15] =$	0.030376
$h[6] = h[14] =$	-0.066090
$h[7] = h[13] =$	-0.085807
$h[8] = h[12] =$	0.070096
$h[9] = h[11] =$	0.311490
$h[10] =$	0.428571

arbitrarily set halfway between the passband and the stopband levels. It is also possible to have more than one sample in the transition band. The methods for optimizing transition-band values are iterative and involve repeatedly computing sets of impulse response coefficients and the corresponding frequency responses. Therefore, before moving to specific optimization approaches we will examine some of the mathematical details and explore some ways for introducing computational efficiency into the process.

13.2 Odd *N* versus Even *N*

Consider the desired response shown in Fig. 13.6 for the case of an odd-length filter with no transition band. If we assume that the cutoff lies midway between

FIR Filter Design: Frequency Sampling Method 249

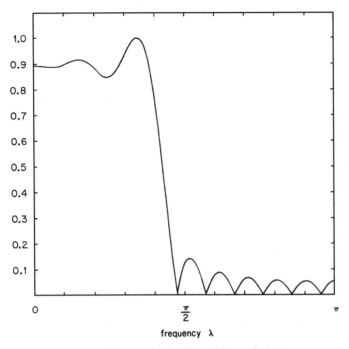

Figure 13.2 Magnitude response for filter of Example 31.1.

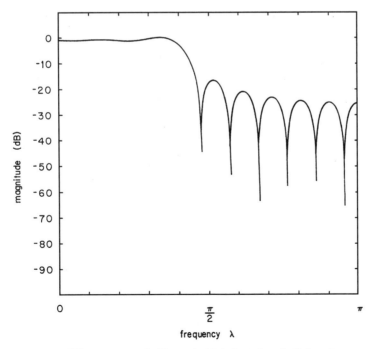

Figure 13.3 Filter response for Example 13.1 plotted on decibel scale.

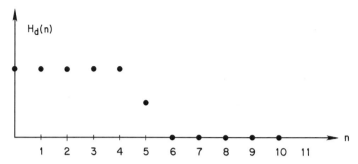

Figure 13.4 Discrete-frequency magnitude response with one transition-band sample midway between the ideal passband and stopband levels.

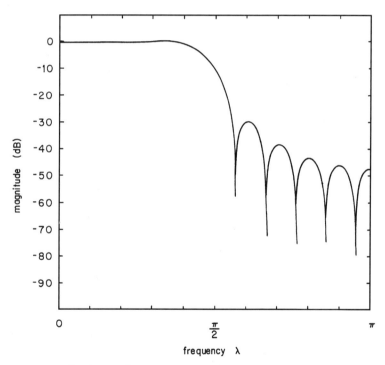

Figure 13.5 Continuous-frequency magnitude response corresponding to the discrete-frequency response of Fig. 13.3.

$n = n_p$ and $n = n_p + 1$ as shown, the cutoff frequency is $2\pi F(n_p + 1/2)$, where F is the interval between frequency domain samples. For the normalized case where $T = 1$, we find $F = 1/N$, so the normalized cutoff is given by

$$\lambda_U = \frac{\pi(2n_p + 1)}{N} \tag{13.1}$$

This equation allows us to compute the cutoff frequency when n_p and N are

TABLE 13.2 Coefficients for the 21-tap Filter with a Single Transition-Band Sample Value of 0.5

$h[0] = h[20] =$	0.002427
$h[1] = h[19] =$	0.008498
$h[2] = h[18] =$	-0.010528
$h[3] = h[17] =$	-0.023810
$h[4] = h[16] =$	0.016477
$h[5] = h[15] =$	0.047773
$h[6] = h[14] =$	-0.020587
$h[7] = h[13] =$	-0.096403
$h[8] = h[12] =$	0.023009
$h[9] = h[11] =$	0.315048
$h[10] =$	0.476190

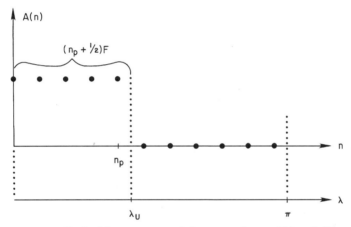

Figure 13.6 Desired frequency-sampled response for an odd-length filter with no transition-band samples.

given. However, in most design situations we will need to start with known (desired) values of N and λ_U and then determine n_p. We can solve Eq. (13.1) for n_p, but for an arbitrary value λ_U, the resulting value of n_p might not be an integer. Therefore, we write

$$n_p = \left\lfloor \frac{N\lambda_{UD}}{2\pi} - \frac{1}{2} \right\rfloor \tag{13.2}$$

where λ_{UD} denotes the desired λ_U and $\lfloor \cdot \rfloor$ denotes the floor function that yields the largest integer less than or equal to the argument. Equation (13.2) yields a value for n_p that guarantees that the cutoff will lie somewhere between n_p and $n_p + 1$, but not necessarily at the midpoint. The difference $\Delta\lambda = |\lambda_U - \lambda_{UD}|$ is an indication of how good the choices of n_p and N are—the smaller $\Delta\lambda$ is, the better the choices are.

It is a common practice to assume that the cutoff frequency lies midway between $n = n_p$ and $n = n_p + 1$ as in the preceding analysis. If the continuous-frequency amplitude response is a straight line between $A(n) = 1$ at $n = n_p$ and $A(n) = 0$ at $n = n_p + 1$, the value of the response midway between these points will be 0.5. However, since $A(n)$ is the *amplitude* response, the attenuation at the assumed cutoff is 6 dB. For an attenuation of 3 dB, the cutoff should be assigned to lie at a point which is 0.293 to the right of n_p and 0.707 to the left of $n_p + 1$.

If we assume that the cutoff lies at $n_p + 0.293$, the cutoff frequency is $2\pi F(n_p + 0.293)$ and the normalized cutoff is given by

$$\lambda_U = \frac{2\pi(n_p + 0.293)}{N} \tag{13.3}$$

The required number of samples in the two-sided passband is $2n_p + 1$ where

$$n_p = \left\lfloor \frac{N\lambda_{UD}}{2\pi} - 0.293 \right\rfloor$$

For convenience we will denote the λ_U given by Eq. (13.1) as λ_6 and the λ_U given by Eq. (13.3) as λ_3.

Even N

Now let us consider the response shown in Fig. 13.7 for the case of an even-length filter with no transition band. If we assume that the cutoff lies midway

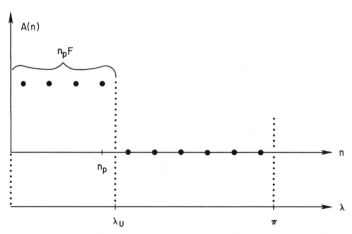

Figure 13.7 Desired frequency-sampled response for an even-length filter with no transition-band samples.

between $n = n_p$ and $n = n_p + 1$, the cutoff frequency is $2\pi F n_p$, and the normalized cutoff is

$$\lambda_6 = \frac{2\pi n_p}{N}$$

Solving for n_p and using the floor function to ensure integer values, we obtain

$$n_p = \left\lfloor \frac{N\lambda_{6D}}{2\pi} \right\rfloor$$

If we assume that the cutoff lies at $n_p + 0.293$, the cutoff frequency is $2\pi F(n_p - 0.207)$ and the normalized cutoff is

$$\lambda_3 = \frac{2\pi(n_p - 0.207)}{N}$$

The required number of samples in the two-sided passband $2n_p$ where

$$n_p = \left\lfloor \frac{N\lambda_{3D}}{2\pi} + 0.207 \right\rfloor$$

If processing constraints or other implementation considerations place an upper limit $N_{\max}$ on the total number of taps that can be used in a particular situation, it sometimes is advantageous to choose between $N = N_{\max}$ and $N = (N_{\max} - 1)$ based on which value of N yields λ_U that is closer to λ_{UD}.

Example 13.2 For $N_{\max} = 21$ and $\lambda_{6D} = 3\pi/7$, determine whether $N = 21$ or $N = 20$ would be the better choice based on values of $\Delta\lambda$.

solution For $N = 20$,

$$n_p = \left\lfloor \frac{20(3\pi/7)}{2\pi} \right\rfloor = \left\lfloor \frac{30}{7} \right\rfloor = 4$$

$$\lambda_6 = \frac{2\pi(4)}{20} = \frac{2\pi}{5}$$

$$\Delta\lambda = \left| \frac{3\pi}{7} - \frac{2\pi}{5} \right| = \frac{\pi}{35}$$

For $N = 21$,

$$n_p = \left\lfloor \frac{21(3\pi/7)}{2\pi} - \frac{1}{2} \right\rfloor = \lfloor 4 \rfloor = 4$$

$$\lambda_6 = \frac{9\pi}{21} = \frac{3\pi}{7}$$

$$\Delta\lambda = \left| \frac{3\pi}{7} - \frac{3\pi}{7} \right| = 0$$

For this contrived case, $N = 21$ is not only the better choice—it is the best choice, yielding $\Delta\lambda = 0$.

Example 13.3 For $N_{\max} = 21$ and $\lambda_{3D} = 2\pi/5$, determine whether $N = 21$ or $N = 20$ would be the better choice based on values of $\Delta\lambda$.

solution For $N = 20$,

$$n_p = \left\lfloor \frac{20(2\pi/5)}{2\pi} + 0.207 \right\rfloor = \lfloor 4.209 \rfloor = 4$$

$$\lambda_3 = \frac{2\pi(4 - 0.207)}{20} = 1.1916$$

$$\Delta\lambda = \left| \frac{2\pi}{5} - 1.1916 \right| = 0.065$$

For $N = 21$,

$$n_p = \left\lfloor \frac{21(2\pi/5)}{2\pi} - 0.293 \right\rfloor = \lfloor 3.907 \rfloor = 3$$

$$\lambda_3 = \frac{2\pi(3.293)}{21} = 0.9853$$

$$\Delta\lambda = \left| \frac{2\pi}{5} - 0.9853 \right| = 0.2714$$

Since $0.065 < 0.2714$, the better choice appears to be $N = 20$.

13.3 Design Formulas

The IDFT can be used as it was in Example 13.1 to obtain the impulse response coefficients $H[n]$ from a desired frequency response that has been specified at uniformly spaced discrete frequencies. However, for the special case of FIR filters with constant group delay, the IDFT can be modified to take advantage of symmetry conditions. In Sec. 13.2, the DTFT was adapted to the four specific types of constant-group-delay FIR filters to obtain the dedicated formulas for $H(\omega)$ and $A(\omega)$ that were summarized in Table 13.2. For the discrete-frequency case, the DFT can be similarly adapted to obtain the explicit formulas for $A(k)$ given in Table 13.3. (The entries in the table are for the normalized case where $T = 1$.) After some trigonometric manipulation, we can arrive at the corresponding inverse relations or *design formulas* listed in Table 13.4. These formulas are implemented by the function FreqSampFilterDesign() provided in file fs_dsgn.cpp.

FIR Filter Design: Frequency Sampling Method

TABLE 13.3 Discrete-Frequency Amplitude Response of FIR Filters with Constant Group Delay

Type	$A[k]$
1 $h[n]$ symmetric N odd	$h[M] + \sum_{n=0}^{M-1} 2h[n]\cos\left[\dfrac{2\pi(M-n)k}{N}\right] = h[M] + \sum_{n=1}^{M} 2h[M-n]\cos\left(\dfrac{2\pi kn}{N}\right)$
2 $h[n]$ symmetric N even	$\sum_{n=0}^{(N/2)-1} 2h[n]\cos\left[\dfrac{2\pi(M-n)k}{N}\right] = \sum_{n=1}^{(N/2)-1} 2h\left[\dfrac{N}{2}-n\right]\cos\left\{\dfrac{2\pi k[n-(1/2)]}{N}\right\}$
3 $h[n]$ antisymmetric N odd	$\sum_{n=0}^{M-1} 2h[n]\sin\left[\dfrac{2\pi(M-n)k}{N}\right] = \sum_{n=1}^{M-1} 2h[M-n]\sin\left(\dfrac{2\pi kn}{N}\right)$
4 $h[n]$ antisymmetric N even	$\sum_{n=0}^{(N/2)-1} 2h[n]\sin\left[\dfrac{2\pi(M-n)k}{N}\right] = \sum_{n=1}^{N/2} 2h\left[\dfrac{N}{2}-n\right]\sin\left\{\dfrac{2\pi k[n-(1/2)]}{N}\right\}$

NOTE: $M = (N-1)/2$.

TABLE 13.4 Formulas for Frequency Sampling Design of FIR Filters with Constant Group Delay

Type	$h[n], \ n = 0, 1, 2, \ldots, N-1$
1 $h[n]$ symmetric N odd	$\dfrac{1}{N}\left\{A(0) + \sum_{k=1}^{M} 2A[k]\cos\left[\dfrac{2\pi(n-M)k}{N}\right]\right\}$
2 $h[n]$ symmetric N even	$\dfrac{1}{N}\left\{A(0) + \sum_{k=1}^{(N/2)-1} 2A[k]\cos\left[\dfrac{2\pi(n-M)k}{N}\right]\right\}$
3 $h[n]$ antisymmetric N odd	$\dfrac{1}{N}\left\{\sum_{k=1}^{M} 2A[k]\sin\left[\dfrac{2\pi(M-n)k}{N}\right]\right\}$
4 $h[n]$ antisymmetric N even	$\dfrac{1}{N}\left\{A[N/2]\sin(\pi(M-n)) + \sum_{k=1}^{(N/2)-1} 2A[k]\sin\left[\dfrac{2\pi(M-n)k}{N}\right]\right\}$

13.4 Frequency Sampling Design with Transition-Band Samples

As mentioned in the introduction to this chapter, the inclusion of one or more samples in a transition band can greatly improve the performance of filters designed via the frequency sampling method. In Sec. 13.1, some improvement was obtained simply by placing one transition-band sample halfway between the passband's unity amplitude and the stopband's zero value. However, even more improvement can be obtained if the value of this single transition-band sample is optimized. Before proceeding, we first need to decide exactly what constitutes an *optimal* value for this sample: we could seek the sample that minimizes passband ripple, minimizes stopband ripple, or minimizes some function that depends on both stopband and passband ripple. The most commonly used approach is to optimize the transition-band value so as to minimize the peak stopband ripple.

For any given set of desired amplitude response samples, determination of the peak stopband ripple entails the following steps:

1. From the specified set of desired amplitude response samples H_d, compute the corresponding set of impulse response coefficients h using the function FreqSampFilterDesign() presented in Sec. 13.3.

2. From the impulse response coefficients generated in Step 1, compute a fine-grained discrete-frequency approximation to the continuous-frequency amplitude response using the ComputeMugResp method belonging to class FreqSampFilterResponse provided in file fs_resp.cpp.

3. Search the amplitude response generated in step 2 to find the peak value in the stopband. This search can be accomplished using the function GetStopbandPeak() belonging to class FreqSampFilterResponse.

In general, we will need five parameters to specify the location of the stopband(s) so that GetStopbandPeak() knows where to search. The first parameter specifies the band configuration—lowpass, highpass, bandpass, or bandstop. The other parameters are indices of the first and last samples in the filter's passbands and stopbands. Lowpass and highpass filters need only two parameters n_1 and n_2, but bandpass and bandstop filters need four: n_1, n_2, n_3, and n_4. The specific meaning of these parameters for each of the basic filter configurations is shown in Fig. 13.8.

To see how this information is used, consider the lowpass case where n_2 is the index of the first stopband sample in the desired response $H_d[n]$. The goal is to find the peak stopband value in the filter's *continuous-frequency* magnitude response. The computer must compute samples of a discrete-frequency approximation to this continuous-frequency response. This approximation should not be confused with the desired response $H_d[n]$, which is also a discrete-frequency magnitude response. The latter contains only N samples, where N is the number of taps in the filter. The approximation to the continuous-frequency response must contain a much larger number of points. For the examples in this chapter, numbers ranging from 120 to 480 have been used. In searching for the peak of a lowpass response, GetStopbandPeak() directs its attention to samples n_s and beyond in the discrete-frequency approximation to the continuous-frequency amplitude response where

$$n_s = \frac{2Ln_2}{N}$$

and

L = number of samples in the one-sided approximation to the continuous response (that is, numPts)

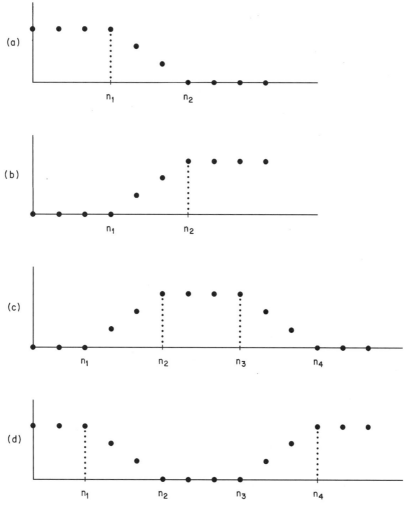

Figure 13.8 Parameters for specifying band configurations: (a) low-pass, (b) high-pass, (c) bandpass, and (d) bandstop.

N = number of taps in the filter

n_2 = index of first sample in the desired positive-frequency stopband

For highpass, bandpass, and bandstop filters, the search is limited to the stopband in a similar fashion.

The approach for finding the peak, as previously outlined in steps 1 through 3, contains some fat (unnecessary computations) that could be eliminated to gain speed at the expense of clarity and modularity. For example, computing the entire amplitude response is not necessary, since only the stopband values are of interest to the optimization procedure. Also, for any given filter, consecutive peaks in the response will be separated by a number of samples that remains more or less constant—this fact could be exploited to compute and examine only those portions of the response falling within areas where stopband ripple peaks can be expected.

Optimization

In subsequent discussions, T_A will be used to denote the value of the single transition-band sample. One simple approach for optimizing the value of T_A is to start with $T_A = 1$ and decrease continually by some fixed increment, evaluating the peak stopband ripple after each decrease. At first, the ripple will decrease each time T_A is decreased, but once the optimal value is passed, the ripple will increase as we continue to decrease T_A. Therefore, once the peak ripple starts to increase, we should decrease the size of the increment and begin *increasing* instead of decreasing T_A. Once peak ripple again stops decreasing and starts increasing, we again decrease the increment and reverse the direction. Eventually, T_A should converge to the optimum value. A slightly more sophisticated strategy for finding the optimum value of T_A is provided by the *golden section search* [1]. This method is based on the fact that the minimum of a function $f(x)$ is known to be bracketed by a triplet of points $a < b < c$ provided that $f(b) < f(a)$ and $f(b) < f(c)$. Once an initial bracket is established, the span of the bracket can be methodically decreased until the three points $a, b,$ and c converge on the abscissa of the minimum. The name golden section comes from the fact that the most efficient search is obtained when the middle point of the bracket is a fractional distance 0.61803 from one endpoint and 0.38197 from the other. A C function GoldenSearch(), provided in file goldsrch.cpp, performs a golden section search tailored for the filter design application.

Example 13.4 For a 21-tap lowpass filter, find the value for the transition-band sample $H_d[5]$ such that the peak stopband ripple is minimized.

solution The program provided in file prog_13a.cpp can be used to solve this example as well as all of the remaining examples in this chapter. The optimal value for $H_d[5]$ is 0.400147, and the corresponding amplitude response is shown in Fig. 13.9. The filter coefficients are listed in Table 13.5. Compared to the case where $H_d[5] = 0.5$, the peak stopband ripple has been reduced by 11.2 dB.

13.5 Optimization with Two Transition-Band Samples

The optimization problem gets a bit more difficult when there are two or more samples in the transition band. Let us walk through the case of a type 1 lowpass filter with 21 taps having a desired response specified by

$$H_d[n] = \begin{cases} 1.0 & 0 \leq |n| \leq 4 \\ H_B & |n| = 5 \\ H_A & |n| = 6 \\ 0.0 & 7 \leq |n| \leq 10 \end{cases}$$

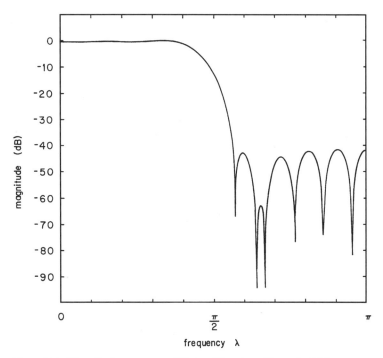

Figure 13.9 Magnitude response of 21-tap filter from Example 13.4.

TABLE 13.5 Coefficients for the Filter in Example 13.4

$h[0] = h[20] =$	0.009532
$h[1] = h[19] =$	0.002454
$h[2] = h[18] =$	−0.018536
$h[3] = h[17] =$	−0.018963
$h[4] = h[16] =$	0.025209
$h[5] = h[15] =$	0.044232
$h[6] = h[14] =$	−0.029849
$h[7] = h[13] =$	−0.094246
$h[8] = h[12] =$	0.032593
$h[9] = h[11] =$	0.314324
$h[10] =$	0.466498

The values of H_A and H_B will be optimized to produce the filter having the smallest peak stopband ripple.

1. Letting $H_B = 1$ and using a stopping tolerance of 0.01 in the single-sample GoldenSearch() function from Sec. 13.4, we find that the peak stopband ripple

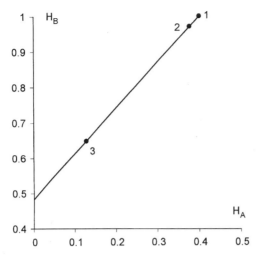

Figure 13.10 Line of steepest descent plotted in the H_A-H_B plane.

is minimized for $H_A = 0.400147$. Thus, we have defined one point in the H_A-H_B plane; specifically $(H_A = 0.400147, H_B = 1.0)$.

2. We define a second point in the plane by setting $H_B = 0.97$ and once again searching for the optimum H_A value that minimizes the peak stopband ripple. This yields a second point at $(0.376941, 0.97)$.

3. The two points $(0.400147, 1)$ and $(0.376941, 0.97)$ then can be used to define a line in the H_A-H_B plane as shown in Fig. 13.10. Our ultimate goal is to determine the ordered pair (H_A, H_B) that minimizes the peak stopband ripple of the filter. In the vicinity of $(H_{A1}, 1)$, the line shown in Fig. 13.10 is the best path to search along, and is called the *line of steepest descent*. On the way to achieving our ultimate goal, a useful intermediate goal is to find the point along the line where the filter's stopband ripple is minimized. To use the single-sample search procedure from Sec. 13.4 to search along this line, we can define positions on the line in terms of their projections onto the H_A axis. To evaluate the filter response for a given value of H_A, we need to have H_B expressed as a function of H_A. The slope of the line is easily determined from points 1 and 2 as

$$m = \frac{1 - 0.97}{0.400147 - 0.376941} = 1.292769$$

Thus we can write

$$H_B = 1.292769 H_A + b \qquad (13.4)$$

where b is the H_B intercept. We can then solve for b by substituting the values for H_A, H_B at point 1 into Eq. (13.4) to obtain

$$b = H_B - 1.292769 H_A$$
$$= 1 - (1.292769)(0.400147) = 0.4827$$

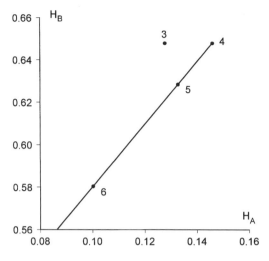

Figure 13.11 Second line of steepest descent.

Thus the line of steepest descent is defined in the H_A-H_B plane as

$$H_B = 1.292769 H_A + 0.4827 \qquad (13.5)$$

The nature of the filter design problem requires that $0 \leq H_A \leq 1$ and $0 \leq H_B \leq 1$. Furthermore, examination of Eq. (13.5) indicates that $H_B < H_A$ for all values of H_A between zero and unity. Thus, the fact that H_B must not exceed unity can be used to further restrict the values of H_A. From step 1, $H_B = 1$ for $H_A = 0.400147$. Therefore, the search along the line is limited to values of H_A such that $0 \leq H_A \leq 0.400147$. The point along the line of Eq. (13.5) where the peak stopband ripple is minimized is found to be (0.127717, 0.647806). The peak stopband ripple at this point is -46.597 dB.

4. The ripple performance of -46.597 is not the best that we can do. The straight line shown in Fig. 13.10 is in fact just an extrapolation from points 1 and 2. Generally, the actual *path* of steepest descent will not be a straight line and will diverge farther from the extrapolated line as the distance from point 1 increases. Thus when we find the optimum point (labeled as point 3) *lying along the straight line*, we really have not found the optimum point *in general*. One way to deal with this situation is to hold H_B constant at the value corresponding to point 3 and then find the optimal value of H_A—without constraining H_A to lie on the line. This results in point 4 as shown in Fig. 13.11. (Figure 13.11 uses a different scale than does Fig. 13.10 so that fine details can be more clearly shown.) The coordinates of point 4 are (0.145898, 0.647806).

5. We now perturb H_B by taking 97 percent of the value corresponding to point 4 [that is, $H_B = (0.97)(0.647806) = 0.62837$]. Searching for the value of H_A that minimizes the peak stopband ripple, we obtain point 5 at (0.132742, 0.62837).

6. The two points (0.127717, 0.647806) and (0.132742, 0.62837) can then be used to define the new line of steepest descent shown in Fig. 13.11. Using the approach discussed above in step 3, we then find the point along the line at

which the stopband ripple is minimized. This point is found to be (0.10022, 0.580328), and the corresponding peak ripple value is -66.6962 dB.

7. We can then continue this process of defining lines of steepest descent and optimizing along the line until the change in peak stopband ripple from one iteration to the next is smaller than some preset limit. Typically, the optimization is terminated when the peak ripple changes by less than 0.1 dB between iterations. Using this criterion, the present design converges after the fifth line of steepest descent is searched to find the point ($H_A = 0.0983006$, $H_B = 0.581637$) where the peak stopband ripple is -70.5209 dB.

Software notes

Optimizing the value of H_A, with H_B expressed as a function of H_A, requires some changes to the way in which the function GoldenSearch() interfaces to the function SetTrans(). In the single-sample case, the search was conducted with H_A as the independent variable. For the two-sample-transition case, the software has been designed to conduct the search in terms of the displacement ρ measured along an arbitrary line. (This approach is more general than it needs to be for the two-sample case, but doing things this way makes extension to three or more samples relatively easy—see Sec. 13.6 for details.) The function GoldenSearch2 provided in file goldsrch.cpp has been modified to include a call to an new overloaded version of SetTrans which takes three parameters. This new function, provided in file fs_spec.cpp, accepts ρ as an input and resolves it into the H_A and H_B components needed for computation of the impulse response and the subsequent estimation of the continuous-frequency amplitude response. The line along which ρ is being measured is specified to SetTrans via the origins and slopes arrays. The values of H_A and H_B corresponding to $\rho = 0$ are passed in origins[1] and origins[2] respectively. The changes in H_A and H_B corresponding to $\Delta\rho = 1$ are passed in slopes[1] and slopes[2], respectively. Setting slopes[1] = 1 and origins[1] = 0 is the correct way to specify $H_A = \rho$. (Note that if we set slopes[1] = 1, origins[1] = 0, slopes[2] = 0, and origins[2] = 0, the single-sample case can be handled as a special case of the two-sample case, since these values are equivalent to setting $H_A = \rho$ and $H_B = 0$.) The iterations of the optimization strategy are mechanized by the function optimize2() provided in file optmiz2.cpp.

Example 13.5 Complete the design of the 21-tap filter that was started at the beginning of this section.

solution As mentioned previously, when GoldenSearch2() is used with a stopping tolerance of 0.01, the example design converges after five lines of steepest descent have been searched. Each line involves 3 points—2 points to define the line plus 1 point where the ripple is minimized. The coordinates and peak stopband ripple levels for the 15 points of the example design are listed in Table 13.6. Each of these points required 8 iterations of GoldenSearch2(). The impulse response coefficients for the filter corresponding to the transition-band values of $H_A = 0.0983006$ and $H_B = 0.581637$ are listed in Table 13.7. The corresponding magnitude response is plotted in Fig. 13.12.

TABLE 13.6 Points Generated in the Optimization Procedure for Example 13.5

Iteration	H_A	H_B	Stopband peak, dB
1	0.400147	1.0	−41.975
2	0.376941	0.97	−42.438
3	0.127717	0.647806	−46.597
4	0.145898	0.647806	−57.384
5	0.132742	0.628372	−60.073
6	0.10022	0.580328	−66.696
7	0.0983006	0.580328	−66.829
8	0.0870643	0.562918	−63.507
9	0.0932756	0.572542	−68.427
10	0.0932756	0.572542	−68.427
11	0.0801199	0.555366	−62.359
12	0.0983006	0.581609	−70.521
13	0.0983006	0.579103	−70.521
14	0.0851449	0.56173	−64.925
15	0.0983006	0.581637	−70.521

TABLE 13.7 Impulse Response Coefficients for the Filter of Example 13.5

$h[0] = h[20] = 0.0026788137$
$h[1] = h[19] = 0.0047456252$
$h[2] = h[18] = -0.0062278465$
$h[3] = h[17] = -0.018152267$
$h[4] = h[16] = 0.0073477613$
$h[5] = h[15] = 0.042014489$
$h[6] = h[14] = -0.0072063371$
$h[7] = h[13] = -0.092181836$
$h[8] = h[12] = 0.0067210347$
$h[9] = h[11] = 0.31350545$
$h[10] = 0.49351023$

Example 13.6 Redesign the filter of Example 13.5 using `tol = 0.001` instead of `tol = 0.01`.

solution The number of iterations required for each point increases from 8 to 15, but the design procedure terminates after only two lines of steepest descent. The coordinates and peak stopband ripple levels for the six points of this design are listed in Table 13.8. The impulse response coefficients are listed in Table 13.9.

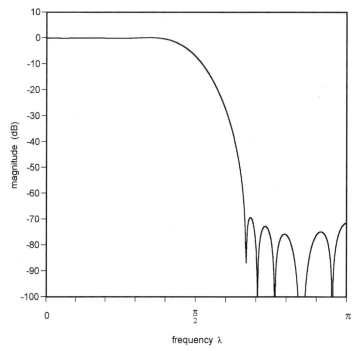

Figure 13.12 Magnitude response for Example 13.5.

TABLE 13.8 Points Generated in the Optimization Procedure for Example 13.6

Iteration	H_A	H_B	Stopband peak, dB
1	0.399413	1.0	−42.080
2	0.377954	0.97	−42.608
3	0.0985806	0.579679	−70.540
4	0.0985806	0.579437	−70.540
5	0.085425	0.562054	−65.168
6	0.0985806	0.579666	−70.540

13.6 Optimization with Three Transition-Band Samples

Just as the two-transition-sample case was more complicated than the single-sample, the three-sample case is significantly more complicated than the two-sample case. Let us consider the case of a type 1 lowpass filter having a desired response as shown in Fig. 13.13. (The following discussion assumes that the three variables H_A, H_B, and H_C are each assigned to one of the three axes in a three-dimensional rectilinear coordinate system.)

TABLE 13.9 Impulse Response Coefficients for the Filter of Example 13.6

$h[0] = h[20] =$	0.0027293958
$h[1] = h[19] =$	0.0047543648
$h[2] = h[18] =$	-0.0063519203
$h[3] = h[17] =$	-0.018198024
$h[4] = h[16] =$	0.0075480155
$h[5] = h[15] =$	0.042071702
$h[6] = h[14] =$	-0.0074728381
$h[7] = h[13] =$	-0.092227119
$h[8] = h[12] =$	0.0070325293
$h[9] = h[11] =$	0.31352249
$h[10] =$	0.49318281

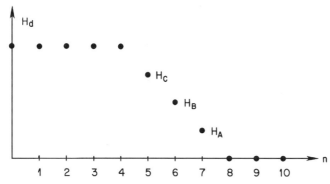

Figure 13.13 Desired response for a 21-tap type 1 filter with three samples in the transition-band.

1. Consider points along the lined defined by $H_C = 1$, $H_B = 1$. (Note: $H_C = 1$ defines a plane parallel to the H_A-H_B plane, and $H_B = 1$ defines a plane that intersects the $H_C = 1$ plane in a line that is parallel to the H_A axis.) Use a single-variable search strategy (such as the golden section search) to locate the point along this line for which the peak stopband ripple is minimized. Denote the value of H_A at this point as H_{A1}.

2. Consider points along the line defined by $H_C = 1$, $H_B = 1 - \epsilon$. Use a single-variable search strategy to locate the point along this line for which the peak stopband ripple is minimized. Denote the value of H_A at this point as H_{A2}.

3. The points $(H_{A1}, 1)$ and $(H_{A2}, 1 - \epsilon)$ define a line in the H_A-H_B plane as shown in Fig. 13.10 for the two-sample case. (Actually the points and the line are in the plane defined by $H_C = 1$, and their *projections* onto the H_A-H_B plane are shown in Fig. 13.10. However, since the planes are parallel, everything looks the same regardless of whether we plot the points in the

$H_C = 1$ plane or their projections in the H_A-H_B, that is, $H_C = 0$, plane.) In the vicinity of $(H_{A1}, 1)$, this line is the line of steepest descent. Search along the line to find the point where the peak stopband ripple is minimized. Denote the values of H_A and H_B at this point as H_{A3} and H_{B3} respectively. As noted previously, the true path of steepest descent is in fact curved, and the straight line just searched is merely an extrapolation based on the two points $(H_{A1}, 1)$ and $(H_{A2}, 1 - \epsilon)$. Thus, the point (H_{A3}, H_{B3}) is not a true minimum. However, this point can be taken as a starting point for a second round of steps 1, 2, and 3, which will yield a refined estimate of the minimum's location. This refined estimate can in turn be used as a starting point for a third round of steps 1, 2, and 3. This cycle of steps 1, 2, and 3 is repeated until the peak ripple at (H_{A3}, H_{B3}) changes by less than some predetermined amount.

13.7 Quantized Coefficients

It is possible to obtain FIR filter coefficients via the methods of the preceeding sections using double precision floating point arithmetic, and then quantize these coefficients to the number of bits required for any particular application.

Example 13.7 Start with the filter design obtained in Example 13.4, and determine how rounding the coefficient magnitudes to 7 bits will impact the frequency response of the filter.

solution The original filter coefficients and the corresponding quantized values are listed in Table 13.10. As shown in Fig. 13.14, the stopband ripple performance is degraded by approximately 10 dB.

TABLE 13.10 Coefficients for the 21-tap Lowpass Filter of Example 13.7

n	Original $h[n]$	Quantized $h[n]$
0, 20	0.00953244	0.0078125
1, 19	0.00245426	0.0000000
2, 18	−0.01853638	−0.0078125
3, 17	−0.01896320	−0.0078125
4, 16	0.02520947	0.0234375
5, 15	0.04423237	0.0468750
6, 14	−0.02984900	−0.0234375
7, 13	−0.09424598	−0.0859375
8, 12	0.03259311	0.0312500
9, 11	0.31432400	0.3125000
10	0.46649783	0.4687500

FIR Filter Design: Frequency Sampling Method 267

The frequency sampling method does provide an opportunity to take quantization into account during the design process and thereby generate only those coefficient sets whose values can be accurately represented in a given number of bits. This quantization is accomplished using the QuantizeCoefficients() function which is a member of the FirFilterDesign class, this class being the base class from which the class FreqSampFilterDesign inherits. This quantization is performed just before the coefficients are used to compute the filter's magnitude response inside the optimization loop.

Example 13.8 Repeat Example 13.4, but this time quantize the coefficients to 7 bits before computing the magnitude responses that are used in the optimization process.

solution The resulting coefficients are listed in Table 13.11. As shown in Fig. 13.15, the stopband ripple performance is degraded by about 5 dB from the unquantized case. Comparison of Figs. 13.14 and 13.15 reveals that, for this particular filter, quantizing *during* the design optimization improves performance by about 5 dB relative to quantizing *after* the design optimization is complete.

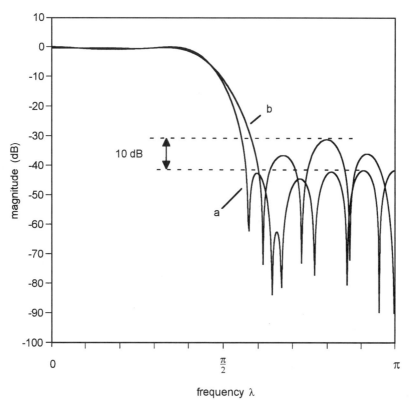

Figure 13.14 Magnitude response of 21-tap filters from (*a*) Example 13.4 and (*b*) Example 13.7.

268 Chapter Thirteen

TABLE 13.11 Coefficients for the 21-tap Lowpass Filter of Example 13.8

n	$h[n]$
0, 20	0.0078125
1, 19	0.0000000
2, 18	−0.0156250
3, 17	−0.0156250
4, 16	0.0234375
5, 15	0.0390625
6, 14	−0.0312500
7, 13	−0.0937500
8, 12	0.0312500
9, 11	0.3125000
10	0.4609375

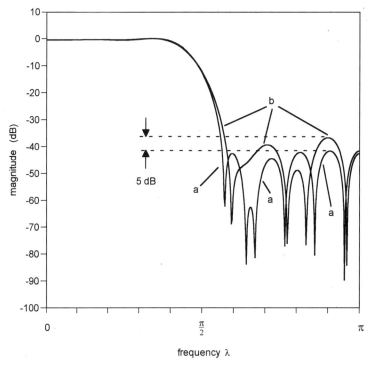

Figure 13.15 Magnitude response of 21-tap filters from (a) Example 13.4 and (b) Example 13.8.

Reference

1. Press W. H., S. A. Teukolsky, W. T. Vetterling, and B. P. Flannery: *Numerical Recipes in C*, 2ed., Cambridge Univ. Press, 1992.

Chapter 14

FIR Filter Design: Remez Exchange Method

In general, an FIR approximation to an ideal lowpass filter will have an amplitude response of the form shown in Fig. 14.1. This response differs from the ideal lowpass response in three quantifiable ways:

1. The pass band has ripples that deviate from unity by $\pm\delta_p$.
2. The stopband has ripples that deviate from zero by $\pm\delta_p$. (Note that Fig. 14.1 shows an *amplitude* response rather than the usual magnitude response, and therefore negative ordinates are possible.)
3. There is a transition-band of finite nonzero width ΔF between the passband and stopband.

The usual design goals are to minimize δ_p, δ_s, and ΔF. As it is generally not possible to minimize simultaneously for three different variables, some compromise is unavoidable. *Chebyshev approximation* is one approach to this design problem.

14.1 Chebyshev Approximation

In the Chebyshev approximation approach, the amplitude response of a type 1 (that is, odd-length, even-symmetric) linear phase lowpass N-tap FIR filter is formulated as a sum of r cosines:

$$A(f) = \sum_{k=0}^{r-1} c_k \cos(2\pi k f) \qquad (14.1)$$

where $r = (N+1)/2$, and the coefficients c_k are chosen so as to yield an $A(f)$ that is optimal (in a sense that will be defined shortly).

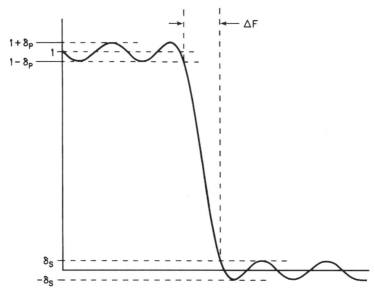

Figure 14.1 Typical amplitude response of an FIR approximation to an ideal lowpass filter.

For a lowpass filter, the passband B_p and stopband B_s are defined as

$$B_p = \{F : 0 \leq F \leq F_p\} \tag{14.2}$$

$$B_s = \{F : F_s \leq F \leq 0.5\} \tag{14.3}$$

where F_p and F_s are, respectively, the edge frequencies for the passband and stopband. [Equation (14.2) is read as "B_p is the set of all F such that F is greater than or equal to zero and less than or equal to F_p.] We then can define a set $\mathcal{F}$ as the union of B_p and B_s:

$$\mathcal{F} = B_p \cup B_s \tag{14.4}$$

In other words, $\mathcal{F}$ is the set of all frequencies between 0 and 0.5 not including the transition frequencies $F : F_p < F < F_s$. In mathematical terms, $\mathcal{F}$ is described as a *compact subset* of [0, 0.5]. The desired response $D(f)$ is the ideal lowpass response given by

$$D(f) = \begin{cases} 1 & F \in B_p \\ 0 & F \in B_s \end{cases} \tag{14.5}$$

Thus, we could define the optimal approach as the one that minimizes the maximum error given by

$$\max_{F \in \mathcal{F}} |D(f) - A(f)| \tag{14.6}$$

However, the maximum error given by Eq. (14.6) treats passband error and stopband error as equally important. A more general approach is to include a weighting function:

$$W(f) = \begin{cases} \frac{1}{K} & F \in B_p \\ 1 & F \in B_s \end{cases} \quad (14.7)$$

which allows stopband errors to be given more importance than passband errors, or vice versa. Thus, we define the maximum approximation error as

$$\|E(f)\| = \max_{F \in \mathcal{F}} W(f)|D(f) - A(f)| \quad (14.8)$$

The crux of the Chebyshev approximation design approach is to identify the coefficients c_k for Eq. (14.1) that minimizes $\|E(f)\|$.

Several examples of FIR design via Chebyshev approximation appear in the early literature [1, 2, 3, 4, 5, 6]. However, the Chebyshev approximation method did not begin to enjoy widespread use until it was shown that the Remez exchange algorithm could be used to design linear phase FIR filters with the Chebyshev error criterion [7]. Use of the Remez exchange algorithm depends on an important mathematical result known as the *alternation theorem*.

Alternation theroem

The response $A(f)$ given by Eq. (14.1) will be the unique, best-weighted Chebyshev approximation to the desired response $D(f)$ if and only if the error function $E(f) = W(f)[D(f) - A(f)]$ exhibits at least $r + 1$ extrema at frequencies in $\mathcal{F}$. The frequencies at which extrema occur are called *extremal frequencies*. Let f_n denote the nth extremal frequency such that

$$f_1 < f_2 < \cdots < f_{n-1} < f_n < f_{n+1} < \cdots < f_r < f_{r+1}$$

Then it can be proven [11] that

$$E(f_n) = -E(f_{n+1}) \quad n = 1, 2, \ldots, r \quad (14.9)$$

and

$$|E(f_n)| = \max_{f \in \mathcal{F}} E(f) \quad (14.10)$$

Together, Eqs. (14.9) and (14.10) mean simply that the error is equal at all the extremal frequencies. Equation (14.9) further indicates that maxima and minima alternate (hence, *alternation* theorem).

14.2 Strategy of the Remez Exchange Method

The alternation theorem given in the previous section tells us how to recognize an optimal set of c_k for Eq. (14.1) when we have one, but it does not tell us how to go about obtaining such c_k. The Remez exchange algorithm provides an approach for finding the FIR filter corresponding to the optimal c_k as follows:

1. Make an initial guess of the $r + 1$ extremal frequencies.

2. Compute the error function corresponding to the candidate set of extremal frequencies (see Sec. 14.3).
3. Search to find the extrema (and therefore the extremal frequencies) of the error function (see Sec. 14.4).
4. Adopt the extremal frequencies found in step 3 as the new set of candidate extremal frequencies and return to step 2.
5. Repeat steps 2, 3, and 4 until the extremal frequencies have converged (see Sec. 14.4).
6. Use the final set of extremal frequencies to compute $P(f)$ and the corresponding impulse response coefficients for the filter (see Sec. 14.5).

The error function mentioned in step 2 is computed as

$$E(f) = W(f)[D(f) - A(f)] \tag{14.11}$$

where $D(f)$ is given by Eq. (14.5) and $W(f)$ is given by Eq. (14.7). Although Eq. (14.1) gives the form of $A(f)$, some other means must be used to evaluate $A(f)$ since the coefficients c_k are unknown. We can obtain $A(f)$ from the extremal frequencies F_k using

$$A(f) = \begin{cases} \gamma_k & \text{for } f = F_0, F_1, \ldots, F_{r-1} \\ \dfrac{\sum_{k=0}^{r-1} \dfrac{\beta_k}{x-x_k} \gamma_k}{\sum_{k=0}^{r-1} \dfrac{\beta_k}{x-x_k}} & \text{otherwise} \end{cases} \tag{14.12}$$

The parameters needed for evaluation of (14.12) are given by

$$\beta_k = \prod_{\substack{i=0 \\ i \neq k}}^{r-1} \frac{1}{x_k - x_i}$$

$$\gamma_k = D(F_k) - (-1)^k \frac{\delta}{W(F_k)}$$

$$\delta = \frac{\sum_{k=0}^{r} \alpha_k D(F_k)}{\sum_{k=0}^{r} \frac{(-1)^k \alpha^k}{W(F_k)}}$$

$$\alpha_k = \prod_{\substack{i=0 \\ i \neq k}}^{r} \frac{1}{x_k - x_i}$$

$$x = \cos(2\pi f)$$

$$x_k = \cos(2\pi F_k)$$

If estimates of the extremal frequencies rather than their *true* values are used in the evaluation of $A(f)$, the resulting error function $E(f)$ will exhibit extrema at frequencies that are different from the original estimates. If the frequencies of these newly observed extrema then are used in a subsequent evaluation of $A(f)$, the new $E(f)$ will exhibit extrema at frequencies that are closer to the true extremal frequencies. If this process is performed repeatedly, the observed extremal frequencies eventually will converge with the true extremal frequencies, which can then be used to obtain $A(f)$ and the filter's impulse response.

Although $A(f)$ is defined over continuous frequency, computer evaluation of $A(f)$ must necessarily be limited to a finite number of discrete frequencies—therefore, $A(f)$ is evaluated over a closely spaced set or *dense grid* of frequencies. The convergence of the observed extremal frequencies will be limited by the granularity of this dense grid, but it has been empirically determined that an average grid density of 16 to 20 frequencies per extremum will be adequate for most designs. Since the maximization of $E(f)$ is only conducted over $f \in \mathcal{F}$, it is not necessary to evaluate $A(f)$ at all within the transition-band. The frequency interval between consecutive points should be approximately the same in both the passband and stopband. Furthermore, the grid should be constructed in such a way that frequency points are provided at $f = 0$, $f = F_p$, $f = F_s$, and $f = 0.5$. An integrated procedure for defining the dense grid and making the initial (equispaced) guesses for the candidate extremal frequencies is provided in Algorithm 14.1.

Algorithm 14.1 Constructing the dense frequency grid.

1. Compute the number of candidate extremal frequencies to be placed in the passband as

$$m_p = \left\lfloor \frac{rF_p}{0.5 + F_p - F_s} - 0.5 \right\rfloor$$

2. Determine the candidate extremal frequencies within the passband as

$$F_k = \frac{kF_p}{m_p} \quad k = 1, 2, \ldots, m_p$$

3. Compute the number of candidate extremal frequencies to be placed in the stopband as

$$m_s = r + 1 - m_p$$

4. Determine the candidate extremal frequencies within the stopband as

$$F_k = F_s + \frac{k(0.5 - F_s)}{m_s - 1} \quad k = 0, 1, \ldots, m_s - 1$$

5. Determine the passband grid frequencies as

$$f_j = jI_p \quad j = 0, 1, 2, \ldots, m_p L$$

where $I_p = \frac{F_p}{m_p L}$
L = average grid density (in points per extremum)

6. Determine the stopband grid frequencies as

$$f_j = F_s + nI_s \quad n = 0, 1, \ldots, (m_s - 1)L$$
$$j = m_p L + n + 1$$

where

$$I_s = \frac{(0.5 - F_s)}{(m_s - 1)L}$$

For computer calculations, the dense grid of frequencies can be implemented by the function `SetupGrid()` which is a member of class `RemezAlgorithm` provided in file `remezalg.cpp`. A grid of actual frequency values is never really created—instead, most of the frequency bookkeeping is done using integers to represent the frequencies' locations within the grid. A call to `GetFrequency()` is used to convert an integer location index into the corresponding floating point frequency value when needed for a calculation.

Generating the desired response and weighting functions

Based on the requirements of the intended application, the desired response function $D(f)$ is defined in accordance with Eq. (14.5) for each frequency $f = f_j$ in the dense grid. For frequency-selective filters, $D(f)$ will usually take on only one of two values—unity in the passband and zero in the stopband. The ideal lowpass response is generated by the member function `DesiredResponse()`.

The passband ripple limit δ_1 and stopband ripple limit δ_2, as shown in Fig. 14.1, are determined by the designer in a manner consistent with the requirements of the intended application. The weight function $W(f)$ is then computed in accordance with Eq. (14.7) with $K = \delta_1/\delta_2$ for each frequency in the dense grid. The member function `WeightFunction()` determines whether the frequency value provided as input lies in the stopband or passband and then returns the appropriate value for $W(f)$.

14.3 Evaluating the Error

Algorithm 14.2 provides a step-by-step procedure for evaluating the error function defined by Eq. (14.11).

Algorithm 14.2 Evaluating the estimation error for the Remez exchange.

1. For $k = 0, 1, \ldots, r-1$, use the candidate extremal frequencies F_k to compute β_k as

$$\beta_k = \prod_{\substack{i=0 \\ i \neq k}}^{r-1} \frac{1}{\cos 2\pi F_k - \cos 2\pi F_i}$$

2. For $k = 0, 1, \ldots, r - 1$, use the β_k from step 1 to compute α_k as

$$\alpha_k = \frac{\beta_k}{\cos 2\pi F_k - \cos 2\pi F_r}$$

3. Use the α_k from step 2 and the extremal frequencies F_k to compute δ as

$$\delta = \frac{\sum_{k=0}^{r} \alpha_k D(F_k)}{\sum_{k=0}^{r} \frac{(-1)^k \alpha_k}{W(F_k)}}$$

where $D(f)$ and $W(f)$ are the desired response and weight functions respectively.

4. For $k = 0, 1, \ldots, r - 1$, use δ from step 3 to compute γ_k as

$$\gamma_k = D(F_k) - (-1)^k \frac{\delta}{W(F_k)}$$

5. Use β_k from step 1, the γ_k from step 4, and the candidate extremal frequencies F_k to compute $A(f)$ for each frequency $f = f_j$ in the dense grid as

$$A(f_j) = \begin{cases} \gamma_k & \text{for } f = F_0, F_1, \ldots, F_{r-1} \\ \frac{\sum_{k=0}^{r-1} \psi_k \gamma_k}{\sum_{k=0}^{r-1} \psi_k} & \text{otherwise} \end{cases}$$

where

$$\psi_k = \frac{\beta_k}{\cos(2\pi f_j) - \cos(2\pi F_k)}$$

6. For each frequency f_j in the dense grid, use $A(f_j)$ from step 5 to compute $E(f_j)$ as

$$E(f_j) = W(f_j)[D(f_j) - A(f_j)]$$

For computer evaluation, the error function is calculated by `RemezError()` which makes use of `ComputeRemezAmplitudeResponse()`. The function `ComputeRemezAmplitudeResponse()` could have been made an integral part of `RemezError()` and designed to automatically generate $A(\ )$ and $E(\ )$ for all frequencies in the dense grid. However, a function in this form would not be useable for generating the uniformly spaced samples of the final $A(\ )$ that are needed to conveniently obtain the impulse response of the filter.

14.4 Selecting Candidate Extremal Frequencies

After Eq. (14.11) has been evaluated, the values of $E(f_j)$ must be checked to determine what the values of F_k should be for the next iteration of the optimization algorithm. Based upon the particular frequencies being checked, the testing can be divided into the five different variations that are described in the following paragraphs. A member function `RemezSearch()` is provided in class `RemezAlgorithm`.

Testing $E(f)$ for $f = 0$

If $E(0) > 0$ and $E(0) > E(f_1)$, then a ripple peak (local maximum) exists at $f = 0$. (Note that f_1 denotes the first frequency within the dense grid after $f = 0$, and due to the way we have defined the frequency spacing within the grid, we know that $f_1 = I_p$.) Even if a peak or valley exists at $f = 0$, it may be a *superfluous* extremum not needed for the next iteration. If a ripple peak does exist at $f = 0$, and $|E(0)| \geq |\rho|$, then the maximum is not superfluous and $f = f_0 = 0$ should be used as the first candidate extremal frequency—in other words, set $F_0 = f_0 = 0$. Similarly, if $E(0) < 0$ and $E(0) < E(f_1)$, a ripple trough (ripple valley, local minimum) exists at $f = 0$. If $|E(0)| \geq |\rho|$, this minimum is not superfluous and we should set $F_0 = f_0 = 0$.

Testing $E(f)$ within the passband and the stopband

The following discussion applies to testing of $E(f)$ for all values of f_j for which $f_0 < f_j < f_p$ or for which $f_s < f_j < 0.5$. A ripple peak exists at f_j if

$$E(f_j) > E(f_{j-1}) \quad \text{and} \quad E(f_j) > E(f_{j+1}) \quad \text{and} \quad E(f_j) > 0 \qquad (14.13)$$

Equation (14.13) can be rewritten as Eq. (14.14) for frequencies in the passband and as Eq. (14.15) for frequencies in the stopband:

$$E(f_j) > E(f_j - I_p) \quad \text{and} \quad E(f_j) > E(f_j + I_p) \quad \text{and} \quad E(f_j) > 0 \qquad (14.14)$$

$$E(f_j) > E(f_j - I_s) \quad \text{and} \quad E(f_j) > E(f_j + I_s) \quad \text{and} \quad E(f_j) > 0 \qquad (14.15)$$

A ripple trough exists at f_j if

$$E(f_j) < E(f_{j-1}) \quad \text{and} \quad E(f_j) < E(f_{j+1}) \quad \text{and} \quad E(f_j) < 0 \qquad (14.16)$$

Equation (14.16) can be rewritten as Eq. (14.17) for frequencies in the passband and as Eq. (14.18) for frequencies in the stopband:

$$E(f_j) < E(f_j - I_p) \quad \text{and} \quad E(f_j) < E(f_j + I_p) \quad \text{and} \quad E(f_j) < 0 \qquad (14.17)$$

$$E(f_j) < E(f_j - I_s) \quad \text{and} \quad E(f_j) < E(f_j + I_s) \quad \text{and} \quad E(f_j) < 0 \qquad (14.18)$$

If either Eqs. (14.13) or (14.16) is satisfied, $f = f_j$ should be selected as a candidate extremal frequency—that is, set $F_k = f_j$ where k is the index of the next extremal frequency due to be specified.

Testing of $E(f)$ at the passband and stopband edges

There is some disagreement within the literature regarding the testing of the passband and stopband edge frequencies f_p and f_s. Some authors (such as [8]) indicate the following testing strategy for f_p and f_s:

If $E(f_p) > 0$ and $E(f_p) > E(f_p - I_p)$, then a ripple peak is deemed to exist at $f = f_p$ regardless of how $E(f)$ behaves in the transition band which lies immediately to the right of $f = f_p$. If a ripple peak exists at $f = f_p$, and if $|E(f_p)| \geq |\rho|$, then this peak is not superfluous and $f = f_p$ should be selected as a candidate extremal frequency—i.e., set $F_k = f_p$ where k is the index of the next extremal frequency due to be specified. Similarly, if $E(f_p) < 0$ and $E(f_p) < E(f_p - I_p)$, a ripple trough exists at $f = f_p$. If $|E(f_p)| \geq |\rho|$, this minimum is not superfluous, and we should set $F_k = f_p$ where k is the index of the next extremal frequency to be specified.

If $E(f_s) > 0$ and $E(f_s) > E(f_s + I_s)$, then a ripple peak is deemed to exist at $f = f_s$ regardless of how $E(f)$ behaves in the transition band which lies immediately to the left of $f = f_s$. If a ripple peak does exist at $f = f_s$, and if $|E(f_s)| \geq |\rho|$, then this peak is not superfluous and $f = f_s$ should be selected as a candidate extremal frequency—i.e., set $F_k = f_s$ where k is the index of the next extremal frequency due to be specified. Similarly, if $E(f_s) < 0$ and $E(f_s) < E(f_s + I_s)$, a ripple trough exists at $f = f_s$. If $|E(f_s)| \geq |\rho|$, this minimum is not superfluous, and we should set $F_k = f_s$ where k is the index of the next extremal frequency to be specified.

Other authors (such as [9]) indicate that f_p and f_s are *always* extremal frequencies. In this author's experience, the testing indicated by [8] is always satisfied, so f_p and f_s are always selected as extremal frequencies. This testing has been omitted from the software provided with this book both to reduce execution time and to avoid the danger of having small numerical inaccuracies cause one of these points to erroneously fail the test and thereby be rejected.

Testing of $E(f)$ for $f = 0.5$

If $E(0.5) > 0$ and $E(0.5) > E(0.5 - I_s)$, then a ripple peak exists at $f = 0.5$. If a ripple peak does exist at $f = 0.5$, and if $|E(0.5)| \geq |\rho|$, then this maximum is not superfluous and $f = 0.5$ should be used as the final candidate extremal frequency. Similarly, if $E(0.5) < 0$ and $E(0.5) < E(0.5 - I_s)$, a ripple trough exists at $f = 0.5$. If $|E(0.5)| \geq |\rho|$, this minimum is not superfluous.

Rejecting superfluous candidate frequencies

The Remez algorithm requires that only $r + 1$ extremal frequencies be used in each iteration. However, when the search procedures just described are used, it is possible to wind up with more than $r + 1$ candidate frequencies. This situation can be remedied easily by retaining only the $r + 1$ frequencies F_k for which $|E(F_k)|$ is the largest. The retained frequencies are renumbered from 0 to r before proceeding. An alternative approach is to reject the frequency corresponding to the smaller of $|E(F_0)|$ and $|E(F_r)|$, regardless of how these two values compare to the absolute errors at the other extrema. Since there is only one solution for a given set of filter specifications, both approaches should lead to the same result. However, one approach may lead to a faster solution or be less prone to numeric difficulties.

Deciding when to stop

There are two schools of thought on deciding when to stop the exchange algorithm. The original criterion [7] examines the extremal frequencies and stops the algorithm when the frequencies do not change from one iteration to the next. This criterion is implemented in the member function RemezStop(). This approach has always worked well for the author, but it does have a potential flaw. Suppose that one of the true extremal frequencies for a particular filter lies at $f = F_T$, and due to the way the dense grid has been defined, F_T lies exactly at the midpoint between two grid frequencies. It is conceivable that on successive iterations, the observed extremal frequency could alternate between f_n and f_{n+1} and thereby never allow the stopping criterion to be satisfied.

A different criterion [8], uses values of the error function rather than the locations of the extremal frequencies. In theory, when the Remez algorithm is working correctly, each successive iteration will produce continually improving estimates of the correct extremal frequencies, and the values of $|E(F_k)|$ will become exactly equal for all values of k. However, due to the finite resolution of the frequency grid, as well as finite precision arithmetic, the estimates may in fact never converge to exact equality. One remedy is to stop when the largest $|E(F_k)|$ and the smallest $|E(F_k)|$ differ by some reasonably small amount. The difference as a fraction of the largest $|E(F_k)|$ is given by

$$Q = \frac{\max |E(F_k)| - \min |E(F_k)|}{\max |E(F_k)|}$$

Typically, the iterations are stopped when $Q \leq 0.01$. This second stopping criterion is implemented in the member function RemezStop2().

14.5 Obtaining the Impulse Response

In Sec. 14.2, the final step in the Remez exchange design strategy consisted of using the final set of extremal frequencies to obtain the filter's impulse response. This can be accomplished by using Eq. (14.12) to obtain $A(f)$ from the set of extremal frequencies, and then using the appropriate formula from Table 13.4 to generate the corresponding impulse response $h[n]$.

14.6 Using the Remez Exchange Method

All of the constituent functions of the Remez method that have been presented in previous sections are called in the proper sequence by the constructor for the RemezAlgorithm class which is provided in file remezalg.cpp.

Deciding on the filter length

To use the Remez exchange method, the designer must specify N, f_p, f_s, and the ratio δ_1/δ_2. The algorithm will provide the filter having the smallest values of $|\delta_1|$ and $|\delta_2|$ that can be achieved under these constraints. However, in many applications, the values specified are f_p, f_s, δ_1, and δ_2 with the designer left free to set N as required. Faced with such a situation, the designer can use f_p, f_s, and $K = \delta_1/\delta_2$ as dictated by the application, then design filters for increasing values of N until the δ_1 and δ_2 specifications are satisfied. An approximation of the required number of taps can be obtained by one of the formulas given in the following. For filters having passbands of *moderate* width, the approximate number of taps required is given by

$$\tilde{N} = 1 + \frac{-20 \log \sqrt{\delta_1 \delta_2} - 13}{14.6(f_s - f_p)} \qquad (14.19)$$

For filters with extremely narrow passbands, Eq. (14.19) can be modified to be

$$\tilde{N} = \frac{0.22 - (20 \log \delta_2)/27}{(f_s - f_p)} \qquad (14.20)$$

For filters with extremely wide passbands, the required number of taps is approximated by

$$\tilde{N} = \frac{0.22 - (20 \log \delta_1)/27}{(f_s - f_p)} \qquad (14.21)$$

Example 14.1 Suppose we wish to design a lowpass filter with a maximum passband ripple of $\delta_1 = 0.025$ and a minimum stopband attenuation of 60 dB or $\delta_2 = 0.001$. The normalized cutoff frequencies for the passband and stopband are, respectively, $f_p = 0.215$ and $f_s = 0.315$.

solution Using Eq. (14.19) to approximate the required filter length N, we obtain

$$N = 1 + \frac{-20 \log \sqrt{(0.001)(0.025)} - 13}{14.6(0.315 - 0.215)}$$
$$= 23.6$$

The next larger odd length would be $N = 25$. If we run RemezAlgorithm() with the following inputs:

$N \stackrel{\triangle}{=}$ filter_length $= 25$

$L \stackrel{\triangle}{=}$ grid_density $= 16$

$K \stackrel{\triangle}{=}$ ripple_ratio $= 25.0$

$f_p \stackrel{\triangle}{=}$ passband_edge_freq $= 0.215$

$f_s \stackrel{\triangle}{=}$ stopband_edge_freq $= 0.315$

we obtain the extremal frequencies listed in Table 14.1 and the filter coefficients listed in Table 14.2. The frequency response of the filter is shown in Figs. 14.2 and 14.3. The actual passband and stopband values of 0.0195 and 0.000780 are significantly better than the specified values of 0.025 and 0.001.

Example 14.2 The ripple performance of the 25-tap filter designed in Example 14.1 exhibits a certain amount of overachievement, and the estimate of the minimum number of taps was closer to 23 than 25. Therefore, it would be natural for us to ask

TABLE 14.1 Extremal Frequencies for Example 14.1

k	f_k
0	0.000000
1	0.042232
2	0.084464
3	0.126696
4	0.165089
5	0.199643
6	0.215000
7	0.315000
8	0.322708
9	0.343906
10	0.372813
11	0.407500
12	0.447969
13	0.500000

TABLE 14.2 Coefficients for the 25-Tap Filter of Example 14.1

$h[0] = h[24] =$	-0.004069
$h[1] = h[23] =$	-0.010367
$h[2] = h[22] =$	-0.001802
$h[3] = h[21] =$	0.015235
$h[4] = h[20] =$	0.003214
$h[5] = h[19] =$	-0.027572
$h[6] = h[18] =$	-0.005119
$h[7] = h[17] =$	0.049465
$h[8] = h[16] =$	0.007009
$h[9] = h[15] =$	-0.096992
$h[10] = h[14] =$	-0.008320
$h[11] = h[13] =$	0.315158
$h[12] =$	0.508810

if we could in fact achieve the desired performance with a 23-tap filter. If we rerun RemezAlgorithm() with filter_length=23, we obtain the extremal frequencies and filter coefficients listed in Tables 14.3 and 14.4. The frequency response of this filter is shown in Figs. 14.4 and 14.5. The passband ripple is approximately 0.034, and the stopband ripple is approximately 0.00138—therefore, we conclude that a 23-tap filter does not satisfy the specified requirements.

FIR Filter Design: Remez Exchange Method   281

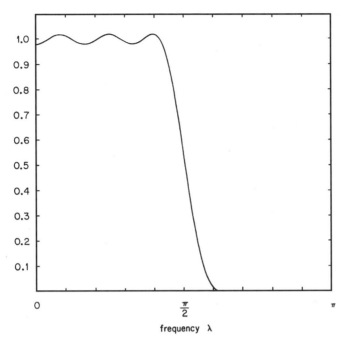

Figure 14.2 Magnitude response (as a fraction of peak) for the filter of Example 14.1.

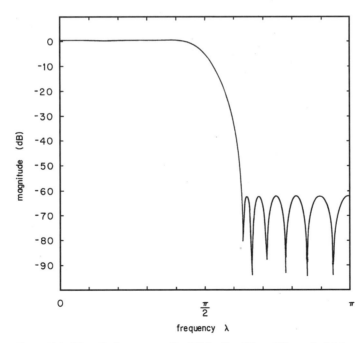

Figure 14.3 Magnitude response (in dB) for the filter of Example 14.1.

TABLE 14.3 Extremal Frequencies for Example 14.2

k	f_k
0	0.000000
1	0.051510
2	0.103021
3	0.152292
4	0.194844
5	0.215000
6	0.315000
7	0.324635
8	0.349688
9	0.382448
10	0.419062
11	0.459531
12	0.500000

TABLE 14.4 Coefficients for the 23-tap Filter of Example 14.2

$h[0] = h[22] =$	-0.000992
$h[1] = h[21] =$	0.007452
$h[2] = h[20] =$	0.018648
$h[3] = h[19] =$	0.002873
$h[4] = h[18] =$	-0.026493
$h[5] = h[17] =$	-0.003625
$h[6] = h[16] =$	0.048469
$h[7] = h[15] =$	0.005314
$h[8] = h[14] =$	-0.096281
$h[9] = h[13] =$	-0.006601
$h[10] = h[12] =$	-0.314911
$h[11] =$	0.507077

When the transition width $f_p - f_s$ is small relative to the sampling rate, an alternative formula for estimating the required number of taps is given by

$$\hat{N} = \frac{D_\infty(\delta_p, \delta_s)}{\Delta F} \tag{14.22}$$

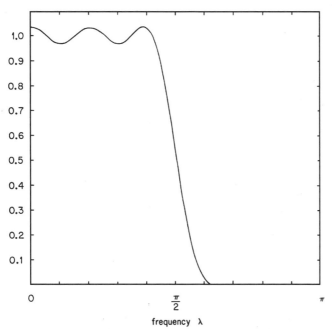

Figure 14.4 Magnitude response (as a fraction of peak) for the filter of Example 14.2.

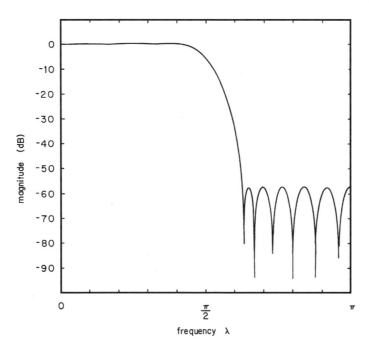

Figure 14.5 Magnitude response (in dB) for the filter of Example 14.2.

where

$$D_\infty(\delta_p, \delta_s) = \log_{10} \delta_s [a_1 (\log_{10} \delta_p)^2 + a_2 \log_{10} \delta_p + a_3]$$
$$+ [a_4 (\log_{10} \delta_p)^2 + a_5 \log_{10} \delta_p + a_6]$$

$$a_1 = 5.309 \times 10^{-3} \quad a_2 = 7.114 \times 10^{-2}$$
$$a_3 = -4.761 \times 10^{-1} \quad a_4 = -2.66 \times 10^{-3}$$
$$a_5 = -5.941 \times 10^{-1} \quad a_6 = -4.278 \times 10^{-1}$$

14.7 Extension of the Basic Method

So far we have considered use of the Remez exchange method for odd-length, linear phase FIR filters having even-symmetric impulse responses (that is, type 1 filters). The Remez method was originally adapted specifically for the design of type 1 filters [7]. However, in a subsequent paper [10], it was noted that the amplitude response of any constant-group-delay FIR filter can be expressed as

$$A(f) = Q(f)P(f)$$

where $P(f) = \sum_{k=0}^{r-1} c_k \cos(2\pi k f)$

$$Q(f) = \begin{cases} 1 & h[n] \text{ symmetric, } N \text{ odd} \\ \cos \pi f & h[n] \text{ symmetric, } N \text{ even} \\ \sin 2\pi f & h[n] \text{ antisymmetric, } N \text{ odd} \\ \sin \pi f & h[n] \text{ antisymmetric, } N \text{ even} \end{cases}$$

Recall that the error $E(f)$ was defined as

$$E(f) = W(f)[D(f) - A(f)] \qquad (14.23)$$

If we substitute $Q(f)P(f)$ and factor out $Q(f)$, we obtain

$$E(f) = W(f)Q(f)\left[\frac{D(f)}{Q(f)} - P(f)\right]$$

We can then define a new weighting function $\hat{W}(f) = W(f)Q(f)$ and a new desired response $\hat{D}(f) = D(f)/Q(f)$, and thereby obtain

$$E(f) = \hat{W}(f)[\hat{D}(f) - P(f)] \qquad (14.24)$$

Equation (14.24) is of the same form as Eq. (14.23) with $\hat{W}(f)$ substituted for $W(f)$, $\hat{D}(f)$ substituted for $D(f)$, and $P(f)$ substituted for $A(f)$. Therefore, the procedures developed in previous sections can be used to solve for $P(f)$ provided that $\hat{W}(f)$ is used in place of $W(f)$ and $\hat{D}(f)$ is used in place of $D(f)$. Once this $P(f)$ is obtained, we can multiply by the appropriate $Q(f)$ to obtain $A(f)$. The appropriate formula from Table 13.4 can then be used to obtain the impulse response coefficients $h[n]$.

References

1. Martin, M. A. "Digital Filters for Data Processing," Tech. Report no. 62-SD484, Missle and Space Division, General Electric Co., 1962.
2. Tufts, D. W., D. W. Rorabacher, and M. E. Mosier. "Designing Simple, Effective Digital Filters," *IEEE Trans. Audio Electroacoust.*, vol. AU-18, pp. 142–158, December 1970.
3. Tufts, D. W., and J. T. Francis. "Designing Lowpass Filters—Comparison of Some Methods and Criteria," *IEEE Trans. Audio Electroacoust.*, vol. AU-18, pp. 487–494, December 1970.
4. Helms, H. D. "Digital Filters with Equiripple or Minimax Responses," *IEEE Trans Audio Electroacoust.*, vol. AU-19, pp. 87–94, March 1971.
5. Herrman, O. "Design of Nonrecursive Digital Filters with Linear Phase," *Electronics Letters*, vol. 6, pp. 328–329, 1970.
6. Hofstetter, E. M., A. V. Oppenheim, and J. Siegel. "A New Technique for the Design og Non-Recursive Digital Filters," *Proc. Fifth Annual Princeton Conf. on Inform. Sci and Syst.*, pp. 64–72, 1971.
7. Parks, T. W., and J. H. McClellan. "Chebyshev Approximation for Nonrecursive Digital Filters with Linear Phase," *IEEE Trans. Circuit Theory*, vol. CT-19, pp. 189–194, March 1972.
8. Antoniou, A. "Accelerated Procedure for the Design of Equiripple Non-recursive Digital Filters," *Proceedings IEE, PART G*, vol. 129, pp. 1–10, 1982.
9. Parks, T. W., and C. S. Burrus. *Digital Filter Design*, Wiley-Interscience, New York, 1987.
10. Parks, T. W., and J. H. McClellan. "A Computer Program for Designing Optimum FIR Linear Phase Digital Filters," *IEEE Trans. Audio Electroacoust.*, vol. AU-21, pp. 506–526, December 1973.
11. Cheyney, E. W., *Introduction to Approximation Theory*, McGraw-Hill, New York, 1966.

Chapter 15

IIR Filter Fundamentals

The general form for an *infinite impulse response* (IIR) filter's output $y[k]$ at time k is given by

$$y[n] = \sum_{n=1}^{N} a_n y[k-n] + \sum_{m=0}^{M} b_m x[k-m] \qquad (15.1)$$

This equation indicates that the filter's output is a linear combination of the present input, the M previous inputs, and the N previous outputs. The corresponding system function is given by

$$H(z) = \frac{\sum_{m=0}^{M} b_m z^{-m}}{1 - \sum_{n=1}^{N} a_n z^{-n}} \qquad (15.2)$$

where at least one of the a_n is nonzero and at least one of the roots of the denominator is not exactly cancelled by one of the roots of the numerator. For a stable filter, all the poles of $H(z)$ must lie inside the unit circle, but the zeros can lie anywhere in the z plane. It is usual for M, the number of zeros, to be less than or equal to N, the number of poles. Whenever the number of zeros exceeds the number of poles, the filter can be separated into an FIR filter with $M - N$ taps in cascade with an IIR filter with N poles and N zeros. Therefore, IIR design techniques are conventionally restricted to cases for which $M \leq N$.

Except for the special case in which all poles lie on the unit circle (in the z plane), it is not possible to design an IIR filter having exactly linear phase. Therefore, unlike FIR design procedures that are concerned almost exclusively with the magnitude response, IIR design procedures are concerned with both the magnitude response and the phase response.

15.1 Frequency Response of IIR Filters

For an IIR filter defined by the difference equation (15.1), the filter's frequency response can be computed from the coefficients a_n and b_m as

$$H[k] = \frac{\sum_{m=0}^{M} b_m \exp\left(j\frac{2\pi mk}{LT}\right)}{\sum_{n=0}^{N} \alpha_n \exp\left(j\frac{2\pi nk}{LT}\right)} \quad (15.3)$$

where

$$\alpha_n = \begin{cases} 1 & n = 0 \\ -a_n & 0 < n \leq N \end{cases}$$

and L is the number of discrete points in the response over the frequency range from $-0.5/T$ to $0.5/T$, T being the sampling interval. Recall that the frequency response of any digital filter is periodic with a period of T^{-1} so the response is completely defined by the response over this range. Typically, normalized responses with $T \equiv 1$ are assumed if a value for T is not explicitly stated. A C++ function that implements Eq. (15.3) is provided in file iir_resp.cpp.

15.2 Structures for IIR Realizations

A number of different structures can be used to realize IIR filters. For any given set of filter coefficients, all of these different structures are equivalent, assuming that they are implemented with infinite-precision arithmetic. However, when implemented using finite precision arithmetic with quantized coefficients and quantized signals, the different structures can exhibit very different behaviors. Therefore, some care needs to be exercised in the selection of a structure for implementing a particular filter. This section explores several of the most common implementation structures used for IIR filters.

Direct form

A direct realization of Eq. (15.1) is shown in Fig. 15.1 using the signal flow graph notation introduced in Sec. 6.4. The structure shown is known as the *direct form 1 realization* or *direct form 1 structure* for the IIR filter represented by Eq. (15.1). Examination of the figure reveals that the system can be viewed as two systems in cascade—the first system using input samples $x[k-M]$ through $x[k]$ to generate an intermediate signal that is labeled as $w[k]$ in the figure, and the second system using $w[k]$ plus output samples $y[k-N]$ through $x[k]$ to generate $y[k]$. Since these two systems are linear time-invariant systems, the order of the cascade can be reversed to yield the equivalent system shown in Fig. 15.2. Examination of this figure reveals that the unit delays in parallel running down the center of the diagram can be paired such that within each pair the two delays each take the same input signal. This fact can be exploited

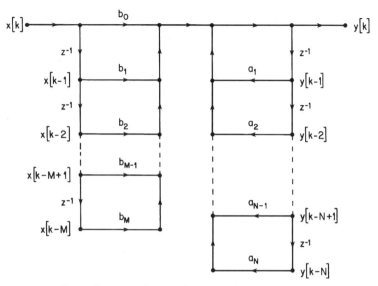

Figure 15.1 Direct form 1 realization for an IIR filter.

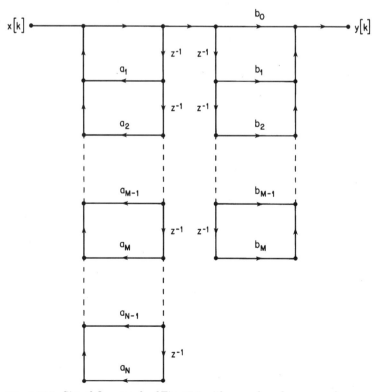

Figure 15.2 Signal flow graph of Fig. 15.1 with cascade order reversed.

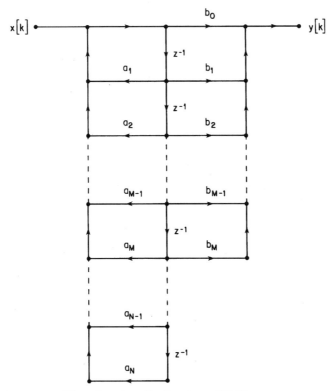

Figure 15.3 Direct form 2 realization for an IIR filter.

to merge the two delay chains into a single chain, as shown in Fig. 15.3. The structure shown in this figure is known as the *direct form 2 realization* of the IIR system represented by Eq. (15.1).

Cascade form

The numerator and denominator of the polynomial system function given by Eq. (15.2) each can be factored into a product of (in general) second-order polynomials with real coefficients:

$$H(z) = \frac{\prod_{k=1}^{M_S} \left(d_{0k} + d_{1k}z^{-1} + d_{2k}z^{-2}\right)}{\prod_{k=1}^{N_S} \left(c_{0k} + c_{1k}z^{-1} + c_{2k}z^{-2}\right)}$$

Each of the second-order factors in the numerator corresponds to a complex-conjugate pair of zeros, and each of the second-order factors in the denominator corresponds to a complex-conjugate pair of poles. If there are an odd number of poles, one of the coefficients c_{2k} will be zero. Likewise, if there are an odd number of zeros, one of the coefficients d_{2k} will be zero. In one widely used approach, numerator factors are paired with denominator factors and the system function is expressed as a product of the ratios formed by these factor pairs:

$$H(z) = \prod_{k=1}^{N_S} \frac{d_{0k} + d_{1k}z^{-1} + d_{2k}z^{-2}}{c_{0k} + c_{1k}z^{-1} + c_{2k}z^{-2}} \qquad (15.4)$$

Each term in the numerator and denominator of Eq. (15.4) can be divided by c_{0k} to yield

$$H(z) = \prod_{k=1}^{N_S} \frac{(d_{0k}/c_{0k}) + (d_{1k}/c_{0k})z^{-1} + (d_{2k}/c_{0k})z^{-2}}{1 + (c_{1k}/c_{0k})z^{-1} + (c_{2k}/c_{0k})z^{-2}} \qquad (15.5)$$

Making the substitutions

$$a_{1k} = \frac{c_{1k}}{c_{0k}}$$

$$a_{2k} = \frac{c_{2k}}{c_{0k}}$$

$$b_{0k} = \frac{d_{0k}}{c_{0k}}$$

$$b_{1k} = \frac{d_{1k}}{c_{0k}}$$

$$b_{2k} = \frac{d_{2k}}{c_{0k}}$$

puts Eq. (15.5) into the more convenient form given by

$$H(z) = \prod_{k=1}^{N_S} \frac{b_{0k} + b_{1k}z^{-1} + b_{2k}z^{-2}}{1 + a_{1k}z^{-1} + a_{2k}z^{-2}} \qquad (15.6)$$

Each one of the N_S factors in Eq. (15.6) corresponds to an IIR section of the form shown in Fig. 15.4. The complete filter represented by Eq. (15.6) can be realized by cascading N_S of these sections.

15.3 Assessing the Impacts of Quantization and Finite-Precision Arithmetic

When it comes to the effects of quantization and finite precision arithmetic, IIR filters share all of the potential problems that FIR filters have, plus a few that are unique to the recursive structures that are used for IIR implementations. As discussed previously for FIR filters, the three manifestations of finite digital word length are *coefficient quantization*, *signal quantization*, and

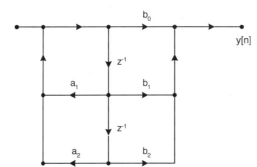

Figure 15.4 Direct form 2 realization for an IIR filter section that implements two poles and two zeros.

finite-precision arithmetic. The first two of these operate in the IIR context similarly to the way they operate in the FIR context. However, due to the feedback that is inherent in recursive implementations of IIR filters, the effects of finite-precision arithmetic can cause problems that are peculiar to the IIR case. These problems include small oscillatory outputs for zero input, and limit-cycle oscillations that are caused by arithmetic overflows.

Direct form realizations

Consider the direct form 1 realization of an IIR filter as depicted in Fig. 15.5. Each of the input samples $x[k]$ and each of the coefficients $a[n]$ and $b[m]$ will be quantized to some finite number of bits. If each $x[k]$ is quantized to B_x bits and each $b[m]$ is quantized to B_b bits, a total of $(B_x + B_b)$ bits will be needed to represent each product term $x[k-m] \cdot b[m]$ without further loss of precision. Still more bits will be needed to precisely represent the sum $s_1[k]$ of these product terms. The sum $s_2[k]$ is formed by adding $s_1[k]$ and all of the product terms $a[n] \cdot s_2[k-n]$. Typically, $s_2[k]$ will need to be accumulated with relatively

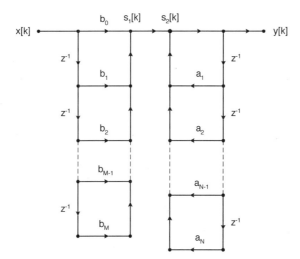

Figure 15.5 Signal flow graph of direct form 1 realization for an IIR filter.

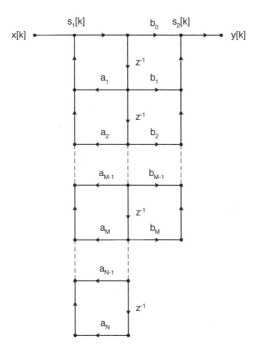

Figure 15.6 Signal flow graph of direct form 2 realization for an IIR filter.

high precision, and this leads to a relatively expensive implementation for the second delay chain. In some schemes, the sum $s_2[k]$ will be accumulated with relatively high precision and then truncated or rounded to a much smaller number of bits before being input to the second delay chain.

Now consider the direct form 2 realization depicted in Fig. 15.6 When implemented with infinite precision, the structures in Figs. 15.5 and 15.6 are equivalent. However, their finite-precision implementations can exhibit significantly different behavior. In the direct form 2 structure, it is the sum $s_1[k]$ that must be delayed rather than $s_2[k]$, and the truncation strategy for reducing the cost of the single delay chain will be different.

15.4 Software Notes

The DirectFormFir class developed in Chap. 11 has been adapted for evaluation of the IIR case and is provided as DirectFormIir in file dir1_iir.cpp. The SweptResponse class from Chap. 11 can be used as is for evaluating IIR designs. The particular instance of FilterImplementation used by SweptResponse is constructed externally and passed in as a parameter. FilterImplementation is a base class that can be externally instantiated as either DirectFormFir or DirectFormIir.

Chapter 16

IIR Filter Design: Invariance and Pole-Zero Placement Methods

This chapter presents three different methods for the design of IIR filters. The *impulse invariance* method is based on setting the unit sample response of the digital filter equal to a sequence of uniformly spaced samples from the impulse response of an analog filter. The *step invariance* method is based on setting the unit step response of the digital filter equal to a sequence of uniformly spaced samples from the step response of an analog filter. The *matched z transformation* method is based upon a direct mapping of s-plane pole and zero locations into the corresponding z-plane locations. A fourth method, *bilinear transformation*, is presented in Chap. 17.

16.1 Impulse Invariance

The basic idea behind the impulse invariance approach is simple: the unit sample response of the digital filter is set equal to a sequence of uniformly spaced samples from the impulse response of an analog filter:

$$h[n] = h_a(nT) \qquad (16.1)$$

(An analog filter used in this context is usually referred to as a *prototype* filter.) This approach is conceptually simple, but from a practical viewpoint, evaluation of Eq. (16.1) is not a straightforward matter. By definition, for an *infinite* impulse response filter, the sequence $h[n]$ will be nonzero over an infinite domain of n. Furthermore, based on the s-plane-to-z-plane mapping discussed in Sec. 7.2, we can conclude that the imposition of Eq. (16.1) will not result in a simple relationship between the frequency response corresponding to $h[n]$ and the frequency response corresponding to $h_a(t)$. In fact, this relationship can be shown to be

$$H(e^{j\lambda}) = \frac{1}{T}\sum_{k=-\infty}^{\infty} H_a\left(j\frac{\lambda+2\pi k}{T}\right) \quad (16.2)$$

where $h[n] \overset{DTFT}{\longleftrightarrow} H(e^{j\lambda})$
$h_a(t) \overset{FT}{\longleftrightarrow} H_a(j\omega)$

Put simply, Eq. (16.2) indicates that $H(e^{j\lambda})$ will be an aliased version of $H_a(j\omega)$. The only way the aliasing can be avoided is if $H_a(j\omega)$ is band limited such that

$$H_a(j\omega) = 0 \quad \text{for} \quad |\omega| \geq \frac{\pi}{T} \quad (16.3)$$

If Eq. (16.3) is satisfied, then

$$H(e^{j\lambda}) = \frac{1}{T}H_a\left(j\frac{\lambda}{T}\right) \quad |\lambda| \leq \pi \quad (16.4)$$

For a practical analog filter, Eq. (16.3) will never be satisfied exactly, but the impulse invariance method can be used to advantage with responses that are nonzero but negligible beyond some frequency.

The transfer function of the analog prototype filter can be expressed in the form of a partial-fraction expansion as

$$H_a(s) = \sum_{k=1}^{N} \frac{A_k}{s - s_k} \quad (16.5)$$

where the s_k are the poles of $H_a(s)$ and the A_k are given by

$$A_k = [(s - s_k)H_a(s)]|_{s=s_k}$$

Based on transform pair 8 from Table 4.2, the impulse response then can be written as

$$h_a(t) = \sum_{k=1}^{N} A_k \exp(s_k t) u(t) \quad (16.6)$$

The unit sample response of the digital filter is then formed by sampling the prototype filter's impulse response to obtain

$$h[n] = \sum A_k (e^{s_k T})^n u(t) \quad (16.7)$$

The corresponding system function for the digital filter $H(z)$ is obtained as the z transform of Eq. (16.7):

$$H(z) = \sum_{k=1}^{N} \frac{A_k}{1 - e^{s_k T} z^{-1}} \qquad (16.8)$$

Based on the foregoing, we can formulate Algorithm 16.1 for impulse-invariant design of an IIR filter.

Algorithm 16.1 Impulse-invariant design of IIR filters.

1. Obtain the transfer function $H_a(s)$ for the desired analog prototype filter. (The material provided in Chap. 5 will prove useful here.)
2. For $k = 1, 2, \ldots, N$, determine the poles s_k of $H_a(s)$ and compute the coefficients A_k using

$$A_k = [(s - s_k) H_a(s)]|_{s=s_k} \qquad (16.9)$$

3. Using the coefficients A_k obtained in step 2, generate the digital filter system function $H(z)$ as

$$H(z) = \sum_{k=1}^{N} \frac{A_k}{1 - \exp(s_k T) z^{-1}} \qquad (16.10)$$

    where T is the sampling interval of the digital filter.
4. The result obtained in step 3 will be a sum of fractions. Obtain a common denominator, and express $H(z)$ as a ratio of polynomials in z^{-1} in the form

$$H(z) = \frac{\sum_{k=0}^{M} b_k z^{-k}}{1 - \sum_{k=1}^{N} a_k z^{-k}} \qquad (16.11)$$

5. Use the a_k and b_k obtained in step 4 to realize the filter in any of the structures given in Sec. 15.2.

Example 16.1 Use the technique of impulse invariance to derive a lowpass IIR digital filter from a second-order Butterworth analog filter with 3-dB cutoff frequency of 3 kHz. The sampling rate for the digital filter is 30,000 samples per second.

solution From Chap. 5 we obtain the normalized transfer function for a second-order Butterworth filter as

$$H(s) = \frac{1}{(s - s_1)(s - s_2)}$$

where $s_1 = \frac{-\sqrt{2}}{2} + j\frac{\sqrt{2}}{2}$
$s_2 = \frac{-\sqrt{2}}{2} - j\frac{\sqrt{2}}{2}$

The specified cutoff frequency of $f = 3000$ yields $\omega_c = 6000\pi$, and the denormalized response is given by

$$H_a(s) = \frac{\omega_c^2}{(s - \omega_c s_1)(s - \omega_c s_2)}$$

$$= \frac{\omega_c^2}{[s + \omega_c(\sqrt{2}/2) - j\omega_c(\sqrt{2}/2)][s + \omega_c(\sqrt{2}/2) + j\omega_c(\sqrt{2}/2)]}$$

The partial fraction expansion of $H_a(s)$ is given by

$$H_a(s) = \frac{A_1}{s + \omega_c(\sqrt{2}/2) - j\omega_c(\sqrt{2}/2)} + \frac{A_2}{s + \omega_c(\sqrt{2}/2) + j\omega_c(\sqrt{2}/2)}$$

where $A_1 = \frac{-j\sqrt{2}}{2\omega_c}$

$A_2 = \frac{j\sqrt{2}}{2\omega_c}$

Using these values for A_1 and A_2 plus the fact that

$$\omega_c T = \frac{6000\pi}{30{,}000} = \frac{\pi}{5}$$

we obtain from Eq. (16.10) the discrete system function $H(z)$ as

$$H(z) = \frac{-j\sqrt{2}/(2\omega_c)}{1 - \exp\left(\frac{-\pi\sqrt{2}}{10} + j\frac{\pi\sqrt{2}}{10}\right)z^{-1}} + \frac{j\sqrt{2}/(2\omega_c)}{1 - \exp\left(\frac{-\pi\sqrt{2}}{10} - j\frac{\pi\sqrt{2}}{10}\right)z^{-1}}$$

$$= \frac{2.06797 \times 10^{-5} z^{-1}}{1 - 1.158045 z^{-1} + 0.4112407 z^{-2}}$$

Programming considerations

Step 1. Butterworth, Chebyshev, and Bessel filters are *all-pole* filters—their transfer functions have no finite zeros. Closed-form expressions are available for the poles of Butterworth and Chebyshev filters. The poles of Bessel filters can readily be obtained by finding the roots of the denominator polynomial as discussed in Chap. 5. The transfer function for an elliptical filter has both poles and zeros. The poles are readily available by using the quadratic formula to find the denominator roots for each factor in Eq. (5.38). The zeros $\pm j\alpha\sqrt{\alpha_i}$ are obtained by inspection of Eq. (5.38). The software for performing the impulse invariance transformation is therefore designed to accept $H_a(s)$ specified as an array of poles and an array of zeros.

Step 2. Evaluation of A_k for step 2 of the algorithm is straightforward. The coefficients A_k can be written as $A_k = N_{Ak}/D_{Ak}$ where the numerator N_{Ak} is obtained as

$$N_{Ak} = \begin{cases} H_0 \prod_{m=1}^{M}(p_k - q_m) & M \neq 0 \\ H_0 & M = 0 \end{cases}$$

and q_m is the mth zero of $H_a(s)$, p_k is the kth pole of $H_a(s)$, and M is the total number of zeros. Equation (16.9) can be evaluated using simple arithmetic—there is no symbolic manipulation needed. The denominator D_{Ak} is obtained as

$$D_{Ak} = \prod_{\substack{n=1 \\ n \neq k}}^{N}(p_k - p_n)$$

Step 3. Evaluation of $H(z)$ is more than plain, straightforward arithmetic. At this point, for each value of k, the coefficient A_k is known and the coefficient $\exp(s_k T)$ can be evaluated. However, z remains a variable and hence will demand some special consideration. To simplify the notation in the subsequent development, let us rewrite $H(z)$ as

$$H(z) = \sum_{k=1}^{N} \frac{A_k}{1 + \beta_k z^{-1}} \qquad (16.12)$$

where $\beta_k = -\exp(s_k T)$.

Step 4. For the summation in (16.12), the common denominator will be the product of each summand's denominator:

$$D(z) = \prod_{k=1}^{N}(1 + \beta_k z^{-1}) \qquad (16.13)$$

To see how Eq. (16.13) can be evaluated easily by computer, let us examine the sequence of partial products $D_k(z)$ encountered in the evaluation:

$$D_1(z) = (1 + \beta_1 z^{-1})$$
$$D_2(z) = (1 + \beta_2 z^{-1})D_1(z) = D_1(z) + \beta_2 z^{-1}D_1(z)$$
$$D_3(z) = (1 + \beta_3 z^{-1})D_2(z) = D_2(z) + \beta_3 z^{-1}D_2(z)$$
$$\vdots$$
$$D(z) = D_N(z) = (1 + \beta_N z^{-1})D_{N-1}(z) = D_{N-1}(z) + \beta_N z^{-1}D_{N-1}(z)$$

Examination of this sequence reveals that the partial product $D_k(z)$ at iteration k can be expressed in terms of the partial product $D_{k-1}(z)$ as

$$D_k(z) = D_{k-1}(z) + \beta_k z^{-1} D_{k-1}(z)$$

The partial product $D_{k-1}(z)$ will be a $(k-1)$-degree polynomial in z^{-1}:

$$D_{k-1}(z) = \delta_0(z^{-1})^0 + \delta_1(z^{-1})^1 + \delta_2(z^{-1})^2 + \cdots + \delta_{k-1}(z^{-1})^{k-1}$$

The product $\beta_k z^{-1} D_{k-1}(z)$ is then given by

$$\beta_k z^{-1} D_{k-1}(z) = \delta_0 \beta_k (z^{-1})^1 + \delta_1 \beta_k (z^{-1})^2 + \delta_2 \beta_k (z^{-1})^3 + \cdots + \delta_{k-1} \beta_k (z^{-1})^k$$

and $D_k(z)$ is given by

$$D_k(z) = \delta_0(z^{-1})^0 + (\delta_1 + \delta_0 \beta_k)(z^{-1})^1 + (\delta_2 + \delta_1 \beta_k)(z^{-1})^2 + \cdots$$
$$+ (\delta_{k-1} + \delta_{k-2} \beta_k)(z^{-1})^{k-1} + \delta_{k-1} \beta_k (z^{-1})^k$$

Therefore, we can conclude that if δ_n is the coefficient for the $(z^{-1})^n$ term in $D_{k-1}(z)$, then the coefficient for the $(z^{-1})^n$ term in $D_k(z)$, is $(\delta_n + \delta_{n-1} \beta_k)$ with the proviso that $\delta_k \triangleq 0$ in $D_{k-1}(z)$. The polynomial $D_{k-1}(z)$ can be represented in the computer as an array of k coefficients, with the array index corresponding to the subscript on δ and the superscript on (z^{-1}). The coefficients for the partial product $D_k(z)$ can be obtained from the coefficients for $D_{k-1}(z)$ as indicated by the following code fragment

```
for( j=k; j>=1; j--){delta[j] = delta[j] + beta * delta[j-1];}
```

The loop is executed in reverse order so that the coefficients can be updated in place without prematurely overwriting the old values. If this fragment is placed within an outer loop with k ranging from 1 to num_poles, the final values in delta[n] will be the coefficients a_n for Eq. (16.11).

For the summation in Eq. (16.10), the numerator can be computed as

$$N(z) = \sum_{k=1}^{N} \left[A_k \prod_{\substack{n=1 \\ n \neq k}}^{N} \left(1 - \beta_n z^{-1}\right) \right] \qquad (16.14)$$

For each value of k, the product in Eq. (16.14) can be evaluated in a manner similar to the way in which the denominator is evaluated. A complete function for computing the coefficients a_k and b_k is provided in file impinvar.cpp.

16.2 Step Invariance

One major drawback to filters designed via the impulse invariance method is their sensitivity to the specific characteristics of the input signal. The digital filter's unit sample response is a sampled version of the prototype filter's impulse response. However, the prototype filter's response to an arbitrary input cannot, in general, be sampled to obtain the digital filter's response to a sampled version of the same arbitrary input. In many applications a filter's step response is of more concern than is the filter's impulse response. In such cases, the impulse invariance technique can be modified to yield Algorithm 16.2, which can be used to design a digital filter based on the principle of step invariance.

Algorithm 16.2 Step invariant design of IIR filters.

1. Obtain the transfer function for the desired analog prototype filter.
2. Multiply $H_a(s)$ by $1/s$ to obtain $G_a(s)$, the Laplace transform of the filter's response to the unit step function.
3. For $k = 1, 2, \ldots, N$, determine the poles s_k of $G_a(s)$ and compute the coefficients A_k using

$$A_k = [(s - s_k)G_a(s)]|_{s=s_k}$$

4. Using the coefficients A_k obtained in step 3, generate the system function $G(z)$ as

$$G(z) = \sum_{k=1}^{N} \frac{A_k}{1 - \exp(s_k T)z^{-1}}$$

5. Multiply $G(z)$ by $(1 - z^{-1})$ to remove the z transform of a unit step and thereby obtain $H(z)$ as

$$H(z) = (1 - z^{-1}) \sum_{k=1}^{N} \frac{A_k}{1 - \exp(s_k T)z^{-1}}$$

6. Obtain a common denominator for the terms in the summation of step 5, and express $H(z)$ as a ratio of polynomials in z^{-1} in the form

$$G(z) = \frac{\sum_{k=0}^{M} b_k z^{-k}}{1 - \sum_{k=0}^{N} a_k z^{-k}}$$

7. Use the a_k and b_k obtained in step 6 to realize the filter in any of the structures in Sec. 15.2.

Programming considerations

The step invariance method is similar to the impulse invariance method, with two important differences. In step 2 of Algorithm 16.2, the transfer function

$H_a(s)$ is multiplied by $1/s$. Assuming that $H_a(s)$ is represented in terms of its poles and zeros, multiplication by $1/s$ is accomplished by simply adding a pole at $s = 0$. (Strictly speaking, if the analog filter has a zero at $s = 0$, multiplication by $1/s$ creates a pole at $s = 0$ which cancels the zero. However, since none of the analog prototype filters within the scope of this book have zeros at $s = 0$, we shall construct the software without provisions for handling at zero at $s = 0$.)

In step 5 of Algorithm 16.2, the system function $G(z)$ is multiplied by $(1 - z^{-1})$ to remove the z transform of a unit step and thereby obtain the system function $H(z)$. Conceptually, this multiplication is appropriately located in step 5. However, for ease of implementation, it makes sense to defer the multiplication until after the coefficients are generated in step 6. A function modified to perform the step invariance technique is provided in file stpinvar.cpp.

16.3 Matched z Transformation

Starting with an analog prototype filter having both poles and zeros, it is possible to derive an IIR digital filter by direct mapping of the s-plane poles and zeros into the z-plane using the replacement relations

$$s + a \to 1 - z^{-1} e^{-aT} \tag{16.15}$$

$$(s + a - jb)(s + a + jb) = 2as + a^2 + b^2 \to 1 - 2z^{-1} e^{-aT} \cos(bT) + z^{-2} e^{-2aT} \tag{16.16}$$

where T is the sampling interval. Equation (16.15) is used for mapping real poles or zeros located at $s = -a$. Equation (16.16) is used for mapping complex conjugate pairs of poles or zeros located at $s = -a \pm jb$. This mapping is sometimes called the matched z transformation and is useful only for analog systems having *both* poles and zeros and with zeros that lie at frequencies of less than half the sampling frequency.

Example 16.2 Apply the matched z transformation to the ninth order elliptical filter having a normalized transfer function given by

$$H(s) = \frac{H_0}{s + p_0} \prod_{i=1}^{4} \frac{s^2 + A_i}{s^2 + B_i s + C_i} \tag{16.17}$$

where A_i, B_i, and C_i are as given in Table 16.1, $H_0 = 0.015317$, and $p_0 = 0.470218$. Denormalize the filter to have a cutoff frequency of 3 kHz, and assume a sampling rate of 30,000 samples per second.

solution The values in Table 16.1 are denormalized using the rules given in Sec. 5.1 to obtain the values for A'_i, B'_i, and C'_i as given in Table 16.2. The zeros of Eq. (16.17) occur in complex conjugate pairs, so we set the numerator of each factor in the product

TABLE 16.1 Normalized Analog Filter Coefficients for Example 16.2

i	a_i	b_i	c_i
1	4.174973	0.6786235	0.4374598
2	1.606396	0.3091997	0.7415493
3	1.182293	0.1127396	0.8988261
4	1.076828	0.0272625	0.9538953

TABLE 16.2 Denormalized Analog Filter Coefficients for Example 16.2

i	a_i	b_i	c_i
1	37574757	2035.8705	3937138
2	14457564	927.5991	6673943
3	10640637	338.2188	8089434
4	9691452	81.7875	8585057

TABLE 16.3 Digital Filter Coefficients for Example 16.2

i	α_i	β_i	γ_i
0	—	0.954067	—
1	0.204328	1.930162	0.934389
2	0.126744	1.962256	0.969553
3	0.108733	1.979858	0.988789
4	0.103770	1.987759	0.997278

equal to the left-hand side of Eq. (16.16) and solve for a_i and b_i in terms of A'_i.

$$s^2 + A'_i = s^2 + 2as + a^2 + b^2$$

$$b_i = \sqrt{A'_i}$$

$$a_i = 0$$

Therefore, the numerator of $H[z]$ can be written as

$$N[z] = \prod_{n=1}^{4}[1 - 2z^{-1}\cos(T\sqrt{A_n}) + z^{-2}]$$

$$= \prod_{n=1}^{4}[1 - \alpha_n z^{-1} + z^{-2}]$$

where the a_n are as given in Table 16.3. The single real pole of Eq. (16.17) lies at

$s = -p_0$. The remaining poles occur in complex conjugate pairs, so we set the denominator of each factor in the product of Eq. (16.17) equal to the left-hand side of Eq. (16.16) and solve for a_i and b_i in terms of B_i' and C_i'.

$$s^2 + B_i' s + C_i' = s^2 + 2a_i s + a_i^2 + b_i^2$$

$$a_i = \frac{B_i'}{2}$$

$$b_i = \sqrt{C_i' - \frac{(B_i')^2}{4}}$$

Therefore the denominator of $H[z]$ can be written as

$$D[z] = [1 - z^{-1} \exp(-p_0 T)]$$

$$\times \prod_{n=1}^{4} \left[1 - 2z^{-1} \exp\left(\frac{-B_n' T}{2}\right) \cos\left(T\sqrt{C_n' - \frac{(B_n')^2}{4}}\right) + z^{-2} \exp(-B_n' T) \right]$$

$$= (1 - \beta_0 z^{-1}) \prod_{n=1}^{4} [1 - \beta_n z^{-1} + \gamma_n z^{-2}]$$

where the β_n and γ_n are as given in Table 16.3.

Chapter 17

IIR Filter Design: Bilinear Transformation

A popular technique for the design of IIR digital filters is the *bilinear transformation method*, which offers several advantages over the techniques presented in the previous chapter.

17.1 Bilinear Transformation

The bilinear transformation converts the transfer function for an analog filter into the system function for a digital filter by making the substitution

$$s \to \frac{2}{T}\frac{1-z^{-1}}{1+z^{-1}}$$

If the analog prototype filter is stable, the bilinear transformation will result in a stable digital filter.

Algorithm 17.1 Bilinear transformation.

1. Obtain the transfer function $H_a(s)$ for the desired analog prototype filter.
2. In the transfer function obtained in step 1, make the substitution

$$s = \frac{2}{T}\frac{1-z^{-1}}{1+z^{-1}}$$

   where T is the sampling interval of the digital filter. Call the resulting digital system function $H(z)$.
3. The analog prototype filter's transfer function $H_a(s)$ will, in general, be a ratio of polynomials in s. Therefore, the system function $H(z)$ obtained in step 2 will, in general, contain various powers of the ratio $(1-z^{-1})/(1+z^{-1})$ in both the numerator and the denominator. Multiply both the numerator and denominator

by the highest power of $(1+z^{-1})$, and collect terms to obtain $H(z)$ as a ratio of polynomials in z^{-1} of the form

$$H(z) = \frac{\sum_{k=0}^{M} b_k z^{-k}}{1 - \sum_{k=0}^{N} a_k z^{-k}} \tag{17.1}$$

4. Use the a_k and b_k obtained in step 3 to realize the filter in any of the structures given in Sec. 15.2.

Example 17.1 Use the bilinear transform to obtain an IIR filter from a second-order Butterworth analog filter with a 3-dB cutoff frequency of 3 kHz. The sampling rate for the digital filter is 30,000 samples per second.

solution The analog prototype filter's transfer function is given by

$$H_a(s) = \frac{\omega_c^2}{s^2 + \sqrt{2}\omega_c s + \omega_c^2}$$

where $\omega_c = 6000\pi$. Making the substitution $s = 2(1-z^{-1})/(T(1+z^{-1}))$ yields

$$H(z) = \frac{\omega_c^2}{\left(\frac{2}{T}\right)^2 \left(\frac{1-z^{-1}}{1+z^{-1}}\right)^2 + \sqrt{2}\omega_c \left(\frac{2}{T}\right)\left(\frac{1-z^{-1}}{1+z^{-1}}\right) + \omega_c^2}$$

where $T = 1/30{,}000$. After the appropriate algebraic simplifications and making use of the fact that

$$\omega_c T = \frac{6000\pi}{30{,}000} = \frac{\pi}{5}$$

we obtain the desired form of $H(z)$ as

$$H(z) = \frac{0.063964 + 0.127929 z^{-1} + 0.063964 z^{-2}}{1 - 1.168261 z^{-1} + 0.424118 z^{-2}} \tag{17.2}$$

Comparison of Eqs. (17.1) and (17.2) reveals that

$$a_1 = 1.168261 \qquad a_2 = -0.424118$$

$$b_0 = 0.063964 \qquad b_1 = 0.127929 \qquad b_2 = 0.063964$$

17.2 Factored Form of the Bilinear Transform

Often, an analog prototype filter will be specified in terms of its poles and zeros—that is, the numerator and denominator of the filter's transfer function will be in factored form. The bilinear transformation can be applied directly to this factored form. An additional benefit of this approach is that the process of finding the *digital* filter's poles and zeros is greatly simplified. Each factor in the numerator of the analog filter's transfer function will be of the form $(s - q_n)$, and each factor in the denominator will be of the form $(s - p_n)$, where q_n and p_n are, respectively, the nth zero and nth pole of the filter. When the bilinear transformation is applied, the corresponding factors become

$$\left(\frac{2(1-z^{-1})}{T(1+z^{-1})} - q_n\right) \quad \text{and} \quad \left(\frac{2(1-z^{-1})}{T(1+z^{-1})} - p_n\right)$$

The zeros of the digital filter are obtained by finding the values of z for which

$$\frac{2(1-z^{-1})}{T(1+z^{-1})} - q_n = 0$$

The desired values of z are given by

$$z_z = \frac{2 + q_n T}{2 - q_n T} \tag{17.3}$$

In a similar fashion, the poles of the digital filter are obtained from the poles of the analog filter using

$$z_p = \frac{2 + p_n T}{2 - p_n T} \tag{17.4}$$

The use of Eqs. (17.3) and (17.4) is straightforward for the analog filter's *finite* poles or zeros. Usually, only the finite poles and zeros of a filter are considered, but in the present context, *all* poles and zeros of the analog filter must be considered. The analog filter's infinite zeros will map into zeros of $z = -1$ for the digital filter.

Algorithm 17.2 Bilinear transformation for transfer functions in factored form.

1. For the desired analog prototype filter, obtain the transfer function $H_a(s)$ in the factored form given by

$$H_a(s) = H_0 \frac{\prod_{m=1}^{M}(s - q_m)}{\prod_{n=1}^{N}(s - p_n)}$$

2. Obtain the poles z_{pn} of the analog filter from the poles p_n of the analog filter using

$$z_{pn} = \frac{2 + p_n T}{2 - p_n T} \quad n = 1, 2, \ldots, N$$

3. Obtain the zeros z_{zm} of the digital filter from the zeros q_m of the analog filter using

$$z_{zm} = \frac{2 + q_m T}{2 - q_m T} \quad n = 1, 2, \ldots, M$$

4. Using the values of z_{pn} obtained in step 2 and the values of z_{zm} obtained in step 3, form $H(z)$ as

$$H(z) = H_0 \frac{T^N}{\prod_{n=1}^{N}(2 - p_n T)} \cdot \frac{(z+1)^{N-M} \prod_{m=1}^{M}(z - z_{zm})}{\prod_{n=1}^{N}(z - z_{pn})} \quad (17.5)$$

The factor $(z+1)^{N-M}$ supplies the zeros at $z = -1$, which correspond to the zeros at $s = \infty$ for analog filters having $M < N$. The first rational factor in Eq. (17.5) is a constant gain factor that is needed to obtain results that exactly match the results obtained via Algorithm 17.1. However, in practice, this factor is often omitted to yield

$$H(z) = H_0 \frac{(z+1)^{N-M} \prod_{m=1}^{M}(z - z_{zm})}{\prod_{n=1}^{N}(z - z_{pn})}$$

Example 17.2 The Butterworth filter of Example 17.1 has a transfer function given in factored form as

$$H_a(s) = \frac{\omega_c^2}{\left[s + \omega_c\left(\frac{\sqrt{2}}{2}\right) - j\omega_c\left(\frac{\sqrt{2}}{2}\right)\right]\left[s + \omega_c\left(\frac{\sqrt{2}}{2}\right) + j\omega_c\left(\frac{\sqrt{2}}{2}\right)\right]}$$

Apply the bilinear transform to this factored form to obtain the IIR filter's system function $H(z)$.

solution The analog filter has poles at

$$s = \omega_c \frac{-\sqrt{2}}{2} \pm j\omega_c \frac{\sqrt{2}}{2}$$

Using (17.4), we obtain the poles of the digital filter as

$$z_{p_1} = \frac{2 + \left(\frac{-\sqrt{2}}{2} + j\frac{\sqrt{2}}{2}\right)\omega_c T}{2 - \left(\frac{-\sqrt{2}}{2} + j\frac{\sqrt{2}}{2}\right)\omega_c T}$$

$$= 0.584131 + 0.28794j$$

$$z_{p_2} = \frac{2 + \left(\frac{-\sqrt{2}}{2} - j\frac{\sqrt{2}}{2}\right)\omega_c T}{2 - \left(\frac{-\sqrt{2}}{2} - j\frac{\sqrt{2}}{2}\right)\omega_c T}$$

$$= 0.584131 - 0.28794j$$

The two zeros at $s = \infty$ map into two zeros at $z = -1$. Thus, the system function is given by

$$H(z) = H_c \frac{(z+1)^2}{(z - 0.584131 + 0.287941j)(z - 0.584131 - 0.287941j)}$$

where
$$\begin{aligned}
H_c &= \frac{H_0 T^2}{(2 - p_1 T)(2 - p_2 T)} \\
&= \frac{(6000\pi)^2}{(30{,}000)^2 \left(2 + \frac{\pi\sqrt{2}}{10} - j\frac{\pi\sqrt{2}}{10}\right)\left(2 + \frac{\pi\sqrt{2}}{10} + j\frac{\pi\sqrt{2}}{10}\right)} \\
&= 0.063964
\end{aligned}$$

If the numerator and denominator factors are multiplied out and all terms are divided by z^2, we obtain

$$H(z) = \frac{0.063964(1 + 2z^{-1} + z^{-2})}{1 - 1.168261z^{-1} + 0.424118z^{-2}} \tag{17.6}$$

which matches the result of Example 17.1.

17.3 Properties of the Bilinear Transformation

Assume that the analog prototype filter has a pole at $s_P = \sigma + j\omega$. The corresponding IIR filter designed via the bilinear transformation will have a pole at

$$\begin{aligned}
z_P &= \frac{2 + sT}{2 - sT} \\
&= \frac{2 + (\sigma + j\omega)T}{2 - (\sigma + j\omega)T} \\
&= \frac{2 + \sigma T + j\omega T}{2 - \sigma T - j\omega T}
\end{aligned}$$

The magnitude and angle of this pole are given by

$$|z_P| = \sqrt{\frac{(2 + \sigma T)^2 + (\omega T)^2}{(2 - \sigma T)^2 + (\omega T)^2}} \tag{17.7}$$

$$\arg(z_P) = \tan^{-1}\left(\frac{\omega T}{2 + \sigma T}\right) - \tan^{-1}\left(\frac{-\omega T}{2 - \sigma T}\right)$$

The poles of a stable analog filter must lie in the left half of the s plane—that is, $\sigma < 0$. When $\sigma < 0$, the numerator of Eq. (17.7) will be smaller than the denominator, and thus $|z_P| < 1$. This means that analog poles in the left half of the s plane map into digital poles inside the unit circle of the z plane—stable analog poles map into stable digital poles. Poles that lie on the $j\omega$ axis of the s plane have $\sigma = 0$ and, consequently, map into z-plane poles which have unity magnitude and, hence, lie on the unit circle. Analog poles at $s = 0$ map into digital poles at $z = 1$, and analog poles at $s = \pm j\infty$ map into digital poles at $z = -1$.

Frequency warping

The mapping of the s plane's $j\omega$ axis into the z plane's unit circle is a highly nonlinear mapping. The analog frequency ω_a can range from $-\infty$ to $+\infty$, but the digital frequency ω_d is limited to the range $\pm\pi$. The relationship between ω_a and ω_d is given by

$$w_d = 2\tan^{-1}\frac{\omega_a T}{2} \qquad (17.8)$$

If an analog prototype filter with a cutoff frequency of ω_a is used to design a filter via the bilinear transformation, the resulting digital filter will have a cutoff frequency of ω_d, where ω_d is related to ω_a via Eq. (17.8).

Example 17.3 A lowpass filter with a 3-dB frequency of 3 kHz is used as the prototype for an IIR filter with a sampling rate of 30,000 samples per second. What will be the 3-dB frequency of the digital filter designed via the bilinear transformation?

solution Equation (17.8) yields

$$\omega_d = 2\tan^{-1}\frac{(6000\pi)(1/30{,}000)}{2}$$
$$= 0.6088$$

Since $\omega_d = \pi$ corresponds to a frequency of $30{,}000/2 = 15{,}000$ Hz, the cutoff frequency of the filter is given by

$$\omega_c = \frac{0.6088}{\pi}(15{,}000) = 2906.8 \text{ Hz}$$

The frequency-warping effects become more severe as the frequency of interest increases relative to the digital filter's sampling rate.

Example 17.4 Consider the case of an analog filter with a 3-dB frequency of 3 kHz used as the prototype for an IIR filter designed via the bilinear transformation. Determine the impact on the 3-dB frequency if the sampling rate is changed from 10,000 samples per second to 30,000 samples per second in steps of 1000 samples per second.

solution The various sampling rates and the corresponding warped 3-dB frequencies are listed in Table 17.1.

Fortunately, it is a simple matter to counteract the effects of frequency warping by prewarping the critical frequencies of the analog prototype filter in such a way that the warping caused by the bilinear transformation restores the critical

TABLE 17.1 Warped Cutoff Frequencies for Example 17.4

Sampling rate	Cutoff frequency, Hz	% error
10,000	2405.8	−19.81
11,000	2480.5	−17.32
12,000	2543.1	−15.23
13,000	2595.8	−13.47
14,000	2640.4	−11.99
15,000	2678.5	−10.72
16,000	2711.1	−9.63
17,000	2739.3	−8.69
18,000	2763.6	−7.88
19,000	2784.9	−7.17
20,000	2803.5	−6.55
21,000	2819.9	−6.00
22,000	2834.4	−5.52
23,000	2847.2	−5.09
24,000	2858.7	−4.71
25,000	2868.9	−4.37
26,000	2878.1	−4.06
27,000	2886.4	−3.79
28,000	2893.8	−3.54
29,000	2900.6	−3.31
30,000	2906.8	−3.11

frequencies to their original intended values. Equation (17.8) can be inverted to yield the equation needed for this prewarping:

$$\omega_a = \frac{2}{T} \tan \frac{\omega_d}{2} \qquad (17.9)$$

Example 17.5 We wish to design an IIR filter with a 3-dB frequency of 3 kHz and a sampling rate of 30,000 samples per second. Determine the prewarped 3-dB frequency required for the analog prototype filter.

solution Since $\omega_d = \pi$ corresponds to a frequency of $30{,}000/2 = 15{,}000$ Hz, a frequency of 3 kHz corresponds to a ω_d of

$$\omega_d = \frac{3000\pi}{15{,}000} = \frac{\pi}{5}$$

The prototype analog frequency ω_a is obtained by using this value of ω_d in Eq. (17.9):

$$\omega_a = \frac{2}{(1/30{,}000)} \tan \frac{\pi}{10} = 19{,}495.18$$

The analog prototype filter must have a 3-dB frequency of $19{,}495.18/(2\pi) = 3102.75$ Hz in order for the IIR filter to have a 3-dB frequency of 3 kHz after warping.

17.4 Programming the Bilinear Transformation

Assume that the transfer function of the analog prototype filter is in the form given by

$$H_a(s) = H_0 \frac{\prod_{m=1}^{M} (s - q_m)}{\prod_{n=1}^{N} (s - p_n)}$$

where p_n and q_n denote, respectively, the filter's poles and zeros. To generate a digital filter via the bilinear transformation, we make the substitution

$$s = \frac{2}{T} \left(\frac{1 - z^{-1}}{1 + z^{-1}} \right)$$

and obtain

$$H(z) = H_0 \frac{\prod_{m=1}^{M} \left[\frac{2}{T} \left(\frac{1-z^{-1}}{1+z^{-1}} \right) - q_m \right]}{\prod_{n=1}^{N} \left[\frac{2}{T} \left(\frac{1-z^{-1}}{1+z^{-1}} \right) - p_n \right]}$$

which, after some algebraic manipulation, can be put into the form

$$H(z) = H_0 \frac{(1 + z^{-1})^{N-M} \prod_{m=1}^{M} \left[\left(\frac{2}{T} - q_m \right) - \left(\frac{2}{T} + q_m \right) z^{-1} \right]}{\prod_{n=1}^{N} \left[\left(\frac{2}{T} - p_n \right) - \left(\frac{2}{T} + p_n \right) z^{-1} \right]}$$

Thus, the denominator of $H(z)$ is given by

$$D(z) = \prod_{n=1}^{N} \left(\gamma_n + \delta_n z^{-1} \right) \tag{17.10}$$

where $\gamma_n = \frac{2}{T} - p_n$
$\delta_n = \frac{-2}{T} - p_n$

To see how Eq. (17.11) can be easily evaluated by computer, let us examine the sequence of partial products $\{D_k(z)\}$ encountered in the evaluation:

$$D_1(z) = (\gamma_1 + \delta_1 z^{-1})$$

$$D_2(z) = (\gamma_2 + \delta_2 z^{-1})D_1(z) = \gamma_2 D_1(z) + \delta_2 z^{-1} D_1(z)$$

$$D_2(z) = (\gamma_3 + \delta_3 z^{-1})D_2(z) = \gamma_3 D_2(z) + \delta_3 z^{-1} D_2(z)$$

$$\vdots$$

$$D(z) = D_N(z) = (\gamma_N + \delta_N z^{-1})D_{N-1}(z) = \gamma_N D_{N-1}(z) + \delta_N z^{-1} D_{N-1}(z)$$

Examination of this sequence reveals that the partial product $D_k(z)$ at iteration k can be expanded in terms of the partial product $D_{k-1}(z)$ as

$$D_k(z) = \gamma_k D_{k-1}(z) + \delta_k z^{-1} D_{k-1}(z)$$

The partial product $D_{k-1}(z)$ will be a $(k-1)$-degree polynomial in z^{-1}:

$$D_{k-1}(z) = \mu_0(z^{-1})^0 + \mu_1(z^{-1})^1 + \mu_2(z^{-1})^2 + \cdots + \mu_{k-1}(z^{-1})^{k-1}$$

The products $\gamma_k D_{k-1}(z)$ and $\delta_k z^{-1} D_{k-1}(z)$ are then given by

$$\gamma_k D_{k-1}(z) = \gamma_k \mu_0(z^{-1})^0 + \gamma_k \mu_1(z^{-1})^1 + \gamma_k \mu_2(z^{-1})^2 + \cdots + \gamma_k \mu_{k-1}(z^{-1})^{k-1}$$

$$\delta_k D_{k-1}(z) = \delta_k \mu_0(z^{-1})^1 + \delta_k \mu_1(z^{-1})^2 + \delta_k \mu_2(z^{-1})^3 + \cdots + \delta_k \mu_{k-1}(z^{-1})^k$$

and $D_k(z)$ is given by

$$D_k(z) = \gamma_k \mu_0(z^{-1})^0 + (\gamma_k \mu_1 - \delta_k \mu_0)(z^{-1})^1 + (\gamma_k \mu_2 - \delta_k \mu_1)(z^{-1})^2 + \cdots$$
$$+ (\gamma_k \mu_{k-1} - \delta_k \mu_{k-2})(z^{-1})^{k-1} - \delta_k \mu_{k-1}(z^{-1})^k$$

Therefore, we can conclude that if μ_n is the coefficient for the $(z^{-1})^n$ term in $D_{k-1}(z)$, then the coefficient for the $(z^{-1})^n$ term in $D(z)$ is $(\gamma_k \mu_n + \delta_k \mu_{n-1})$ with the proviso that $\mu_k \triangleq 0$ in $D_{k-1}(z)$. The polynomial $D_{k-1}(z)$ is represented in the computer as an array of k coefficients, with the array index corresponding to the subscript on μ and the superscript (exponent) on (z^{-1}). Thus, array element mu[0] contains μ_0, array element mu[1] contains μ_1, and so forth. The coefficients for the partial product $D_k(z)$ can be obtained from the coefficients for $D_{k-1}(z)$, as indicated by the following fragment:

```
for( j=k; j >=1; j--)
        {mu[j] = gamma * mu[j] + beta * mu[j-1];}
```

The loop is executed in reverse order so that the coefficients can be updated in place without prematurely overwriting the old values. If this fragment is placed within an outer loop with k ranging from 1 to num_poles, the final values in mu[n] will be the coefficients a_n for Eq. (17.1).

A similar loop can be developed for the numerator product $N(z)$ given by

$$N(z) = \prod_{m=1}^{M} \left(\alpha_m - \beta_m z^{-1} \right) \tag{17.11}$$

where $\alpha_m = \frac{-2}{T} + q_m$
$\beta_m = \frac{-2}{T} - q_m$

A function for computation of the bilinear transformation is provided in file bilinear.cpp. A main program for exercising this function, and which can be used for the remaining examples in this chapter, is provided in prog_17a.cpp.

17.5 Computer Examples

In this section we will apply the computer methods from the previous section to a number of different examples.

Example 17.6 Use the bilinear transform design software to obtain an IIR filter from a second-order Butterworth analog filter with a 3-dB cutoff frequency of 3 kHz. The sampling rate for the digital filter is 30,000 samples per second. Compare the frequency responses of the IIR filter and then the corresponding ideal Butterworth filter.

solution The magnitude response of the IIR filter is shown in Fig. 17.1, and the phase response is shown in Fig. 17.2. The solid trace shows the response of the IIR filter, which is virtually identical to the response of the analog filter. These plots have been normalized so that the 3-dB cutoff frequency is 1.0 Hz. Because the actual sampling rate is 10 times the unnormalized cutoff frequency, the normalized sampling rate becomes 10 samples per second. The response of the digital filter is shown only for frequencies up to the normalized folding frequency of 5 Hz. The analog filter's normalized response characteristic continues as shown by the dashed traces in the figures.

Example 17.7 Repeat Example 17.6 for the case of a fourth-order Chebyshev filter with 0.1-dB ripple in the passband.

solution The magnitude response of the IIR filter is shown in Fig. 17.3, and the phase response is shown in Fig. 17.4. Up to the normalized folding frequency of 5.0 Hz, the responses are virtually identical to the responses of the corresponding analog filter.

Example 17.8 Use the bilinear transform design software to obtain an IIR filter from the ninth-order elliptical filter that was designed in Example 5.7. Compare the responses of the IIR and analog filters.

solution The magnitude response of the IIR filter is shown in Fig. 17.5, and the phase response is shown in Fig. 17.6. Up to the normalized folding frequency of 5.0 Hz, the responses are virtually identical to the responses of the corresponding analog filter.

IIR Filter Design: Bilinear Transformation 315

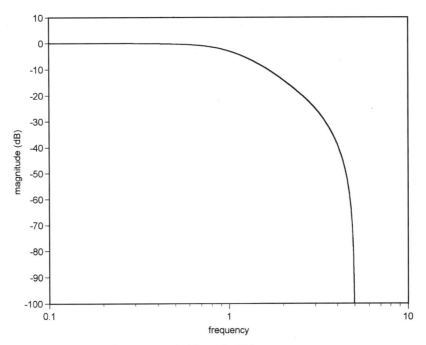

Figure 17.1 Magnitude responses for Example 17.6.

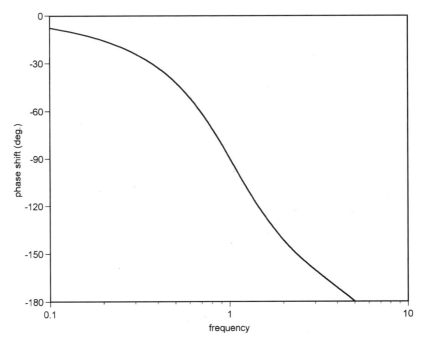

Figure 17.2 Phase responses for Example 17.6.

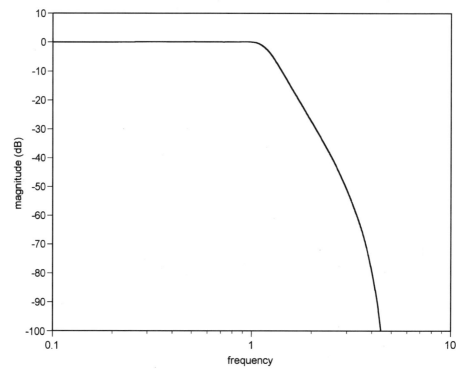

Figure 17.3 Magnitude responses for Example 17.7.

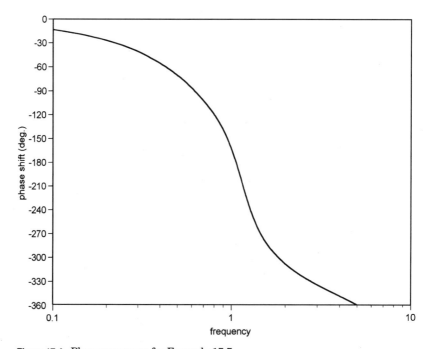

Figure 17.4 Phase responses for Example 17.7.

IIR Filter Design: Bilinear Transformation 317

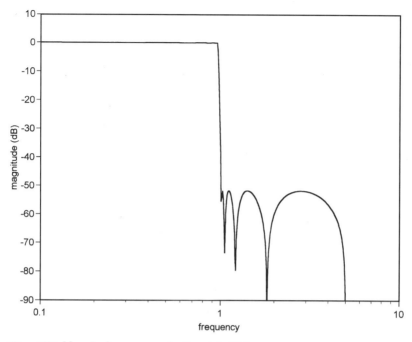

Figure 17.5 Magnitude responses for Example 17.8.

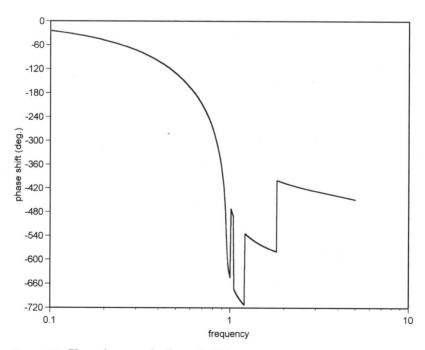

Figure 17.6 Phase responses for Example 17.8.

17.6 Quantization in IIR Filters Obtained via Bilinear Transformation

This section explores the impacts of quantization upon the various IIR filters that were designed in Sec. 17.5.

Example 17.9 Assume that the IIR filter obtained from a fourth-order Chebyshev response in Example 17.7 is to be implemented using 8-bit, 10-bit, or 12-bit coefficients. Assume that arithmetic results are not truncated or rounded. Compare the frequency response of the three quantized filters with the response of the unquantized filter.

solution The magnitude responses of the filters are compared in Fig. 17.7, and their phase responses are compared in Fig. 17.8. The response of the filter with 12-bit quantization is very close to the response of the unquantized filter.

For the next example we will use the swept-tone method that was described in Sec. 11.5 to see how the combination of coefficient quantization and signal quantization might further degrade performance.

Example 17.10 Repeat Example 17.9 using the swept-tone method to determine how the filter will perform when the coefficients are quantized to 12 bits and the input signal is quantized to 12 bits.

solution The magnitude response of the filter with quantized inputs is nearly identical to the 12-bit trace for unquantized inputs shown in Fig. 17.7.

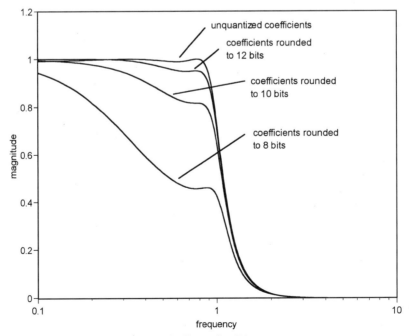

Figure 17.7 Magnitude responses for Example 17.9.

IIR Filter Design: Bilinear Transformation 319

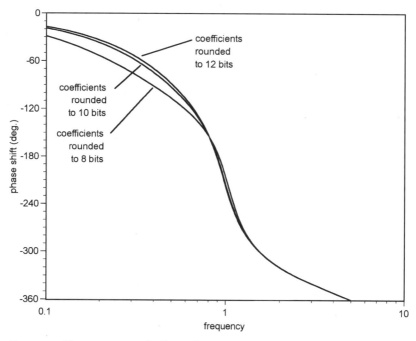

Figure 17.8 Phase responses for Example 17.9.

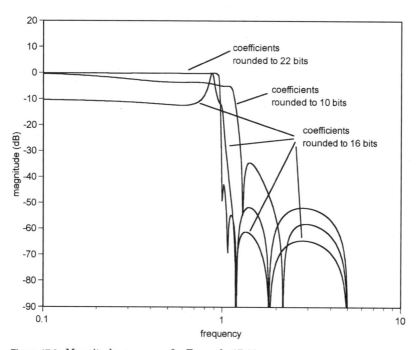

Figure 17.9 Magnitude responses for Example 17.11.

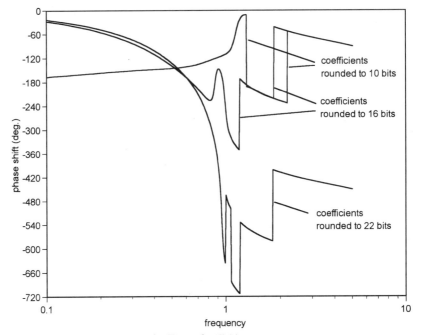

Figure 17.10 Phase responses for Example 17.11.

Example 17.11 Assume that the IIR filter obtained from a ninth-order elliptical response in Example 17.8 is to be implemented using 10-bit, 16-bit, or 22-bit coefficients. Assume that arithmetic results are not truncated or rounded. Compare the frequency response of the three quantized filters with the response of the unquantized filter.

solution The magnitude responses of the filters are compared in Fig. 17.9, and the phase responses are compared in Fig. 17.10.

Chapter 18

Multirate Signal Processing: Basic Concepts

There are many situations in which it is advantageous to be able to change the sampling rate within a DSP system. *Multirate signal processing* is the subarea of DSP concerned with the techniques that can be used to efficiently change the sampling rates within a system. Examples of multirate processing can be found in virtually every applications area. Consider a DSP implementation of a wireless data communications system in which the data starts out as a bit sequence, with one sample per bit. To represent this sequence as a sampled-data baseband waveform suitable for input to a modulator, it will be necessary to increase the sampling rate to have several samples per bit. The modulator will produce an *intermediate frequency* (IF) signal that will contain frequencies that are much greater than the highest frequencies in the baseband waveform. Because of its increased frequency content, such an IF signal must be sampled at a rate that is at least twice the highest frequency contained in the signal. Multirate techniques are used extensively in modern DSP-based audio systems such as CD and DAT players.

In addition to uses driven by the sampling rates dictated by the application, a large body of multirate technology also is concerned with more efficient ways to implement traditional DSP functions in intrinsically single-rate systems. For example, narrow band FIR filters typically require a very large number of taps to meet their stringent frequency response specifications. Multirate techniques can be used to perform this filtering at a lower rate with a large decrease in the number of required taps. Multirate techniques also are widely used for implementing filter banks and digital spectrum analyzers.

Interpolation and decimation are the fundamental operations of interest in multirate processing. Interpolation is the process of increasing the sampling rate of a discrete-time signal, and decimation is the process of decreasing the sampling rate. The basic concepts involved in interpolation and decimation will be covered in this chapter. Practical structures for implementing interpolators

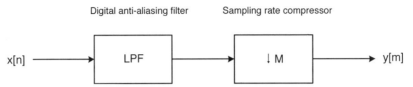

Figure 18.1 Block diagram of system for decimation by a factor of M.

and decimators will presented in Chap. 19, and advanced multirate techniques will be covered in Chap. 20.

18.1 Decimation by Integer Factors

We could conceivably reduce a signal's sampling rate by an integer factor M simply by discarding $M-1$ samples out of every M samples, retaining only every Mth sample from the original sequence. However, the sampling theorem is still in force, and before we can safely reduce the sampling rate, we must first ensure that the signal has no significant spectral content at frequencies greater than half the new sampling rate. This is usually accomplished by passing the signal through a digital lowpass antialiasing filter prior to downsampling, as shown in Fig. 18.1. If the original sampling rate was F_S, the filter must effectively band limit the signal to less than $F_S/(2M)$.

18.2 Interpolation by Integer Factors

The basic idea of using interpolation to produce new sample values between existing samples is straightforward. However, the way in which this interpolation is accomplished is not particularly intuitive. Furthermore, several published mathematical explanations of the usual approach to interpolation are confusing, misleading and, in at least one case, incorrect.

Consider the sampled signal and its spectrum shown in Fig. 18.2. Suppose we wish to triple the sampling rate. Following the usual approach, we proceed by inserting two zero-valued samples between each pair of original samples and obtain the sequence shown in Fig. 18.3. We then use a lowpass filter to limit the signal to a bandwidth equal to half the original sampling rate. The resulting signal and its spectrum is depicted in Fig. 18.4. The confusion surrounds the explanation of why the filtering is needed. Several references state or imply that the insertion of $L-1$ zero-valued samples causes the signal's spectrum to be compressed by a factor of L. In other words, if the original signal had a bandwidth of $F_S/2$, insertion of the zeros would compress that bandwidth down to $F_S/(2L)$. If this were true, it would not be possible to recover the original signal by lowpass filtering. A lowpass filter only attenuates signal components outside of the passband; it would not be able to decompress a compressed spectrum.

Multirate Signal Processing: Basic Concepts   323

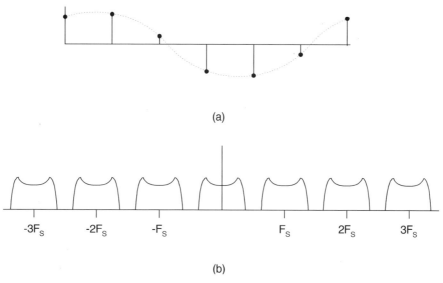

Figure 18.2 (a) Discrete-time signal and (b) its spectrum.

Figure 18.3 Discrete-time signal after zero-valued samples have been inserted.

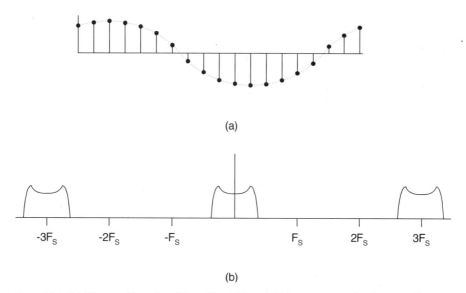

Figure 18.4 (a) Discrete-time signal from Fig. 18.3 and (b) its spectrum after lowpass filtering.

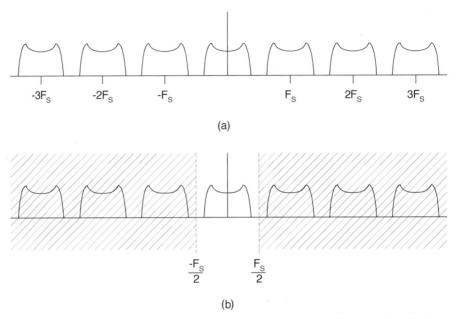

Figure 18.5 Spectrum of a discrete-time signal showing (a) spacing of images along the frequency axis, and (b) relationship between images and system bandwidth.

As discussed in Sec. 6.1, the spectrum of any discrete-time signal contains images of the original spectrum that are periodically repeated along the frequency axis, as shown in Fig. 18.5(a). The center-to-center spacing of the images is equal to the sampling rate F_S. To avoid aliasing, the original spectrum must fit within the frequency interval from $f = -F_S/2$ to $f = F_S/2$. Therefore, we could take the viewpoint that the discrete-time system in question has a bandwidth of either $F_S/2$ (one-sided) or F_S (two-sided). As shown in Fig. 18.5(b), all of the images fall outside of the original system bandwidth. What really happens when the two zero-valued samples are inserted between each pair of original samples is that the system bandwidth becomes three times wider, as shown in Fig. 18.6. This wider bandwidth will include two images along with the original spectrum. The lowpass filtering operation is needed to remove these images from the system passband.

The confusion about compression of the spectrum arises from the widespread practice of using the sampling rate to normalize the frequencies in a DSP system. If the frequencies in Fig. 18.2(b) are normalized by F_S, the bandwidth of the baseband image is confined to the normalized frequency range of $\pm 1/2$. On the other hand, if we say that after interpolation the sampling rate is $3F_S$, and we normalize the frequencies accordingly, the baseband image is confined to the normalized frequency range of $\pm 1/6$, and appears, therefore, to have been compressed by a factor of 3. Graphically, the same comparison could be made by redrawing Fig. 18.6 so that its unshaded interval is the same size as the unshaded interval in Fig. 18.5(b). It would then appear as though the insertion of zero-valued samples caused the signal's spectrum to be compressed.

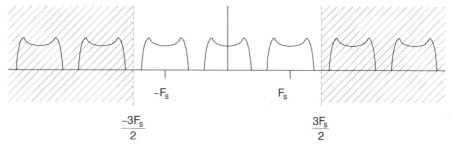

Figure 18.6 Spectrum of discrete-time signal showing relationship between images and system bandwidth after two zero-valued samples have been inserted between each pair of original time samples.

18.3 Decimation and Interpolation by Non-Integer Factors

There are many situations in which the sampling rate needs to be changed by a factor that is not an integer. The sampling rate can be changed by a rational factor L/M by first interpolating by a factor of L and then decimating by a factor of M. If $L > M$, the net effect is interpolation by a factor of L/M. If $M > L$, the net effect is decimation by a factor of M/L. A classic example of non-integer rate conversion can be found in the conversion of compact disc (CD) signals to digital audio tape (DAT) signals. The sampling rate in CD systems is 44.1 kHz, and the sampling rate in DAT systems is 48 kHz. It is widely believed that the rates have been made different intentionally to make direct digital copying from CD to DAT difficult, and thereby discourage illegal copying of copyrighted CDs onto DAT tapes. The different rates only discourage copying, because it is still possible to connect the analog output of a CD player to the analog input of a DAT recorder and make a copy that has only a small decrease in audio quality. Furthermore, using multirate processing techniques it is possible to convert the 44.1 kHz sampling rate to a 48 kHz sampling rate without having to convert the signal back to analog. Interpolation by a factor of 160 can be used to convert the CD sample rate from 44.1 kHz to 7056 kHz. Decimation by a factor of 147 then can be used to convert the 7056 kHz sample rate down to the DAT rate of 48 kHz.

18.4 Decimation and Interpolation of Bandpass Signals

When applied to the case of bandpass signals, the terms decimation and interpolation take on slightly different meanings than for lowpass signals. When a lowpass signal is to be decimated, it is assumed that the useful spectral content of the signal is already sufficiently band-limited so as to permit a reduction of the sampling rate without introducing significant aliasing. The lowpass filtering performed prior to the actual downsampling is meant to attenuate any noise or interference which lies outside of the useful signal's bandwidth but which would be aliased into the signal bandwidth by the downsampling. For the case of bandpass signals, it is usually assumed that, prior to downsampling, the signal bandwidth is to be reduced by moving the signal's passband to a lower center frequency. Conversely, for interpolation of bandpass signals, it is usually assumed that, after upsampling, the signal bandwidth is to be increased by moving the signal's passband to a higher center frequency. Several different approaches for frequency shifting the signal's passband will be examined in the following sections.

Quadrature modulation of bandpass signals

Consider a real-valued signal $x[n]$ having the DTFT spectrum $X(e^{j\omega})$ shown in Fig. 18.7(a). Let X^+ denote the component of $X(e^{j\omega})$ for $\omega > 0$ and let X^- denote the component of $X(e^{j\omega})$ for $\omega < 0$. The band edges are $|\omega_L|$ and $|\omega_L + \omega_\triangle|$, and the center frequency of X^+ is $\omega_0 = \omega_L + \omega_\triangle/2$. Because $x[n]$ is real-valued, X^+ and X^- must be conjugate symmetric about $\omega = 0$, that is

$$X^+(e^{j\omega}) = [X^-(e^{-j\omega})]^*$$

Consequently, either X^+ or X^- alone will uniquely define the signal because X^+ can be constructed from X^-, and X^- can be constructed from X^+.

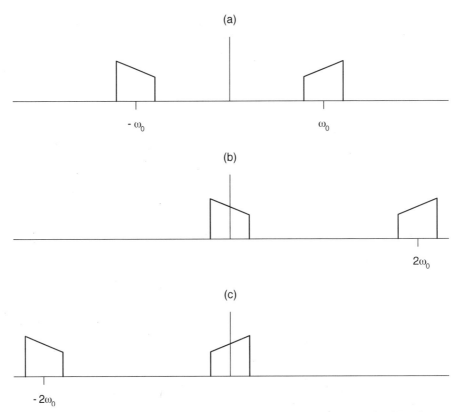

Figure 18.7 Spectral interpretation of quadrature demodulation: (a) real-valued bandpass signal, (b) shifted spectrum for $x[n]\exp(j\omega_0 n)$, (c) shifted spectrum for $x[n]\exp(-j\omega_0 n)$.

If the signal $x[n]$ is multiplied by the complex signal $e^{j\omega_0 n}$, the signal spectrum will be shifted right by ω_0 moving the center of the X^- band to $\omega = 0$ and the center of the X^+ band to $\omega = 2\omega_0$, as shown in Fig. 18.7(b). Alternatively, if the signal $x[n]$ is multiplied by the complex signal $e^{-j\omega_0 n}$, the signal spectrum will be shifted left by ω_0 moving the center of the X^- band to $\omega = 2\omega_0$ and the center of the X^+ band to $\omega = 0$, as shown in Fig. 18.7(c). Because either X^+ or X^- alone will uniquely define the signal, we can lowpass filter the signal of Fig. 18.7(b) to remove the X^+ component at $\omega = 2\omega_0$, or we can lowpass filter the signal of Fig. 18.7(c) to remove the X^- component at $\omega = -2\omega_0$. In either case, decimation can be performed on the resulting baseband signal to reduce the sampling rate to a value equal to or greater than ω_Δ.

Although X^+ and X^- are originally conjugate symmetric relative to each other, there is no guarantee that either X^+ or X^- alone will be conjugate symmetric when moved to $\omega = 0$. This means that the corresponding time signal will, in general, be complex-valued. For the case in which $x[n]$ is multiplied by $e^{j\omega_0 n}$ to shift X^- to $\omega = 0$ and then filtered to remove X^+, the corresponding complex-valued time domain signal is called the *complex envelope* of $x[n]$ and denoted as $\tilde{x}[n]$. The complex envelope can be expressed in terms of an *inphase* component $x_I[n]$ and *quadrature* component $x_Q[n]$ as

$$\tilde{x}[n] = x_I[n] + jx_Q[n] \qquad (18.1)$$

The process of obtaining $x_I[n]$ and $x_Q[n]$ from $x[n]$ is called *quadrature demodulation*. A block diagram of a quadrature demodulator is shown in Fig. 18.8. Equation (18.1) is the *rectangular form* of the complex envelope. The *polar form* of the complex envelope is given by

$$\tilde{x}[n] = a(t)\exp[j\phi(t)] \qquad (18.2)$$

where $a(t)$ is the *envelope* and $\phi(t)$ is the *phase* of the signal $x[n]$. The envelope and phase can be obtained from the inphase and quadrature components by

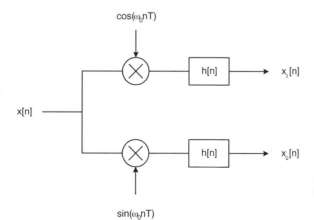

Figure 18.8 Block diagram of a quadrature demodulator.

using

$$a(t) = \sqrt{x_I^2(t) + x_Q^2(t)}$$

$$\phi(t) = \tan^{-1}\left[\frac{x_I(t)}{x_Q(t)}\right]$$

Conversely, the inphase and quadrature components can be obtained from the envelope and phase by using

$$x_I(t) = a(t)\cos[\phi(t)]$$

$$x_Q(t) = a(t)\sin[\phi(t)]$$

Given a complex envelope signal $\tilde{x}[n]$ that has been obtained via quadrature demodulation of a real-valued bandpass signal, it is a straightforward matter to perform *quadrature modulation* to reconstruct the original bandpass signal. The signal $x_I[n] + jx_Q[n]$ can be modulated by $e^{-j\omega_0 n}$ to obtain the signal corresponding to X^-, and the signal $x_I[n] - jx_Q[n]$ can be modulated by $e^{j\omega_0 n}$ to obtain the signal corresponding to X^+. The two signals corresponding to X^+ and X^- then can be summed to form $x[n]$. However, because of some cancellations that occur in forming the sum, it is not necessary to completely generate the signals corresponding to X^+ and X^-. The sum of the two signals can be expressed as follows:

$$x[n] = (x_I[n] + jx_Q[n])\,e^{-j\omega_0 n} + (x_I[n] - jx_Q[n])\,e^{j\omega_0 n} \qquad (18.3)$$

By using Euler's identities, Eq. (18.3) can be simplified to

$$x[n] = x_I[n]\cos(\omega_0 n) + x_Q[n]\sin(\omega_0 n) \qquad (18.4)$$

A block diagram of a quadrature modulator is shown in Fig. 18.9. It should be noted that some texts (such as [1]) refer to the system depicted in Fig. 18.8 as a quadrature *modulator*, and refer to the system depicted in Fig. 18.9 as a quadrature *demodulator*. This might be due to the fact that the term modulation is sometimes used to refer to frequency-shifting of a signal's spectrum in either direction in much the same way that physicists use the term *acceleration* to refer to any change in velocity. However, calling Fig. 18.8 the demodulator and Fig. 18.9 the modulator is more intuitive and consistent with the way these processes are actually used in communications applications.

Single sideband modulation

As stated in Sec. 18.4, when quadrature modulation is used to frequency shift a real-valued bandpass signal down to baseband, the resulting baseband signal (complex envelope) will, in general, have complex values. Processing such a complex-valued signal may be highly inconvenient in some applications. It is

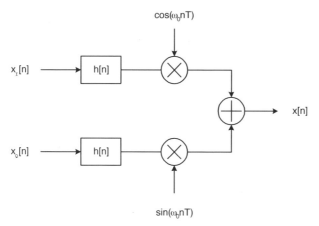

Figure 18.9 Block diagram of a quadrature modulator.

possible to modify the complex envelope of a signal to produce a baseband signal that is real-valued. Consider the spectrum of a complex envelope signal shown in Fig. 18.10(a). This signal can be modulated by $\exp(-j\omega_\triangle n/2)$ to obtain the spectrum X_{SS}^- shown in Fig. 18.10(c). Similarly, modulation by $\exp(j\omega_\triangle n/2)$ can be used to shift the spectrum of the conjugate complex envelope, shown in Fig. 18.10(b), to obtain the spectrum X_{SS}^+ shown in Fig. 18.10(d). The signals corresponding to X_{SS}^+ and X_{SS}^- then can be summed to obtain a signal having the spectrum shown in Fig. 18.10(e). This signal is sometimes referred to as the *single sideband* modulated form of the original bandpass signal [1, 2].

Decimation via integer band sampling

Both the quadrature modulation and single sideband approaches to converting a bandpass signal to a baseband signal involve modulation of the signal with complex exponentials. In many applications, this modulation can become quite expensive in terms of required processing power. In some cases there may be a less expensive alternative. When there is a particular relationship between the sampling rate, decimation rate, and band-edge frequencies of the bandpass signal, a technique called *integer band sampling* can be used to obtain the corresponding baseband signal without the expense of modulation. Specifically, the band-edge frequencies must lie at consecutive integer multiples of f_S/M, that is

$$f_L = \frac{k f_S}{2M} \quad (18.5a)$$

$$f_H = \frac{(k+1) f_S}{2M} \quad (18.5b)$$

where f_S is the sampling rate, M is the decimation rate, and k is a positive integer.

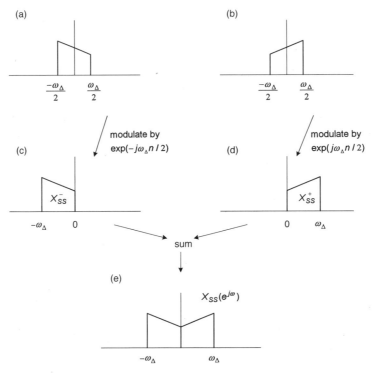

Figure 18.10 Constructing the single-sideband signal from the complex envelope signal: (a) spectrum of complex envelope, (b) spectrum of conjugate complex envelope, (c) spectrum of complex envelope after modulation by $\exp(-j\omega_\Delta n/2)$, (d) spectrum of complex envelope after modulation by $\exp(j\omega_\Delta n/2)$, (e) spectrum of single-sideband signal obtained by summing (c) and (d).

Consider the case of a bandpass signal with $f_L = 1000$ Hz and $f_H = 1200$ Hz. The original sampling rate is $f_S = 2400$ sps, and the decimation factor is $M = 6$. The spectrum of the original signal is shown in Fig. 18.11(a). Only the first image bands immediately past the folding frequencies of $\pm f_S/2$ are shown to keep the drawing down to a reasonable size. If we now downsample the signal by keeping only every sixth sample, the new sampling rate will be 400 sps. The additional images due to the lower sampling rate will be arranged as shown in Fig. 18.11(b). A lowpass filter can then be used to eliminate all but the two images on either side of zero and obtain the spectrum shown in Fig. 18.11(c). Notice that the positive part of this spectrum is *inverted* such that it is the mirror image of the positive part of the original spectrum. The negative part is similarly inverted. This inversion will occur whenever the value of k used in Eq. (18.5) is odd. For the example shown in Fig. 18.11, $k = 5$.

Inversion will not occur for k even. If we consider the case of $k = 4$ while keeping $f_S = 2400$ and $M = 6$, Eq. (18.5) is satisfied if the band edges are $f_L = 800$ and $f_H = 1000$. As shown in Fig. 18.12, the baseband images are not inverted relative to the original spectrum.

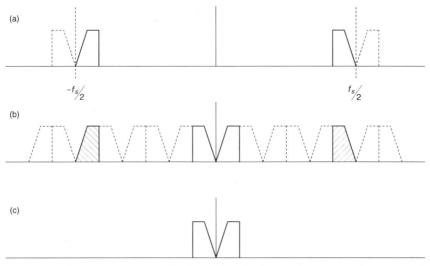

Figure 18.11 Illustration of integer-band sampling for $k = 5$: (a) spectrum of original bandpass signal sampled at $f_S = 2400$, (b) spectrum showing images introduced by downsampling to $f_S = 400$, (c) baseband spectrum produced by lowpass filtering.

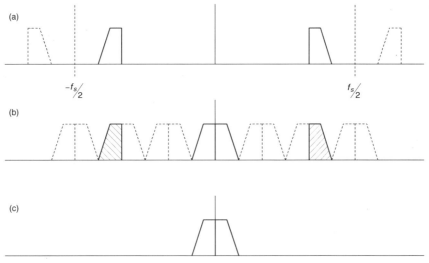

Figure 18.12 Illustration of integer-band sampling for $k = 4$: (a) spectrum of original bandpass signal sampled at $f_S = 2400$, (b) spectrum showing images introduced by downsampling to $f_S = 400$, (c) baseband spectrum produced by lowpass filtering.

Suppose that we are faced with the case previously examined in Fig. 18.11 where $f_L = 1000$ Hz and $f_H = 1200$ Hz. However, this time, spectrum inversion is deemed to be unacceptable. Our goal is to make minor adjustments in the values of f_S and M such that we can use an even value of k in Eq. (18.5) and thereby avoid inversion. Sometimes it can be difficult or even impossible to

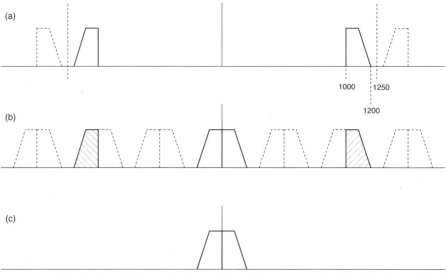

Figure 18.13 Illustration of integer-band sampling: (a) spectrum of original bandpass signal sampled at $f_S = 2500$, (b) spectrum showing images introduced by downsampling to $f_S = 500$, (c) baseband spectrum produced by lowpass filtering.

find values that satisfy both parts of Eq. (18.5). There is one extra degree of flexibility that can be exploited to make this task easier. The bandpass signal has $f_H = 1200$, but if we increase the sampling rate beyond 2400, the folding frequency will be greater than 1200. Once we find values of f_S, M, and k that satisfy Eq. (18.5a), we can then adjust f_H to any value between 1200 and $f_S/2$ that satisfies Eq. (18.5b).

If we make $f_S = 2500$ and $M = 5$, we can obtain $f_L = 1000$ with $k = 4$. These values will yield $f_H = 1250$. The original bandpass signal's spectrum with images due to sampling at $f_S = 1250$ is shown in Fig. 18.13(a). After decimation by $M = 5$, the images will be as shown in Fig. 18.13(b) and it is a simple matter to use a lowpass filter to obtain the baseband signal shown in Fig. 18.13(c). Spectral inversion has been avoided, but there is a price. The original sampling rate had to increased from 2400 to 2500, and the decimated sampling rate increased from 400 to 500. Unlike the spectrum in Fig. 18.11(b), the spectrum in Fig. 18.13(b) does have gaps of 50 Hz between adjacent images. These gaps are due to the fact that the downsampling was designed to accommodate a signal band from 1000 Hz to 1250 Hz, but the actual signal band extends to only 1200 Hz. As we will see in Chap. 25, the presence of these gaps will simplify the design of the lowpass filter needed to extract the baseband signal.

References

1. Crochiere, R. E. and L. R. Rabiner. *Multirate Digital Signal Processing*, Prentice-Hall, Englewood Cliffs, NJ, 1983.
2. Darlington, S. "On Digital Single-Side-Band Modulators," *IEEE Trans. Circuit Theory*, Vol. CT-17, pp. 409–414, August 1970.

Chapter 19

Structures for Decimators and Interpolators

Based on the material presented in the previous chapter, it should be clear that, in principle, decimators and interpolators can be implemented using any of the digital filter types discussed in Chaps. 11 through 17. However, when the choice of digital filter type is limited to FIR filters, several interpolator and decimator structures can be exploited to achieve more efficient implementations than might first be obvious from the basic theory presented in Chap. 18.

19.1 Decimator Structures

Consider the basic decimator shown in Fig. 19.1. The filter can be realized in a direct form structure to yield the decimator structure shown in Fig. 19.2. To produce each output sample, this structure must process M input samples, performing N *multiply-accumulate* (MAC) operations for each one. This is a total of MN MAC operations for each output sample. It is possible to commute the order of the multiplications and sampling rate compression to obtain the structure shown in Fig. 19.3. This structure is much more efficient than Fig. 19.2, requiring only N MAC operations per output sample. The equation implemented by Fig. 19.3 is given by

$$y[m] = \sum_{n=0}^{N-1} h[n]x[Mm - n]$$

19.2 Interpolators

The basic block diagram for a 1-to-L interpolator is shown in Fig. 19.4. The filter can be realized using a transposed direct form structure to yield the interpolator structure shown in Fig. 19.5. This structure produces L output samples

334 Chapter Nineteen

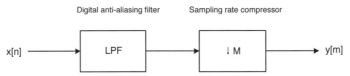

Figure 19.1 Block diagram of basic decimator.

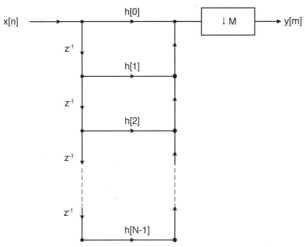

Figure 19.2 Direct form decimator structure obtained by direct expansion of the FIR filter in Fig. 19.1.

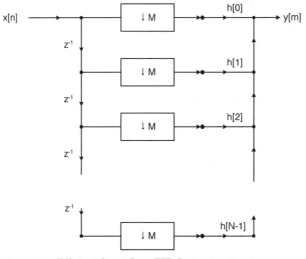

Figure 19.3 Efficient direct form FIR decimator structure.

Figure 19.4 Block diagram of basic interpolator.

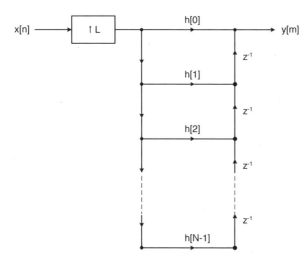

Figure 19.5 FIR interpolator with filter realized in transposed direct form.

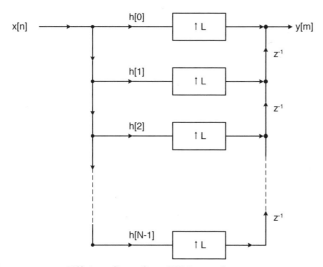

Figure 19.6 Efficient direct form FIR interpolator structure.

for each input sample and performs LN MAC operations per input sample or N MAC operations per output sample. It is possible to commute the order of the multiplications and sampling rate expansion to obtain the structure shown in Fig. 19.6. This structure requires only N MAC operations per input sample or N/L MAC operations per output sample.

19.3 Polyphase FIR Interpolator Structures

A polyphase structure for a 1-to-L interpolator is shown in Fig. 19.7. Each of the L filter functions $p_0[n], p_1[n], \ldots p_{L-1}[n]$ is obtained using

$$p_\rho[n] = h[nL + \rho]$$

where $h[n]$ is the response of the filter when the interpolator is realized in the form of Fig. 19.4. In other words, each of the filters $p_\rho[n]$ is a decimated version of the prototype filter $h[m]$.

The sampling rate expanders (*upsamplers*), delays, and summing nodes on the right hand side of the interpolator can be replaced by a multiplexer or commutator as shown in Fig. 19.8. The equivalence of the structures in Fig. 19.6 and Fig. 19.7 can be verified by examining the operation of the interpolator in Fig. 19.6. When a single sample is provided as input, a sequence consisting of one nonzero sample followed by $L-1$ zero-valued samples will be produced at the output of each of the L upsamplers. The delay elements shift the different sequences such that the nonzero sample produced by any one of the upsamplers aligns with and is added to zero-valued samples in all of the other upsamplers as shown in Fig. 19.9. The resulting overall sequence $y_0[n], y_1[n], y_2[n], \ldots y_{L-1}[n]$ is exactly the sequence produced by the structure in Fig. 19.8 if the commutator starts at branch zero and advances counterclockwise by one branch for each subsequent output sample. The entire commutator cycle repeats for each subsequent input sample. Each branch is connected to the output for only 1 out of every L output samples. The $L-1$ sample times for which the branch is unconnected effectively take the place of the $L-1$ zeros that would be inserted by the upsamplers of Fig. 19.6. As detailed in [1], the term polyphase is used to describe the structure in Fig. 19.7 because each of the filters $p_\rho[n]$ will exhibit a

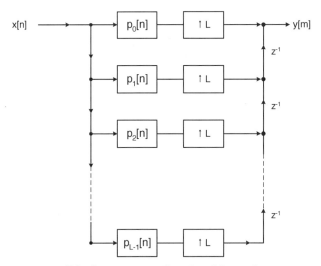

Figure 19.7 Polyphase structure for a 1-to-L interpolator.

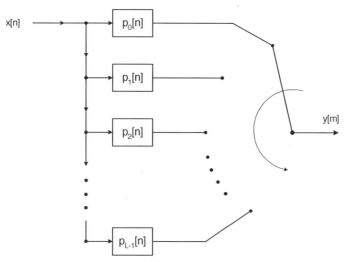

Figure 19.8 Commutator model for a 1-to-L interpolator.

$$\begin{array}{cccccc} y_0[k] & 0 & 0 & 0 & y_0[k+1] & 0 \\ 0 & y_1[k] & 0 & 0 & 0 & y_1[k+1] \\ 0 & 0 & y_2[k] & 0 & 0 & 0 \\ 0 & 0 & 0 & y_3[k] & 0 & 0 \\ \hline y_0[k] & y_1[k] & y_2[k] & y_3[k] & y_0[k+1] & y_1[k+1] \end{array}$$

Figure 19.9 Alignment of constituent sequences in a polyphase implementation of a 1-to- interpolator.

different linear phase response (assuming of course that the original $h[n]$ from which the $p_\rho[n]$ are derived has a linear phase response).

19.4 Polyphase FIR Decimator Structures

A polyphase structure for an M-to-1 decimator is shown in Fig. 19.10. Each of the M filter functions $p_0[n], p_1[n], \ldots p_{M-1}[n]$ is obtained using

$$p_\rho[n] = h[nM + \rho]$$

where $h[n]$ is the response of the filter when the decimator is realized in the form of Fig. 19.1. The sampling rate compressors (*down samplers*) and delays elements on the left hand side of the decimator can be replaced by a demultiplexing commutator as shown in Fig. 19.11.

19.5 Half-Band FIR Filters

Half-band filters have a frequency response of the form shown in Fig. 19.12. Notice that this is a plot of the real-valued response $H(f)$, not a plot of the magnitude response $|H(f)|$; there are some negative values in the stopband.

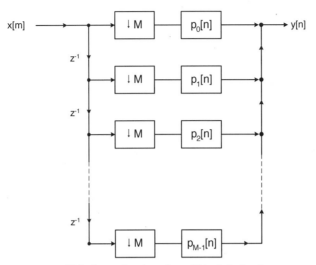

Figure 19.10 Polyphase structure for a M-to-1 decimator.

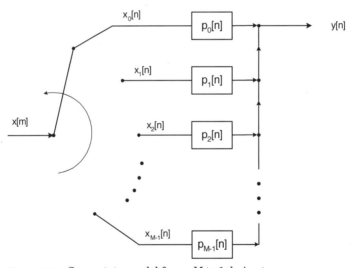

Figure 19.11 Commutator model for an M-to-1 decimator.

The passband ripple and stopband ripple each are equal to δ, and the widths of the passband and stopband each are equal to f_1. The response is symmetrical around the point $(f_S/4, 0.5)$ with

$$H(e^{j2\pi f/f_S}) = 1 - H(e^{j2\pi(0.5-f)/f_S}) \qquad (19.1)$$

This filter gets its name from the fact that it divides the maximum available bandwidth of $f_S/2$ in half, with the passband plus the first half of the transition band extending from $f = 0$ to the half-band frequency $f = f_S/4$, and the second

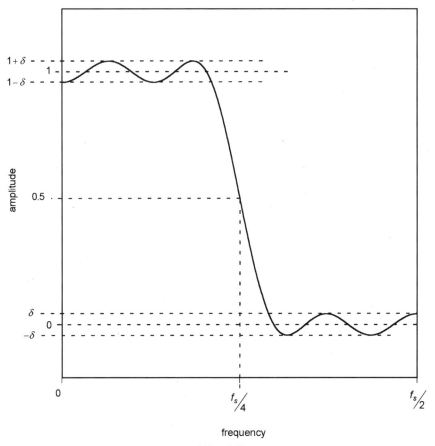

Figure 19.12 Frequency response of a half-band filter.

half of the transition band plus the stopband extending from the half-band frequency to the folding frequency $f = f_S/2$. FIR half-band filters have odd length, and their coefficients satisfy the constraint given by

$$h[k] = \begin{cases} 1, & k = 0 \\ 0, & k = \pm 2, \pm 4, \ldots \end{cases} \quad (19.2)$$

for coefficients indexed from $k = -(N-1)/2$ through $k = (N-1)/2$, or

$$h[k] = \begin{cases} 1, & k = \frac{N-1}{2} \\ 0, & k = \frac{N-1}{2} \pm 2, \frac{N-1}{2} \pm 4, \ldots \end{cases} \quad (19.3)$$

for coefficients indexed from $k = 0$ through $k = N - 1$. This property is one of the primary reasons for the popularity of half-band filters. Because of the zero-valued coefficients, a direct form realization of an N-tap half-band filter

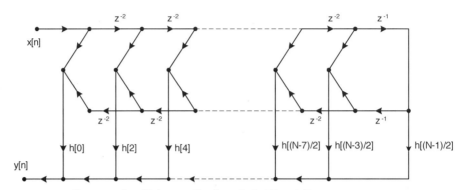

Figure 19.13 Structure for efficient realization of a half-band filter.

can be accomplished with only $(N+3)/2$ multiplications. Because the filter will always have even symmetry, the structure shown in Fig. 19.13 can be used to further reduce the multiplication count to $(N+5)/4$. It is possible to construct half-band filters only for values of N for which $N+1$ divides evenly by 4. For values of N for which $N+1$ does not divide evenly by 4, the first and last coefficient each will be zero, thus reducing the actual length of the filter from N to $N-2$.

References

1. Crochiere, R. E. and L. R. Rabiner. *Multirate Digital Signal Processing*, Prentice-Hall, Englewood Cliffs, NJ, 1983.
2. Haddad, R. A. and T. W. Parsons. *Digital Signal Processing: Theory, Applications, and Hardware*, Computer Science Press, New York, 1991.

Chapter 20

Advanced Multirate Techniques

The basic ideas of multirate processing can be applied in a myriad of ways that are limited only by the insight and creativity of the designer. This chapter presents a few of the more widely used techniques.

20.1 Multistage Decimators

When large changes in sampling rates are required, the design of adequate interpolation or decimation filters becomes more difficult, and the number of taps required for the FIR implementation of these filters can become very large, resulting in a prohibitively large computational burden for real-time applications. In practical systems, large changes in sampling rate usually are accomplished as a sequence of several smaller changes, each implemented in a separate interpolator or decimator stage. Compared to single-stage approaches, multistage approaches usually will result in reduced total computational burden, reduced storage requirements, simpler filter designs, and reduced sensitivity to finite word-length effects in the implementation of the filters. This section will explore multistage approaches for efficient implementation of high-order decimators.

The Basic Idea

When the desired decimation factor M can be expressed as a product of two or more positive integer factors (i.e., $M = M_1 M_2 \cdots M_I$), the decimation can be realized as a cascade of I decimators, each decimating by one of the factors M_i. The filters in the multistage implementation usually will have greatly relaxed transition band requirements, which will allow the total number of filter taps to be significantly smaller than the number of taps required for an equivalent single-stage decimator.

Consider the case of a signal $x[n]$ sampled at a rate of 7056 kHz, which is to be decimated by a factor of 147 to produce a signal $y[m]$ with a sample rate

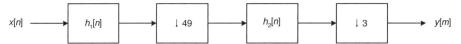

Figure 20.1 Structure for implementing a 147-to-1 decimator as a cascade of a 49-to-1 decimator and a 3-to-1 decimator.

of 48 kHz. Assume that the passband of the signal extends from 0 to 20 kHz. Consequently, the transition band extends from 20 kHz to 24 kHz. Using the approximate design formula (14.22) for equiripple FIR filters, we can estimate the required number of taps as

$$N = \left\lceil \frac{D_\infty(\delta_P, \delta_S)}{\Delta F/F} \right\rceil \qquad (20.1)$$

where $\Delta F = 2400 - 2000 = 4$ kHz
$F = 7056$ kHz
$\delta_P = 0.01$
$\delta_S = 0.001$
$D_\infty(\delta_P, \delta_S) = 2.541$

On evaluating Eq. (20.1), we obtain $N = 4482$. The number of multiplications per second required to implement this filter can be obtained using:

$$R_T^* = \frac{NF}{2M}$$

$$= \frac{(4482)(7.056 \times 10^6)}{2(147)} = 107{,}568{,}000 \text{ mult/sec}$$

Now suppose that the decimation by 147 is accomplished as a decimation by 49 followed by a decimation by 3, using the decimator structure shown in Fig. 20.1. Since the final output signal will have passed through the passbands of two filters, the total amount of passband ripple experienced by the signal could equal the sum of the passband ripples of the two separate filters. To ensure that the total ripple experienced by the signal does not exceed the original specification of $\delta_P = 0.01$, it will be necessary to halve the passband ripple allowance for the two filters, i.e., set $\delta_{P1} = \delta_{P2} = 0.005$. For this value of δ_P we find $D_\infty(\delta_P, \delta_S) = 2.76$.

Important aspects of the frequency plan for this decimator are shown in Fig. 20.2. As shown in part (c) of the figure, all frequency content above 72 kHz remaining in the signal after the first stage filter will be aliased into frequencies below 72 kHz once the downsampling by 49 is performed. Frequencies from 0 to 24 kHz (which comprise the ultimate output spectrum) must be protected from aliasing. However, we do not care about aliasing in frequencies from $F_S = 24$ kHz up to the folding frequency of 72 kHz because these frequencies will all be removed by the second-stage filter. Therefore, to avoid aliasing below 24 kHz, but allow it above 24 kHz, the stopband edge for the first stage filter must be set

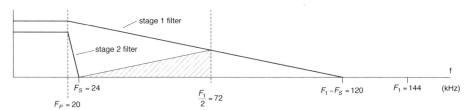

Figure 20.2 Frequency plan for two stage implementation of the 147-to-1 example decimator.

at $F_1 - F_S = 120$ kHz. Thus the transition width $\triangle F$ is $F_1 - F_S - F_P = 100$ kHz. We can estimate the number of taps N_1 required for the first stage filter as

$$N_1 = \left\lceil \frac{(2.76)(7.056 \times 10^6)}{10^5} \right\rceil = 195$$

The number of multiplications per second required to implement this filter is obtained as:

$$R^*_{T1} = \frac{(195)(7.056 \times 10^6)}{2(49)} = 14{,}040{,}000 \text{ mult/sec}$$

In the second stage, $\triangle F = F_S - F_P = 4$ kHz, and the input sampling rate is 144 kHz. We can estimate the number of taps for the second stage filter as

$$N_2 = \left\lceil \frac{(2.76)(1.44 \times 10^5)}{4 \times 10^3} \right\rceil = 100$$

The total multiplication rate needed for this filter is $R^*_{T2} = 2{,}400{,}000$, bringing the total requirement for the complete decimator to 16,440,000 multiplications per second. In many practical implementations the number of taps in each filter needs to be odd, so we might need to increase N_2 to 101, and thereby increase the total multiplication requirement to 16,464,000 multiplications per second.

We could also perform the decimation by 147 in 3 stages with $M_1 = 7$, $M_2 = 7$, and $M_3 = 3$. In this case, $F_1 = F/7 = 1.008$ MHz, and the first-stage transition band extends from $F_P = 20$ kHz to $F_1 - F_S = 984$ kHz. Since we are using 3 stages, we must re-evaluate $D_\infty(\delta_P, \delta_S)$ for $\delta_P = 0.01/3$ and $\delta_S = 0.001$. The result is 2.887. The number of taps for the first stage is then obtained as:

$$N_1 = \left\lceil \frac{(2.887)(7.056 \times 10^6)}{9.64 \times 10^5} \right\rceil = 22$$

with $R^*_{T1} = 11{,}088{,}000$. For the second stage, the output sampling rate is $F_2 = F_1/7 = 144$ kHz, and the transition band extends from $F_P = 20$ kHz to $F_2 - F_S = 120$ kHz. The number of taps for the second stage is obtained as:

$$N_2 = \left\lceil \frac{(2.887)(1.008 \times 10^6)}{10^5} \right\rceil = 30$$

with $R^*_{T_2} = 2{,}160{,}000$. For the third stage, the output sampling rate is $F_3 = F_2/3 = 48$ kHz, and the transition band extends from 20 kHz to 24 kHz. The number of taps for the third stage is obtained as:

$$N_3 = \left\lceil \frac{(2.887)(1.44 \times 10^5)}{4 \times 10^3} \right\rceil = 104$$

with $R^*_{T_3} = 2{,}496{,}000$, bringing the total for this implementation to 15,744,000 multiplications per second. If we force the number of filter taps always to be odd, the values of N_1, N_2, and N_3 would increase to 23, 31, and 105, respectively, resulting in a total multiplication requirement of 16,344,000 multiplications per second. Notice that the two-stage implementation requires a multiplication rate only 1.0073 times the multiplication rate required for a three-stage implementation. In many cases, most of the efficiency gains will be obtained in going from one stage to two, and it will not be worth the additional complexity to use three or more stages.

Further Insight

In a single-stage decimator, the filter must pass all of the signal frequencies with little or no attenuation, while attenuating to negligible levels all frequencies above half the output sampling rate. If the signal is sampled at near critical rates, accomplishing these goals will require that the filter have a narrow transition band. The relaxed transition band requirements for the filters in a multistage decimator can be traced to the fact that transition bands of filters in the early stages can extend from the ultimate passband edge, all the way up to half the output sampling rate of the stage in question. Frequencies above half the final sampling rate do not need to be attenuated to negligible levels by the early stage filter, because these frequencies will fall into the stopband of the filter in at least one of the subsequent sections. In terms of *absolute* frequencies, the filter in the final stage will have the same transition band as the filter in a single-stage implementation, but the filter in the multistage approach will require fewer taps, because it is implemented at a much lower sampling rate making its transition band much wider in terms of *normalized* frequencies.

Algorithm 20.1 Estimating the relative cost for a multistage implementation of a decimator.

1. Consistent with the intended application, determine the required values for passband ripple δ_P, stopband ripple δ_S, decimation factor M, initial sampling rate F, and final passband edge F_P.
2. Select the particular factorization of M that will be used as the basis for a multistage design. Let I denote the number of stages in the design.
3. Compute $D_\infty(\delta_P/I, \delta_S)$.
4. For stage n, compute $\triangle F$ as $\triangle F = F_n - F_S - F_P$ where F_n is the sampling rate at the output of stage n.

5. Estimate the number of taps N required for the stage n filter using Eq. (20.1) with F being the sampling rate at the input to stage n.

Software notes

File `msdcost.cpp` contains the function `MultistageDecimCost`, which can be used to estimate the total number of multiplications per second that will be needed for any specified multistage decimator implementation. The arguments passed into this function include the number of stages, an ordered vector of the per-stage decimation factors, the final passband edge frequency, the overall passband ripple limit, the stopband ripple limit, and a flag indicating whether the filter lengths should always be made odd. The file `prog_20a.cpp` contains a main program that provides some user interface and calls `MultistageDecimCost` with the proper arguments.

20.2 Multistage Interpolators

The Basic Idea
When the desired interpolation factor L can be expressed as a product of two or more positive integer factors (i.e., $L = L_1 L_2 \cdots L_I$), the interpolation can be realized as a cascade of I interpolators, each interpolating by one of the factors L_i. The filters in the multistage implementation will have relaxed transition band requirements which will allow the total number of filter taps to be significantly smaller than the number of taps required for an equivalent single-stage interpolator.

Consider the case of a signal $x[n]$ sampled at a rate of 44.1 kHz, which is to be interpolated by a factor of 160 to produce a signal $y[m]$ with a sample rate of 7.056 MHz. Assume that the signal of interest occupies frequencies from 0 to 20 kHz. If the sample rate is increased by inserting 159 zeros after each sample of the original signal, the system bandwidth will open up to include images of the original spectrum spaced at intervals of 44.1 kHz, as shown in Fig. 20.3. Since the signal of interest extends to 20 kHz, the anti-imaging filter for this interpolator must be designed for a transition band extending from 20 kHz to 22.05 kHz. If we assume a passband ripple of $\delta_P = 0.01$ and a stopband ripple of $\delta_S = 0.001$, we can use the approximate design formula, Eq. (20.1), for equiripple FIR filters to estimate the number of taps needed for the filter as

$$N = \left\lceil \frac{(2.54)(7.056 \times 10^6)}{2050} \right\rceil = 8743$$

The required number of multiplications per second is

$$R_T^* = \frac{(8743)(7.056 \times 10^6)}{2(160)} = 192{,}783{,}150$$

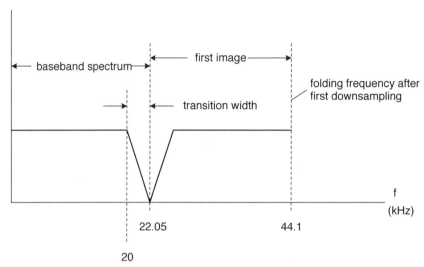

Figure 20.3 Relationship between baseband signal and the first image after the first stage upsampler.

Now suppose that the interpolation by 160 is accomplished as an interpolation by 2 followed by an interpolation by 80. It is customary to number the stages of a multistage interpolator in reverse order, so the first stage in the processing sequence would be denoted as stage 2 for this particular configuration. In order to eliminate the first image without attenuating any of the baseband signal spectrum, the first stage must have an anti-imaging filter that has a transition band of 2050 Hz, just as the single stage interpolator did. However, this filter will be much easier to implement in the multistage case because the output sample rate is only 88.2 kHz, instead of 7.056 MHz. The number of taps needed for this filter can be estimated as:

$$N_2 = \left\lceil \frac{(2.76)(8.82 \times 10^4)}{2050} \right\rceil = 119$$

with $R^*_{T2} = 2{,}623{,}950$. Now the second stage will upsample by 80, and the system bandwidth will open up to include images of the first stage's output spectrum, as depicted in Fig. 20.4. The transition band requirements for the second-stage filter are greatly relaxed due to the wide spacing of images out of the first stage. The transition band can extend from 20 kHz all the way to the lower stopband edge of the first image at 66.15 kHz. The number of taps needed for this filter can be estimated as:

$$N_1 = \left\lceil \frac{(2.76)(7.056 \times 10^6)}{4.615 \times 10^4} \right\rceil = 422$$

with $R^*_{T1} = 18{,}610{,}200$, bringing the total to 21,234,150 multiplications per

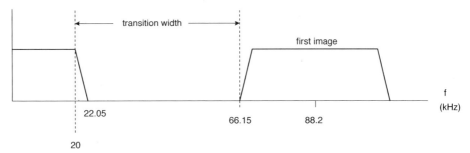

Figure 20.4 Relationship between the first stage output spectrum and its first image after the second stage upsampler.

second. This is a considerable savings over the 192,783,150 MPS needed for a single-stage implementation.

Further Insight

In a single-stage interpolator, the filter must pass all of the signal frequencies with little or no attenuation, while attenuating to negligible levels all of the image frequencies that fall within the system bandwidth after upsampling. The filter's stopband edge must fall below the low frequency edge of the first image. In terms of *absolute* frequencies, the filter in the first stage will have the same transition band as the filter in a single-stage implementation, but the filter in the multi-stage approach will require fewer taps, because it is implemented at a much lower sampling rate making its transition band much wider in terms of *normalized* frequencies. In later stages, the filter's stopband edge still must fall below the low frequency edge of the first image. However, because this image is centered on a frequency equal to the sampling rate at the stage's input, prior stages of interpolation will have moved this first image to a higher frequency, and thus increased the permissible transition width.

The relative cost for any particular multistage implementation of a decimator can be estimated using Algorithm 20.2.

Algorithm 20.2 Estimating the relative cost for a multistage implementation of an interpolator.

1. Consistent with the intended application, determine the required values for passband ripple δ_P, stopband ripple δ_S, interpolation factor L, initial sampling rate F, and final passband edge F_P.
2. Select the particular factorization of L that will be used as the basis for a multi-stage design. Let I denote the number of stages in the design.
3. Compute $D_\infty(\delta_P/I, \delta_S)$.
4. For stage n, compute $\triangle F$ as $\triangle F = F_n - F_S - F_P$ where F_n is the sampling rate at the input to stage n.
5. Estimate the number of taps N required for the stage n filter using Eq. (20.1) with F being the sampling rate at the output of stage n.

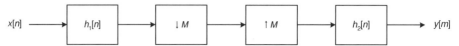

Figure 20.5 Block diagram of a structure for multirate implementation of a lowpass filter.

Software notes

Comparing Algorithms 20.1 and 20.2, it is clear that the transpose relationship between decimators and interpolators will allow the software that was developed for multistage decimators to be used for multistage interpolators with a few simple adjustments. This transpose relationship is the reason why the stages of a multistage interpolator are numbered in reverse order. File msmrcost.cpp contains the function MultistageMultirateCost, which is a straightforward generalization of the function MultistageDecimCost discussed previously. The file prog_20b.cpp contains a main program that provides some user interface and calls MultistageMultirateCost with the proper arguments.

20.3 Multirate Implementation of Lowpass Filters

When a digital filter is to have a bandwidth that is small relative to half the sampling rate, multirate techniques can be used to implement the filter with considerable reductions in computation. Efficient implementations of narrowband lowpass filters can be accomplished using an M-to-1 decimator and a 1-to-M interpolator in cascade as shown in Fig. 20.5. For this structure to be equivalent to a conventional lowpass filter, certain constraints must be satisfied. The desired bandwidth (passband plus transition band) of the filter must be less than $F/(2M)$ because, after decimation, any signal content above this frequency will be aliased into frequencies in the transition band, and possibly the passband. Even when this constraint is satisfied, frequencies in the stopband of the decimation filter $h_1[n]$ will be aliased into the desired passband and transition band. Therefore the filter $h_1[n]$ must have stopband attenuation sufficient to make this aliasing negligible. Similarly, the upsampler will create images that fall in the stopband of the interpolation filter $h_2[n]$, and therefore this filter must have stopband attenuation sufficient to make the imaging negligible.

Assuming that these constraints can be satisfied, how much efficiency will be provided by a multirate implementation? If we neglect possible symmetry in its response, an N-tap conventional filter will require FN MAC operations per second, where F is the sampling rate. Analysis of the multirate structure is easiest if we consider the polyphase implementation shown in Fig. 20.6. To compute each sample of the low rate signal $v[m]$, the decimation section of this structure must sum one output sample from each of the M different polyphase branch filters $p_\rho[m]$. To produce one output sample in each branch requires N/M MAC operations per branch. Thus each sample in $v[m]$ requires $M(N/M) = N$ MAC operations. These samples occur at a rate of F/M, so the required MAC

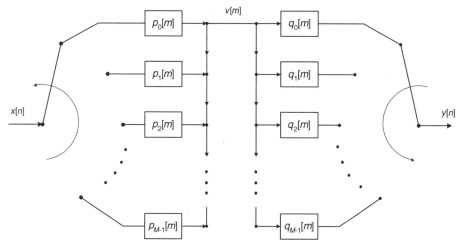

Figure 20.6 Polyphase implementation of a multirate lowpass filter.

rate for the decimation is FN/M. To make each sample of the output signal $y[n]$, the interpolation section selects the output from a single branch $p_\rho[m]$. To produce one output sample requires N/M MAC operations. These samples occur at a rate of F, so the required MAC rate for the interpolation is FN/M, and the total required MAC rate is $2FN/M$. Thus the multirate structure requires $2/M$ times the MAC rate of the corresponding conventional implementation. Larger values of M result in greater computational savings. However, the constraint that $f_S < F/(2M)$ places an upper limit on M for any desired bandwidth f_S of

$$M < \frac{F}{2 f_S}$$

Thus we conclude that the structure shown in Fig. 20.5 yields the greatest efficiency for desired bandwidths which are very narrow relative to the sampling rate.

Specifying the filters $h_1[n]$ and $h_2[n]$

Assume that a lowpass filter is to be implemented using the structure shown in Fig. 20.5. The desired overall equivalent filter characteristic is to have a passband ripple limit of δ_P and a stopband ripple limit of δ_S. In order to guarantee that the multirate implementation satisfies these constraints, the filters $h_1[n]$ and $h_2[n]$ must each have ripple limits $\tilde{\delta}_P$ and $\tilde{\delta}_S$ that satisfy

$$\tilde{\delta}_P \leq \frac{\delta_P}{2}$$

$$\tilde{\delta}_S \leq \delta_S$$

350   Chapter Twenty

Software notes

The class `MultirateLowpass` provided in file `mr_lpf.cpp` provides a multirate implementation of a lowpass filter. Like many of the other filter implementations in this book, this is a derived class that inherits from the virtual base class `FilterImplementation`. The project `prog_20c` generates a program that can be used to exercise `MultirateLowpass`. The following example uses `prog_20c`.

Example 20.1 Suppose we have a DAT recording that includes human voice plus some music and high frequency background noise. To remove as much music and background noise as possible, we wish to digitally lowpass filter the DAT signal to have a passband of 3 kHz and a stopband that begins at 3465 Hz. The passband ripple limit is $\delta_P = 0.05$ and the stopband ripple limit is $\delta_S = 0.005$. Using the software in `prog_20c`, we find that the required number of taps is 169, making the required MAC rate approximately 8.42×10^6 MAC operations per second. The frequency response of an equiripple FIR filter that meets the stated specifications is shown in Fig. 20.7. To implement this filter using the structure of Fig. 20.6, we need to design a prototype filter from which to extract the polyphase filters $p_\rho[m]$ and $q_\rho[m]$. As shown previously,

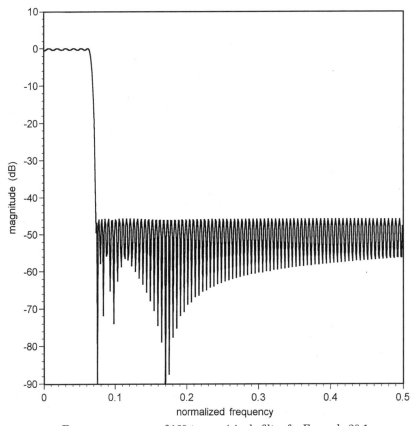

Figure 20.7 Frequency response of 169-tap equiripple filter for Example 20.1.

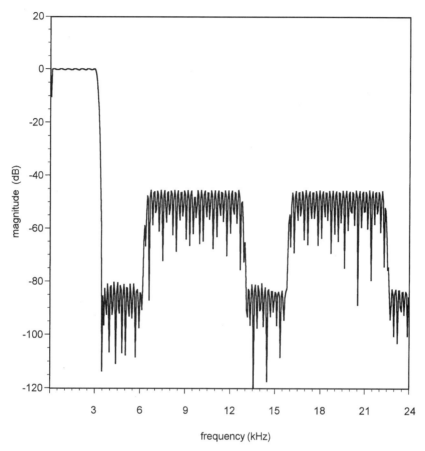

Figure 20.8 Swept response of multirate implementation of lowpass filter for Example 20.1.

we need to design this filter using a passband ripple spec that is half the desired ripple. When this is done, we find that the more stringent ripple requirement raises the minimum number of taps from 169 to 191. For each of the polyphase filters to have the same length, the decimation factor M must be a factor of the prototype length. Furthermore, for $F = 48,000$ and $f_S = 3465$, the largest permissible value for M is 6. Both 191 and 193 are prime numbers, but the prime factorization of 195 is $3 \cdot 5 \cdot 13$, so we can increase the prototype length to 195 and select $M = 5$. The frequency response of the resulting unquantized multirate implementation can be obtained most conveniently by using the swept response techniques discussed in Chap. 11. The capability to perform the swept response measurement is included in prog_20c. The swept response for the multirate implementation with $M = 5$ is shown in Fig. 20.8.

Chapter 21

Random Signals and Sequences

Random process theory is indispensible in the mathematical characterization of noise and interference, and therefore central to the study of statistical signal processing. This chapter introduces all of the fundamental concepts needed to understand and use the statistical processing methods presented in Chaps. 22–24, and the spectral estimation methods in Chaps. 25–26.

21.1 Random Sequences

A discrete-time random signal can be viewed as a *ransom sequence*, but exactly what is a random sequence? This is a difficult concept to pin down precisely. Knuth writes [1]:

> The mathematical theory of probability and statistics carefully sidesteps the question; it refrains from making absolute statements, and instead expresses everything in terms of how much *probability* is to be attached to statements involving random sequences of events.

Knuth goes on to devote 27 pages to attempt to describe random sequences. Readers with some mathematical agility and a slightly philosophical bent will enjoy this material. However, for our purposes, a simpler, more notional description of random sequences provided by Lehmer [2] will suffice:

> A random sequence is a vague notion embodying the idea of a sequence in which each term is unpredictable to the uninitiated and whose digits pass a certain number of tests, traditional with statisticians and depending somewhat on the uses to which the sequence is to be put.

Random sequences will most often be treated mathematically as *realizations* or *sample functions* of a random process. Before we begin investigating random processes and their properties in Secs. 21.9–21.12, we first need to establish the concept of a *random variable*.

21.2 Randomness and Probability

Consider the space $\mathbb{S}$ comprising a (possibly infinite) number of events. An example of such a space would be the set of all possible outcomes from rolling a single die. In this space there would be six events—one corresponding to each face of the die. The space $\mathbb{S}$ in its entirety can be called the *certain event* since it is certain that one of the included events must occur. (Let us assume that a die must land on one of its faces—we will neglect the pathological cases that require the die to remain stably balanced on an edge or on a vertex.) Each event can also be called an *experimental outcome*. Actually, any subset of $\mathbb{S}$ is an event, with those subsets containing a single element being further distinguished as *elementary events*. Some texts such as (Proakis 1983) refer to elementary events as *sample points* of the experiment. In our example, a nonelementary event might be the rolling of a 5. The subset of $\mathbb{S}$, which is the empty set (denoted as $\emptyset$), is called the *impossible event*. A numeric value $P(A)$ can be assigned to every event A in the space $\mathbb{S}$. In probability theory, this value is called the *probability of event A* and is assigned such that

1. The probability of the certain event is unity

$$P(\mathbb{S}) = 1$$

2. All probabilities of events within S are nonnegative

$$P(A) \geq 0$$

3. If events A and B share no common outcomes (i.e., if A and B are *mutually exclusive*), then the probability of either A or B occurring is equal to the probability of event A occurring plus the probability of event B occurring.

$$[A \cap B = \emptyset] \Rightarrow P(A + B) = P(A) + P(B)$$

A single execution of an experiment that selects an event from $\mathbb{S}$ is called a *trial*. A collection of events is called a *field* if it contains the certain event and is closed under finite union and complementation. A collection of events is called a *sigma field* (or σ field) if it contains the certain event and is closed under countable union and complementation. A σ field is also called a *Borel field* in honor of Emile Borel (1871–1956), a French mathematician.

Joint and conditional probabilities

The probability that event A and event B both occur is called the *joint probability* of A and B and is denoted as $P(AB)$ or $P(A \cap B)$. The *conditional probability* that event A will occur given that event B has occurred is denoted by $P(A \mid B)$ and is defined as:

$$P(A \mid B) = \frac{P(AB)}{P(B)} = \frac{P(A \cap B)}{P(B)} \qquad (21.1)$$

If B is a subset of A (that is, $B \subset A$), then

$$P(A \mid B) = \frac{P(AB)}{P(B)} = \frac{P(B)}{P(B)} = 1 \qquad (21.2)$$

If A is a subset of B (that is, $A \subset B$), then

$$P(A \mid B) = \frac{P(AB)}{P(B)} = \frac{P(A)}{P(B)} \geq P(A) \qquad (21.3)$$

since $P(B) \leq 1$. If events A and B are mutually exclusive [that is, $P(A \cap B) = 0$] and exhaustive [that is, $P(A \cup B) = 1$], then

$$P(A \mid B) = \frac{P(B \mid A)}{P(B)} \qquad (21.4)$$

The relationship expressed in Eq. (21.4), called *Bayes rule*, can be extended to the case of more than two events as follows: If the events A_n ($n = 1, 2, \ldots, N$) are mutually exclusive and exhaustive, that is,

$$A_n \cap A_m = \emptyset \quad m, n \in \{1, 2, \ldots, N\}$$

$$m \neq n$$

$$\bigcup_{n=1}^{N} A_n = \mathbb{S}$$

and if B is any event having nonzero probability, then

$$P(A_n \mid B) = \frac{P(A_n \cap B)}{P(B)}$$

$$= \frac{P(B \mid A_n) P(A_n)}{\sum_{m=1}^{N} P(B \mid A_m) P(A_m)} \qquad (21.5)$$

Independent events

Two events A and B are called *independent*, or *statistically independent*, if $P(A \cap B) = P(A) P(B)$. Furthermore, if A and B are independent, then

$$P(A \mid B) = P(A) \quad \text{and} \quad P(B \mid A) = P(B)$$

For the N events $A_1, A_2, \ldots, A_N$ to be independent, their probabilities must satisfy

$$P(A_1 \cap A_2 \cap \cdots \cap A_N) = P(A_1) P(A_2) \cdots P(A_N)$$

21.3 Bernoulli Trials

In many applications, a single experiment can have only two possible outcomes. An electronic component can be tested and found either defective or not defective. A coin can be tossed and land either on heads or tails. If n experiments are performed—if n coins are tossed or if a single coin is tossed n times—the probability that a particular outcome will be observed exactly k times is given by

$$\binom{n}{k} p^k (1-p)^{n-k} = \frac{n!}{k!(n-k)!} p^k (1-p)^{n-k} \tag{21.6}$$

where p is the probability of the desired outcome in a single experiment.

Example 21.1 Suppose that 10 percent of the resistors produced by a particular machine are defective. If we examine a sample of 20 resistors, what is the probability that less than 3 will be defective?

solution To obtain the probability of finding at most 2 defective units, we must sum together the probability of finding no defects, the probability of finding exactly 1 defective unit, and the probability of finding exactly 2 defective units.

$$P(0 \text{ defects}) = \binom{20}{0}(0.10)^0(0.90)^{20} = 0.12158$$

$$P(1 \text{ defect}) = \binom{20}{1}(0.10)^1(0.90)^{19} = 0.27017$$

$$P(2 \text{ defects}) = \binom{20}{2}(0.10)^2(0.90)^{18} = 0.28518$$

$$P(\text{less than 3 defects}) = 0.6769$$

21.4 Random Variables

If we conduct an experiment (such as rolling a die), we can assign a numeric value to each possible outcome of the experiment. The rule for assigning values to outcomes is called a *random variable* (RV), although it is not really a variable in the commonly understood sense. It is more like a function that maps each experimental outcome in the domain into the corresponding numeric value in the range. The outcomes of rolling a single die can be denoted $f_1, f_2, \ldots, f_6$. We can define an RV $X(f_i)$, which assigns a value to each outcome. The obvious choice would be for $f_1 = 1$, $f_2 = 2$, and so on. However, we are not limited to this option—we can define the mapping in many different ways, such as:

$$X(f_i) = 5i \qquad X(f_i) = i^2 \qquad X(f_i) = i - 1$$

Remember: f_i represents the possible outcomes of the experiment, while $X(f_i)$ is the RV that assigns numeric values to each of these outcomes. (This numeric value should not be confused with the numeric probability value that is assigned to each outcome. In the case of a fair die, the probability of each face landing up is 1/6 regardless of the numeric score that may be printed on or assigned to the face.)

Cumulative distribution functions

The *cumulative distribution function* $F_x(x)$ associated with an RV x yields the probability that X does not exceed x:

$$F_x(x) \equiv P\{X \leq x\}$$

Since $F_x(x)$ is a probability, we know that $0 \leq F_x(x) \leq 1$. Even though F_x is a function of x and not X, it is often referred to as the *distribution function* of the RV X. In some of the literature, it is also called the *probability distribution function*, but this is best avoided because the apparent abbreviation "pdf" could also be interpreted as *probability density function*. Although some authors use lowercase *pdf* to denote probability density function, and uppercase *PDF* to denote *probability distribution function*, most authors avoid confusion by using *cdf* to indicate cumulative distribution function and *pdf* to indicate probability density function.

In cases where the independent variable and the subscript on F differ, the subscript indicates the RV of interest. Thus $F_x(y)$ represents the distribution function (evaluated at y) of the RV X. This is equal to the probability that the value of the RV X is less than or equal to y. In cases where the independent variable and the subscript are the same, the subscript is often omitted.

Properties of distribution functions

If $F(x)$ is the distribution function of the RV X, then it will exhibit the following properties:

$$F(-\infty) = 0 \tag{21.7}$$

$$F(+\infty) = 1 \tag{21.8}$$

$$\text{If } x_1 < x_2, \text{ then } F(x_1) \leq F(x_2) \tag{21.9}$$

$$\text{If } F(x_1) = 0, \text{ then } F(x) = 0 \text{ for all } x \leq x_1 \tag{21.10}$$

$$P\{X > x\} = 1 - F(x) \tag{21.11}$$

$$P\{x_1 < X < x_2\} = F(x_2) - F(x_1) \tag{21.12}$$

$$P\{X = x\} = F(x) - \lim_{0 < \epsilon \to 0} F(x - \epsilon) \tag{21.13}$$

$$P\{x_1 \leq X \leq x_2\} = F(x_2) - \lim_{0 < \epsilon \to 0} F(x - \epsilon) \tag{21.14}$$

Probability density function

The derivative of the distribution function is called the *probability density function* (pdf), *density function*, or *frequency function* of the RV X.

21.5 Moments of a Random Variable

Mean

The *mean* $E(X)$ of a *discrete* random variable X is defined as

$$E(X) = \sum_{i=-\infty}^{\infty} x_i p_i \qquad (21.15)$$

where p_i is the probability of the event $X = x_i$. The mean of a *continuous* RV is defined as

$$E(X) = \int_{-\infty}^{\infty} x f(x) \, dx \qquad (21.16)$$

where $f(x)$ is the pdf of the RV X.

- The mean is also called the *expected value, expectation, ensemble average,* or *statistical average*, and is typically denoted by an overbar, or by η_x, η, or μ.

$$\bar{X} = \eta_x = \eta = \mu = E(X)$$

- The mean of the sum of two RVs is equal to the sum of their means.

$$E(Y + Y) = E(X) + E(Y) \qquad (21.17)$$

- In general, the mean of a product of two RVs is not equal to the product of their individual means. However, the mean of the product will equal the product of the means if X and Y are uncorrelated.
- The mean of a constant times a RV is equal to the constant times the mean

$$E(aX) = aE(X) \qquad (21.18)$$

- Taken together, Eqs. (21.17) and (21.18) indicate that expectation is a linear operation.

Mean of a function of an RV

The mean $E\{g(X)\}$ of a function g of a discrete RV X is defined as

$$E\{g(X)\} = \sum_{i=1}^{n} p_i g(x_i) \qquad (21.19)$$

where p_i is the probability of the event $X = x_i$. If $g(X)$ is a function of a continuous RV X, the mean is defined as

$$E\{g(X)\} = \int_{-\infty}^{\infty} g(x) f(x) \, dx \tag{21.20}$$

where $f(x)$ is the pdf of the RV X. This result is known as the *fundamental theorem of expectation*.

Moments

The *n*th *moment* of the discrete RV X is defined as

$$m_n = E(X^n) = \sum_{i=-\infty}^{\infty} x_i^n p_i \tag{21.21}$$

where p_i is the probability of the event $X = x_i$. The *n*th *moment* of the *continuous* RV X is defined as

$$m_n = E\{X^n\} = \int_{-\infty}^{\infty} x^n f(x) \, dx \tag{21.22}$$

where $f(x)$ is the pdf of X. Note that the moment m_1 is the same as the mean.

Central moments

The *k*th *central moment* μ_k of a discrete RV X is defined by

$$\mu_k = \sum_{i=-\infty}^{\infty} (x_i - \bar{X})^k p_i \tag{21.23}$$

where p_i is the probability of the event $X = x_i$. The *k*th central moment of a continuous RV X is defined by

$$\mu_k = E\{(X - \bar{X})^k\} = \int_{-\infty}^{\infty} (x - \bar{X})^k f(x) \, dx \tag{21.24}$$

where $f(x)$ is the pdf of X.

- The second central moment is often called the *variance* and denoted by σ^2 rather than μ_n.
- The positive square root of variance is called the *standard deviation*.
- The third central moment is called *skew*.
- The fourth central moment is called the *kurtosis*.

Properties of variance

$$\text{var}\{cX\} = c^2 \text{ var}\{X\}$$

We can relate the mean and variance of a random variable by expanding the definition of variance

$$\begin{aligned}
\sigma^2 &= E\{(X-\mu)^2\} \\
&= E\{(X^2 + 2X\mu + \mu^2)\} \\
&= E\{X^2\} - 2\mu E\{X\} + \mu^2 \\
&= E\{X^2\} - 2\mu^2 + \mu^2 \\
&= E\{X^2\} - [E\{X\}]^2
\end{aligned}$$

If X and Y are independent, then

$$\text{var}\{X+Y\} = \text{var}\{X\} + \text{var}\{Y\}$$

21.6 Relationships between RVs

If X and Y are continuous RVs, their *joint distribution* $F_{xy}(x, y)$ yields the probability that X does not exceed x and Y does not exceed y.

$$F_{xy}(x, y) = P\{X \leq x, Y \leq y\}$$

The *joint density function*, $p_{xy}(x, y)$, of two continuous RVs x and y is defined as the mixed partial derivative of F_{xy}:

$$p_{xy}(x, y) = \frac{\partial^2}{\partial x\, \partial y} F_{xy}(x, y)$$

The individual density functions for X and Y can be obtained from $p_{xy}(x, y)$ by integration:

$$p_x(x) = \int_{-\infty}^{\infty} p_{xy}(x, y)\, dy$$

$$p_y(y) = \int_{-\infty}^{\infty} p_{xy}(x, y)\, dx$$

In this context, $p_x(x)$ and $p_y(y)$ are referred to as *marginal* pdf's to emphasize the distinction between them and the joint density function.

If X and Y are discrete RVs, their joint probability function p_{ij} is the probability that $X = x_i$ and $Y = y_j$.

$$p_{ij} = P\{X = x_i, Y = y_j\}$$

The marginal probability functions for X and Y can be obtained by summing the joint probability function over all y_j or all x_i

$$P(X = x_i) = \sum_{j=-\infty}^{\infty} P(X = x_i, Y = y_j)$$

$$P(Y = y_j) = \sum_{i=-\infty}^{\infty} P(X = x_i, Y = y_j)$$

Statistical independence

Two RVs are called *statistically independent* if

$$P\{X \in A, Y \in B\} = P\{X \in A\}P\{Y \in B\}$$

where A and B are arbitrary subsets of the ranges of X and Y, respectively. This is equivalent to saying that $\{X \in A\}$ and $\{Y \in B\}$ are independent events. If the RVs X and Y are statistically independent, then the joint density $p_{xy}(x, y)$ equals the product of the marginal densities:

$$p_{xy}(x, y) = p_x(x)\, p_y(y)$$

If the RVs X and Y are not independent, then the joint density cannot be synthesized from the marginal densities alone—either the conditional densities or sufficient other *a priori* information that characterizes the dependencies will be needed.

21.7 Correlation and Covariance

The *correlation* R_{xy} of two RVs X and Y is defined as

$$R_{xy} = E[XY] = \int_{-\infty}^{\infty}\int_{-\infty}^{\infty} xy\, p_{xy}(x, y)\, dx\, dy \qquad (21.25)$$

The *covariance* C_{xy} of two RVs X and Y is defined as

$$C_{xy} = E\{(X - \mu_x)(Y - \mu_y)\} \qquad (21.26)$$

It can be shown that $|C_{xy}| \leq \sigma_x \sigma_y$. In some applications, it is more convenient to use a normalized measure called the *correlation coefficient*, which is defined as

$$r_{xy} = \frac{C_{xy}}{\sigma_x \sigma_y} \qquad (21.27)$$

where C_{xy} = covariance of X and Y
σ_x = standard deviation of X
σ_y = standard deviation of Y

It can be shown that $|r_{xy}| \leq 1$.

Two random variables are called *uncorrelated* if their covariance and correlation coefficients are both zero. (Either one being zero is a sufficient condition for uncorrelatedness, and implies that the other is also zero.) If X and Y are uncorrelated, then

$$E\{XY\} = E\{X\}E\{Y\} \tag{21.28}$$

and
$$\sigma^2_{x+y} = \sigma^2_x + \sigma^2_y \tag{21.29}$$

If two RVs X and Y are statistically independent, then they are also uncorrelated. However, two RVs may be uncorrelated but not independent. In the case of normal RVs, uncorrelatedness is sufficient to establish statistical independence.

Two random variables X and Y are *orthogonal* if and only if $E\{XY\} = 0$. The orthogonality between X and Y can be denoted as $X \perp Y$. If X and Y are *uncorrelated*, then

$$(X - \bar{X}) \perp (Y - \bar{Y}) \tag{21.30}$$

If X and Y are uncorrelated and either or both has zero mean, then X is orthogonal to Y.

21.8 Probability Densities for Functions of an RV

Given:

X is a continuous RV with pdf $p_x(X)$.

Y is a monotonically increasing or monotonically decreasing function of X that is differentiable for all values of X.

Then the pdf of Y is given by

$$p_y(Y) = p_x(Y[X]) \cdot \left|\frac{dY}{dX}\right| \tag{21.31}$$

where $Y[X] \triangleq f^{-1}(y)$ given that $y = f(x)$.

Example 21.2 Let X be a random variable uniformly distributed between 0 and 1:

$$p_x(X) = \begin{cases} 1 & 0 \leq X \leq 1 \\ 0 & \text{otherwise} \end{cases}$$

Find the pdf for $y = -\ln x$.

solution It follows directly that

$$Y[X] = \exp(-Y) \tag{21.32}$$

$$\left|\frac{dY}{dX}\right| = -\exp(-Y) \tag{21.33}$$

Substituting Eqs. (21.32) and (21.33) into Eq. (21.31) yields

$$p_y(Y) = p_x[\exp(-Y)] \cdot |-\exp(-Y)| \tag{21.34}$$

We note that $0 \leq \exp(-Y) \leq 1$ for all $Y \geq 0$; thus

$$p_x[\exp(-Y)] = \begin{cases} 1 & Y \geq 0 \\ 0 & Y < 0 \end{cases}$$

Therefore Eq. (21.34) simplifies to

$$p_y(Y) = \begin{cases} \exp(-Y) & Y \geq 0 \\ 0 & Y < 0 \end{cases}$$

21.9 Random Processes

Similar to the way in which a random variable assigns numeric values to experimental outcomes, a random process assigns functions to experimental outcomes. The functions are called *sample functions*, *member functions*, or *realizations* of the random process. In most of the engineering literature, sample functions predominate; however, in works such as this one that deal with signal processing, we prefer to use the term *realization* to avoid confusion between sample function and either *sampling function* or *sampled function*. In the majority of engineering applications, each realization is a function of time (usually) or space (sometimes). For the specific case of a discrete-time random process, we can say that the random process assigns *sequences* to experimental outcomes.

Consider the collection, or *ensemble*, of realizations from a discrete-time random process shown in Fig. 21.1. Each realization $x_i[n]$ has a time average given by

$$\langle x_i[n] \rangle \stackrel{\triangle}{=} \lim_{M \to \infty} \frac{1}{2M+1} \sum_{n=-M}^{M} x_i[n]$$

This average is an RV that takes on different values as different realizations $x_1[n]$, $x_2[n]$, etc. are substituted for $x_i[n]$.

Conceptually, at any instant of time we could measure the instantaneous values of each realization in the ensemble. The set of values so obtained forms an RV that can be statistically characterized apart from the characterization of the process itself. For clarity we will sometimes refer to such an RV as an *instantaneous* RV, even though such terminology is not widespread. Thus a random process can be viewed as a time-indexed family of RVs, which, in the case of a discrete-time process, becomes a time sequence of random variables. In general, the statistics of the instantaneous RVs are unrelated to the time statistics of the realizations. However, in *ergodic* random processes, the ensemble statistics and time statistics are interchangeable.

If the realizations themselves are random in nature (such as shown above), the process is called a *regular random process*. A regular random process could

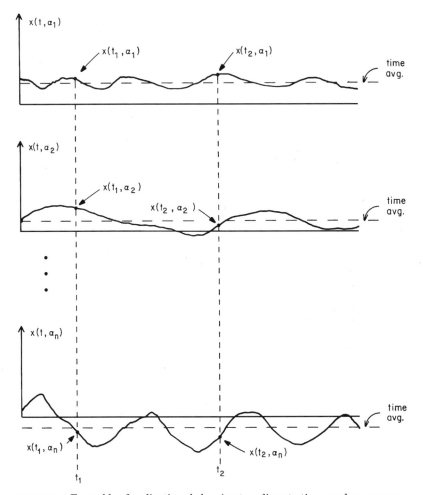

Figure 21.1 Ensemble of realizations belonging to a discrete-time random process.

be used to model the position of sand grains in a sandstorm. The position of each grain is described by a realization of the random process. Since the future position of a grain cannot be expressed exactly in terms of its past position, each sample function is random. Likewise, knowing the position of a particular grain at a specific instant will not give us the position of any other grains. Thus the ensemble of positions at a fixed instant forms a random variable.

If the realizations are deterministic, the process is called a *predictable random process*. An example of a predictable random process would be a group of RC oscillators assembled from randomly selected components. The amplitude, phase, and frequency of the oscillators will be RVs mapped from the specific characteristics of the selected components. However, unlike the motion of a sand particle, the future output of a particular oscillator (once it has been assembled) can be determined from its past output.

It should be noted that an ensemble usually is not a collection of actual realizations; rather, it is an abstract concept that proves useful in the statistical characterization of random phenomena. Instead of having a large collection of oscillators, we may actually have only one oscillator. Prior to switching it on, we still can characterize its behavior based on the statistics of a hypothetical collection of oscillators, which could be built on from the available set of components. However, once the oscillator is turned on and observed for a while, it may be more appropriate to discard the statistical characterization and instead describe the oscillator output as a deterministic function of time.

21.10 Autocorrelation and Autocovariance

The *autocorrelation function* (acf) of a random process is different from, but in many cases closely related to, the time autocorrelation discussed in Chap. 3. It is a common practice to denote both the time (or time-averaged) autocorrelation and the process (or ensemble-averaged) autocorrelation by the letter R and leave it to the reader to determine from context which autocorrelation is meant. This often is a major source of confusion for readers who are not already experts in the areas of RVs and random processes. In this book we use the script $\mathcal{R}$ to denote the process autocorrelation so that it can be distinguished more readily from the time autocorrelation. Harry Urkowitz [3] is one of the few other authors who use different character styles to distinguish between time and process autocorrelations, but he uses the opposite convention: $\mathcal{R}$ for time and R for process. Since this book deals almost exclusively with ergodic processes and their time autocorrelations, we assign the more unusual character style to the relatively few instances in which the process autocorrelation must be specifically indicated. The process autocorrelation of the continuous-valued discrete-time random process x is denoted by $\mathcal{R}_x[n_1, n_2]$, and is obtained by forming the cross-correlation of the random variables at time n_1 and n_2:

$$\mathcal{R}_x[n_1, n_2] = \mathcal{E}[x_1 x_2^*] = \int_{-\infty}^{\infty} \int_{-\infty}^{\infty} x_1 x_2^* \, p_{x_1 x_2}(x_1, x_2) \, dx_1 \, dx_2$$

where x_1 represents $x[n_1]$, x_2 represents $x[n_2]$ and $p_{x_1 x_2}(x_1, x_2)$ is the joint density function of x_1 and x_2. In general, $x[n]$ can be complex-valued, and the superscript asterisk denotes complex conjugation. Many texts assume that $x[n]$ is real-valued and define the autocorrelation without the conjugation operator shown here. The acf of a discrete-valued discrete-time random process is defined by

$$\mathcal{R}_x[n_1, n_2] = \sum_{i=1}^{N} \sum_{j=1}^{N} x[n_1] x^*[n_2] \, P(x[n_1] = a_i, x[n_2] = a_j)$$

Stationarity

The instantaneous values of all the realizations comprising a random process will constitute an RV. This RV will have statistics that are, in general, unrelated to the statistics over time of the individual realizations.

Furthermore, the statistics across the ensemble at one instant may be quite different from the statistics at a different instant. In fact, at each instant an RV with potentially different statistics could be formed. Joint statistics can be determined for any group of two or more RVs across the ensemble at different instants.

Many applications of practical interest will involve random processes in which the RVs across the ensemble exhibit identical statistics from instant to instant. The term *stationary* is used to describe a random process in which the statistics at individual instants, as well as all possible joint statistics, are time invariant. Knowing that a particular process is stationary can be useful; unfortunately, proving the time invariance of all possible joint statistics is usually impossible.

Wide-sense stationarity is a weaker form of invariance that is easier to establish than strict stationarity, yet almost as useful. A discrete-time random process is wide-sense stationary (often abbreviated as wss) if its mean is constant over time and the process autocorrelation $\mathcal{R}_x$ satisfies

$$\mathcal{R}_x(n+k, n) = \mathcal{R}_x(k, 0) \quad \text{for all } n \text{ and } k$$

Sometimes (especially in British literature) a wss process is called *weakly stationary*. A discrete-time random process is called *covariance stationary* if its autocovariance function $\mathcal{K}_x$ satifies

$$\mathcal{K}_x(n+k, n) = \mathcal{K}_x(k, 0) \quad \text{for all } n \text{ and } k$$

Strict stationarity implies wide-sense stationarity, and wide-sense stationarity implies covariance stationarity.

If the process x is wide-sense stationary, $\mathcal{R}_x[n_1, n_2]$ is a function only of the difference $n_2 - n_1$ and not of the specific values of n_1 and n_2. In this case, the process autocorrelation can be written as a function of a single variable

$$\mathcal{R}_x[k] = \mathcal{E}\{x[n]x[n+k]\}$$

Ergodicity

A *ergodic* random process is one in which the statistical properties of the entire process can be determined from the statistical properties of any single realization of the process. Or, loosely speaking, the ensemble statistics equal the time statistics. An ergodic process is stationary, but a stationary process is not necessarily ergodic.

For an ergodic process $x[n]$, the autocorrelation properties of the process are captured completely by the time autocorrelation

$$\mathcal{R}_x[k] = \mathcal{E}\{x[n]x^*[n+k]\}$$
$$= \langle x[n]x^*[n+k]\rangle$$
$$= R_x[k]$$

Properties of autocorrelation functions

1. The acf is nonnegative.
2. The power in the process x is given by

$$\text{Power} = E\{x^2[n]\} = \mathcal{R}_x[n,n] = \mathcal{R}_x[0] = \sigma_x^2$$

3. A second-order process is one for which $\mathcal{R}_x[n,n] < \infty$ for all t.
4. $\mathcal{R}_x[n,m] = \mathcal{R}_x^*[m,n]$ for all n and m.
5.

$$|\mathcal{R}_x[n,m]| \leq \sqrt{\mathcal{R}_x[n,n]\mathcal{R}[m,m]}$$

6. For wss processes, $\mathcal{R}_x[k] = \mathcal{R}_x^*[-k]$.
7. For wss processes, $|\mathcal{R}_x[k]| \leq \mathcal{R}_x[0]$.

Autocovariance

The *autocovariance function* (acvf) of a random process is denoted by $\mathcal{K}_x[n_1, n_2]$ and is obtained by forming the covariance of the instantaneous RVs at time n_1 and n_2.

$$\mathcal{K}_x[n_1, n_2] = E\{(x[n_1] - m_x[n_1])(x[n_2] - m_x[n_2])\} \tag{21.35}$$

Expansion of Eq. (21.35) yields an expression that relates autocovariance, autocorrelation, and mean:

$$\mathcal{K}_x[n_1, n_2] = \mathcal{R}_x[n_1, n_2] - m_x[n_1]m_x[n_2] \tag{21.36}$$

Uncorrelated random processes

A pair of discrete-time random processes $x[n]$ and $y[n]$ is described as *uncorrelated* if for all possible n and m:

$$E[x[n]y[m]] = E[x[n]]E[y[m]]$$

or equivalently if

$$\mathcal{R}_{xy}[n,m] = \mu_x[n]\mu_y[m]$$

A single random process is described as uncorrelated if its instantaneous RVs are uncorrelated with each other. A random process is described as *independent* if its constituent instantaneous RVs are independent.

Autocorrelation matrix

The $N \times N$ autocorrelation matrix of a random process is defined as

$$\bar{\mathcal{R}}_x = E\{\mathbf{xx}^{*T}\} \tag{21.37}$$

$$= \begin{bmatrix} E\{x[0]x^*[0]\} & E\{x[0]x^*[1]\} & \cdots & E\{x[0]x^*[N-1]\} \\ E\{x[1]x^*[0]\} & \ddots & & \vdots \\ \vdots & & \ddots & \\ E\{x[N-1]x^*[0]\} & \cdots & & E\{x[N-1]x^*[N-1]\} \end{bmatrix}$$

$$= \begin{bmatrix} \mathcal{R}_x[0,0] & \mathcal{R}_x[0,1] & \cdots & \mathcal{R}_x[0,N-1] \\ \mathcal{R}_x[1,0] & \ddots & & \\ \vdots & & \ddots & \vdots \\ \mathcal{R}_x[N-1,0] & & \cdots & \mathcal{R}_x[N-1,N-1] \end{bmatrix}$$

For the case of a stationary process, the correlation matrix is always a [Hermitian] symmetric Toeplitz matrix of the form

$$\bar{\mathcal{R}}_x = \begin{bmatrix} \mathcal{R}_x[0] & \mathcal{R}_x[-1] & \cdots & \mathcal{R}_x[-N+1] \\ \mathcal{R}_x[1] & \ddots & \ddots & \vdots \\ \vdots & \ddots & \ddots & \mathcal{R}_x[-1] \\ \mathcal{R}_x[N-1] & \cdots & \mathcal{R}_x[1] & \mathcal{R}_x[0] \end{bmatrix} \tag{21.38}$$

21.11 Power Spectral Density of Random Processes

The *power spectral density* (psd) $S_x(f)$ and the acf $R_x(\tau)$ of a wide-sense stationary random process $X(t)$ comprise a Fourier transform pair.

$$S_x(f) = \mathcal{F}\{R_x(\tau)\} = \int_{-\infty}^{\infty} R_x \exp(-j2\pi ft)\, dt \tag{21.39}$$

$$R_x(\tau) = \mathcal{F}^{-1}\{S_x(f)\} = \int_{-\infty}^{\infty} S_x \exp(j2\pi ft)\, df \tag{21.40}$$

In some of the literature, the acf and psd are developed separately, and then the fact that they form a Fourier transform pair is demonstrated and dubbed the *Wiener-Khintchine theorem*. Some texts take the alternative approach of developing the acf and then defining the psd as the Fourier transform of the acf.

In this case, Eqs. (21.39) and (21.40) are called the *Wiener-Khintchine relations* since they are defined rather than derived.

(Note: There is some disagreement concerning the proper spelling of *Khintchine*. The spelling used here agrees with [4] and [5]. Other observed spellings include *Kinchin* [3], *Khinchin* [6], *Kinchine* [7, 8], and *Khinchine* [9]. This is somewhat understandable due to various transliterations from Cyrillic to Latin alphabets, but there is even some disagreement over *Wiener*, which in at least two texts [9, 5] appears as *Weiner*. At least one text [10] avoids the issue completely by presenting the relationship but not giving it a name.)

21.12 Linear Filtering of Random Processes

Application of a random process $x(t)$ to the input of a linear time-invariant filter will produce a different random process $y(t)$ at the filter output. Some statistical properties of the output can be determined directly from the filter's impulse response and the statistical properties of the input. For an input process having a mean of $\mu_x(t)$, the output will have a mean given by

$$\mu_y(t) = \int_{-\infty}^{\infty} h(\tau)\mu_x(t-\tau)\,d\tau \tag{21.41}$$

For the special case of $x(t)$ being wide-sense stationary, the mean of the output can be simplified to $\mu_y = \mu_x H(0)$.

For the general case of an input process with a process autocorrelation of $\mathcal{R}_x(t_1, t_2)$, the acf of the output will be given by

$$\mathcal{R}_y(t_1, t_2) = \int_{-\infty}^{\infty} h(\tau_1)\,d\tau_1 \int_{-\infty}^{\infty} h(\tau_2)\,\mathcal{R}_x(t_1-\tau_1, t_2-\tau_2)\,d\tau_2 \tag{21.42}$$

For the special case of $x(t)$ being wide-sense stationary, the acf can be simplified to

$$\mathcal{R}_y(\tau) = \int_{-\infty}^{\infty}\int_{-\infty}^{\infty} h(\tau_1)h(\tau_2)\,\mathcal{R}_x(\tau-\tau_1+\tau_2)\,d\tau_1\,d\tau_2 \tag{21.43}$$

For an input with power spectral density of $S_x(f)$, the output will have a power spectral density given by

$$S_y(f) = |H(f)|^2 S_x(f) \tag{21.44}$$

When the input noise process to a linear filter is a Gaussian random process, the resulting output noise process also is Gaussian.

21.13 Estimating the Moments for a Random Process

The mean and correlation or covariance functions of a random process frequently are needed to apply various statistical signal processing techniques.

Often these functions are not known and must be estimated from observed samples of the process in question. There are an infinite number of possible rules for estimating the parameters of an RV or of a random process. Only a few of these rules yield good results, and are therefore useful. A number of properties can be used to describe more precisely the *quality* of a particular estimation rule.

Any estimate $\hat{\theta}$ of a parameter obviously is a function of the observations $\mathbf{x} = \{x_i\}, i = 1, 2, \ldots, N$. The estimate also will depend on the number of observations, so we will use the notation $\hat{\theta}_N$ to indicate an estimate that is based on N observations.

Unbiased estimate. An estimate $\hat{\theta}_N$ is *unbiased* if the expected value of the estimate equals the true value of the parameter θ

$$\mathcal{E}\{\hat{\theta}_N\} = \theta$$

Asymptotically unbiased estimate. An estimate $\hat{\theta}_N$ is *asymptotically unbiased* if the expected value of the estimate approaches θ in the limit as N becomes infinite.

$$\lim_{N \to \infty} \mathcal{E}\{\hat{\theta}_N\} = \theta$$

Consistent estimate. An estimate $\hat{\theta}_N$ is a *consistent estimate* if

$$\lim_{N \to \infty} \Pr[|\hat{\theta}_N - \theta| < \epsilon] = 1$$

for any arbitrarily small number ϵ.

Efficient estimate. An estimate $\hat{\theta}_N$ is a *efficient* with respect to some other estimate if it has lower variance.

Mean

For a stationary, ergodic random process, the mean is usually estimated as the sample mean of N observations:

$$\mu_x = \frac{1}{N} \sum_{n=0}^{N-1} x[n] \tag{21.45}$$

The sample mean is a consistent estimator of the process mean.

Correlation function

The correlation function of a random process is estimated using the *sample correlation function*. There are two different definitions for the sample correlation function. Each has advantages and disadvantages. The first definition, denoted by $\hat{R}'_x$ is given by

$$\hat{R}'_x[k] = \frac{1}{N-k} \sum_{n=0}^{N-1-k} x[n+k]\, x^*[n] \quad 0 \le k < N \qquad (21.46a)$$

$$\hat{R}'_x[k] = \frac{1}{N-|k|} \sum_{n=0}^{N-1-|k|} x[n]\, x^*[n+|k|] \quad -N < k \le 0 \qquad (21.46b)$$

Notice that $\hat{R}'_x[-k] = \hat{R}'^*_x[k]$. For real-valued random processes, this definition simplifies to

$$\hat{R}'_x[k] = \frac{1}{N-k} \sum_{n=0}^{N-1-k} x[n+|k|]\, x[n] \quad |k| < N \qquad (21.47)$$

The sample correlation functions given by Eqs. (21.46) and (21.47) are unbiased, consistent estimators of the correlation function. However, these estimates are not guaranteed to be *positive semidefinite*. The correlation function of a random process is positive semidefinite, and in many applications it is desirable to use an estimate that also is positive semidefinite. This brings us to the second definition of the sample correlation function, denoted by $\hat{R}_x$

$$\hat{R}_x[k] = \frac{1}{N} \sum_{n=0}^{N-1-k} x[n+k] x^*[n] \quad 0 \le k < N \qquad (21.48a)$$

$$\hat{R}_x[k] = \frac{1}{N} \sum_{n=0}^{N-1-|k|} x[n] x^*[n+|k|] \quad -N < k < 0 \qquad (21.48b)$$

The sample correlation function given by Eq. (21.48) is an asymptotically unbiased, consistent estimator of the correlation function. Furthermore, this estimate is always positive semidefinite. Because it is only asymptotically unbiased, this form of the sample correlation function is sometimes referred to as the *biased* estimate of the acf.

Cross-correlation

The sample cross-correlation function can be defined analogous to either Eq. (21.46) or Eq. (21.48). The form corresponding to Eq. (21.46) is

$$\hat{R}_{xy}[k] = \frac{1}{N} \sum_{n=0}^{N-1-k} x[n+k]\, y^*[n] \quad 0 \le k < N \qquad (21.49)$$

$$\hat{R}_{xy}[k] = \frac{1}{N} \sum_{n=0}^{N-1-|k|} x[n]\, y^*[n+|k|] \quad -N < k < 0$$

21.14 Estimating the Correlation Matrix

The straightforward approach to estimating the correlation matrix is to replace every occurrence of $\mathcal{R}_x[k]$ in Eq. (21.38) with the corresponding estimate $\hat{R}_x[k]$ as given by Eq. (21.48). The $P \times P$ estimated correlation matrix (for $P \leq N$) is given by

$$\hat{\mathbf{R}}_x = \begin{bmatrix} \hat{R}_x[0] & \hat{R}_x[-1] & \cdots & \hat{R}_x[-P+1] \\ \hat{R}_x[1] & \ddots & \ddots & \vdots \\ \vdots & \ddots & \ddots & \hat{R}_x[-1] \\ \hat{R}_x[P-1] & \cdots & \hat{R}_x[1] & \hat{R}_x[0] \end{bmatrix} \qquad (21.50)$$

The estimate formed in this manner will be a Hermitian symmetric Toeplitz matrix as is Eq. (21.38). This estimated correlation matrix can be expressed as the matrix product

$$\hat{\mathbf{R}}_x = \frac{1}{N} \mathbf{X}^H \mathbf{X} \qquad (21.51)$$

where

$$\mathbf{X} = \begin{bmatrix} \mathbf{L} \\ \mathbf{T} \\ \mathbf{U} \end{bmatrix}$$

and the lower triangular $(P-1) \times P$ matrix $\mathbf{L}$, rectangular $(N-P+1) \times P$ matrix $\mathbf{T}$, and upper triangular $(P-1) \times P$ matrix $\mathbf{U}$ are defined as

$$\mathbf{L} = \begin{bmatrix} x[0] & 0 & \cdots & \cdots & 0 \\ x[1] & x[0] & 0 & \cdots & 0 \\ \vdots & \vdots & \ddots & \ddots & \vdots \\ x[P-2] & x[P-3] & \cdots & x[0] & 0 \end{bmatrix} \qquad (21.52)$$

$$\mathbf{T} = \begin{bmatrix} x[P-1] & x[P-2] & \cdots & x[0] \\ x[P] & x[P-1] & \cdots & x[1] \\ \vdots & \vdots & & \vdots \\ x[N-1] & x[N-2] & \cdots & x[N-P] \end{bmatrix} \qquad (21.53)$$

$$\mathbf{U} = \begin{bmatrix} 0 & x[N-1] & \cdots & x[N-p+2] & x[N-P+1] \\ \vdots & 0 & \ddots & \vdots & \vdots \\ 0 & \cdots & \ddots & x[N-1] & x[N-2] \\ 0 & \cdots & \cdots & 0 & x[N-1] \end{bmatrix} \quad (21.54)$$

The rows in **L** end with a number of zeros, and the rows in **U** begin with a number of zeros. An alternative approach that avoids using these zeros uses **T** in place of **X**. The correlation matrix is then estimated as

$$\hat{\mathbf{R}}_x = \frac{1}{N-P+1}\mathbf{T}^\mathbf{H}\mathbf{T} \quad (21.55)$$

The estimate formed in this manner will not be a Toeplitz matrix. Within much of the signal processing literature, the estimate in Eq. (21.51) is often referred to as the *autocorrelation method* estimate and that in Eq. (21.55) is referred to as the *covariance method* estimate—even though these names do not correspond to the way the terms *autocorrelation* and *covariance* are used in statistics. Either method could be used to estimate either a correlation matrix or a covariance matrix.

21.15 Markov Processes

Markov processes play a role in the analysis of certain random signals, such as sampled speech, that exhibit correlation from sample to sample. Within the literature, there appears to be some confusion and disagreement concerning the terminology used to describe *Markov processes* and *Markov chains*. The terminology used in [11] appears to be the most detailed and explicitly descriptive of any of the common variations. Except where otherwise noted, the terminology that is presented and used in this book is similar to the terminology of Papoulis.

A *Markov process* can be described as a stochastic process whose future depends (in a probabilistic sense) only upon its present and very recent past. For a given present condition of the process, the distant past behavior of the process will have no impact on its future behavior.

Markov processes can be divided into four categories: (1) continuous-time, continuous-valued; (2) discrete-time, continuous-valued; (3) continuous-time, discrete-valued; and (4) discrete-time, discrete-valued. Discrete-valued Markov processes are usually referred to as *Markov chains*. Kemeny and Snell [12] refer to discrete-valued Markov processes as *finite Markov processes*; they reserve the term *Markov chain* for something we will discuss shortly that is more precisely called a *homogeneous Markov process*.

In this book we are concerned primarily with Markov chains. General Markov processes are mentioned only as a matter of general interest since they are the parent structure within which Markov chains form a specific subset. Markov processes are named after the Russian mathematician A. A. Markov (1856–1922). (Note: Papoulis and some others prefer to spell it Markoff.)

Markov chains

Markov chains can be divided into two categories: *discrete-time* and *continuous-time*. However, as we will see shortly, continuous-time chains implicitly involve discrete-time chains and many authors do not bother to make the distinction. A discrete-time discrete-valued Markov process will have a countable number of possible outcomes. In general, the process will exhibit new outcomes at uniform intervals since a new outcome will be exhibited each time the discrete-time index is incremented. A continuous-time, discrete-valued Markov process will have a countable number of different possible outcomes, with transitions to new outcomes occurring at random points t_n in *continuous* time. However, if the outcomes, $x(t_n^+)$, are considered as a function of n rather than of t_n, the result is a discrete-time Markov chain that is said to be *imbedded* in the continuous-time chain. [Note that $x(t_n^+)$ represents the outcome just after the transition at t_n, and that $x(t_n^-)$ represents the outcome just before the transition.]

In discussions of Markov chains, outcomes of the process are usually referred to as *states*. A discrete-time Markov chain can be specified in terms of its *state probabilities* and *state transition probabilities*. The *state probability* $p_i[n]$ is the probability that the chain will be in state i at time n.

$$p_i(n) = P\{s(n) = s_i\}$$

Usually the state probabilities for a Markov chain are given as *initial state probabilities* $p_i(0)$, which are the absolute probabilities that the chain will start in state i at time zero. The *transition probability* $p_{ij}(n_1, n_2)$ is the probability that the chain will enter state j at time n_2 given that the chain is in state i at time n_1.

$$p_{ij}(n_1, n_2) \equiv P\{s(n_2) = s_j | s(n_1) = s_i\} \qquad (21.56)$$

In cases where $n_2 = n_1 + 1$, the transition probability is called the *one-step transition probability*, and usually is denoted as $p_{ij}(n_2)$. In common usage the unqualified term *transition probability* refers to one-step transition probabilities, and the term *n-step transition probability* distinguishes the multistep case. The transition probabilities of a Markov chain are often represented as a stochastic matrix called the *determining matrix* of the chain.

Classification of Markov chains

Simple discrete-time Markov chains are chains in which the probability distribution of states at time k is fully determined by the state of the chain at the single instant of time $k - 1$. *Complex* discrete-time Markov chains are chains in which the probability distribution of states at time k depends on the states of the chain at two or more time instants prior to time k.

A *homogeneous Markov chain* is a Markov chain in which the transition probabilities in Eq. (21.56) depend only on the difference between n_1 and n_2, rather than on the actual values of n_1 and n_2 themselves. (In other words, the transition probabilities are invariant under a shift of the origin.)

$$p_{ij}(n_1, n_2) = p_{ij}(n + n_2 - n_1) \quad \text{for all } n, n_1, n_2 \in \{0, 1, 2, \ldots\}$$

Most of the elementary literature deals primarily with homogeneous chains rather than with the more general case. Reference [13] refers to homogeneous chains as "chains with stationary probabilities p_{ij}."

A *stationary* Markov chain is a homogeneous Markov chain in which the state probabilities as well as the transition probabilities are invariant under a shift of the origin.

$$[p_{ij}(n_1, n_2) = p_{ij}(n + n_2 - n_1)] \wedge [p_i(n) = p_i(0)] \quad \text{for all } n, n_1, n_2 \in \{0, 1, 2, \ldots\}$$

Thus the initial state probabilities equal the final state probabilities.

A Markov chain C_n with determining matrix $\mathbf{P}$ is called *regular* when the maximal eigenvalue of $\mathbf{P}$ (remember: $\lambda_{\max} = 1$ since $\mathbf{P}$ is stochastic) is a simple root of $P(\lambda)$ and all other eigenvalues of $\mathbf{P}$ have magnitudes strictly less than 1. If any of the final state probabilities p_j for a regular Markov chain C_n are equal to zero, the chain is described as *nonnegatively regular*. If all p_j are nonzero, the C_n is described as *positively regular*, or *normal*. If all the p_j are equal ($p_j = 1/n$), the chain is described as *completely regular*.

The vast majority of existing Markov chain theory deals with simple homogeneous discrete-time Markov chains with finite numbers of states. Kolmogorov calls such chains "Markov chains in the restricted sense of the word", [14]. For brevity, C_n is used in the remainder of this section to denote a simple homogeneous discrete-time Markov chain with n states.

State diagrams of Markov chains

The relationship between the various states and transition probabilities in a homogeneous Markov chain are often depicted as a state diagram. The transition probability from state A to state B is depicted as a directed path from state A to state B. State diagrams could be used to depict nonhomogeneous Markov chains, but this may become quite cumbersome, since in general a different set of transition probabilities will be needed for each time n.

Since each state must always have a next state, the probabilities for all paths exiting a state must sum to unity:

$$\forall i, \quad \sum_j p_{ij} = 1$$

Often the probabilities for all the paths exiting a state as shown in a diagram will sum to something less than 1. In these cases there is an assumed or implied

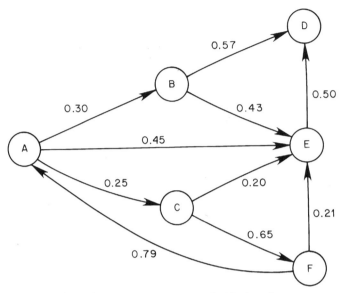

Figure 21.2 State diagram representation of a Markov chain.

path which leaves and then immediately reenters the same state. The transition probability associated with this implied path is equal to the difference between unity and the sum of all the explicitly shown exiting transition probabilities. If there are no exiting pathways, the state is called an *absorbing state*, since the chain will become trapped or absorbed in the state.

Example 21.3 Consider the Markov chain depicted in Fig. 21.2. Inspection of the state diagram reveals that there must be an implied loop at state C having a transition probability $p_{CC} = 0.15$. There are also implied loops at D ($p_{DD} = 1.0$) and at E ($p_{EE} = 0.5$). State D is an absorbing state. If we assign $s_1 \equiv A$, $s_2 \equiv B$ and so on, then we can represent the depicted Markov chain via the stochastic matrix

$$\begin{bmatrix} 0 & 0.30 & 0.25 & 0 & 0.45 & 0 \\ 0 & 0 & 0 & 0.57 & 0.43 & 0 \\ 0 & 0 & 0.15 & 0 & 0.20 & 0.65 \\ 0 & 0 & 0 & 1.00 & 0 & 0 \\ 0 & 0 & 0 & 0.5 & 0.5 & 0 \\ 0.79 & 0 & 0 & 0 & 0.21 & 0 \end{bmatrix}$$

References

1. Knuth, D. E. *The Art of Computer Programming: Vol. 2 Seminumerical Algorithms*, 2nd ed., pp. 142–169, Addison-Wesley Pub. Co., Reading, MA, 1981.

2. Lehmer, D. H. *Proc. 2nd Symp. on Large-Scale Digital Calculating Machinery*, pp. 141–146, Harvard University Press, Cambridge, MA, 1951.
3. Urkowitz, H. *Signal Theory and Random Processes*, Artech House, Dedham, Mass., 1983.
4. Haykin, S. *Communication Systems*, 2nd ed., Wiley, New York, 1983.
5. Simpson, R. S. and R. C. Houts. *Fundamentals of Analog and Digital Communication Systems*, Allyn and Bacon, Boston, 1971.
6. Blachman, M. M. *Noise and Its Effect on Communication*, McGraw-Hill, New York, 1966.
7. Whalen, A. D. *Detection of Signals in Noise*, Academic Press, New York, 1971.
8. Carlson, A. B. *Communication Systems: An Introduction to Signals and Noise in Electrical Communication*, McGraw-Hill, New York, 1968.
9. Stein, S. and J. J. Jones. *Modern Communication Principles*, McGraw-Hill, New York, 1967.
10. Taub, H. and D. L. Schilling. *Principles of Communication Systems*, 2nd ed., McGraw-Hill, New York, 1986.
11. Papoulis, A. *Probability, Random Variables, and Stochastic Processes*, 2nd ed., McGraw-Hill, New York, 1984.
12. Kemeny, J. G. and J. L. Snell. *Finite Markov Chains*, Springer-Verlag, New York, 1976.
13. Bharucha-Reid, A. T. *Elements of the Theory of Markov Processes and Their Applications*, McGraw-Hill, New York, 1960.
14. Romanovsky, V. I. *Discrete Markov Chains*, Volters-Noordhoff Publishing, The Netherlands, 1970.

Chapter 22

Parametric Models of Random Processes

Many discrete-time random processes can be modeled as the output of a shift-invariant system that is driven by a white noise sequence. The output sequence of this model can thus be characterized by the statistics of the input noise and the parameters of the shift-invariant system. Consequently such models are referred to as *parametric models* of random processes. It is most often the case that the system in question has a transfer function that can be expressed as a ratio of polynomials in z. Parametric models employing such filters are further identified as *rational transfer function models* and comprise the specific types of *autoregressive*, *moving average*, and *autoregressive-moving average* models. This chapter is devoted to an exploration of these three specific models and their properties. These models have several uses in signal processing: (1) They are used in simulations to generate pseudorandom sequences that possess particular autocorrelation and spectral properties; (2) These models and variations on them are used in the spectral estimation techniques that are discussed in Chaps. 25 and 26; and (3) They play a central role in *linear prediction*, which is discussed in Chap. 23.

22.1 Autoregressive-Moving Average Model

Many discrete-time random processes can be modeled as the output of a digital filter that is driven by a white noise sequence $w[n]$

$$x[n] = \sum_{k=1}^{p} a_k x[n-k] + \sum_{k=0}^{q} b_k w[n-k] \qquad (22.1)$$

A model of a random process that makes use of an IIR filter as in Eq. (22.1) is called an *autoregressive-moving average* (ARMA) model for the time-series $x[n]$. The first summation constitutes the *autoregressive* portion of the model, and the second summation constitutes the *moving average* portion. An ARMA model of order p autoregressive parameters and order q moving average parameters is denoted as ARMA(p, q).

Software notes

The class `ArmaProcess`, provided in file `armaproc.cpp`, implements an ARMA model of a random process. There are two subclasses that inherit from `ArmaProcess`. The first of these subclasses, `ArmaSource` which is provided in file `arma_src.cpp`, is for the case where the vectors of ARMA parameters **a** and **b** as well as the variance ρ_w of the driving white noise process are known or assumed and provided to the subclass constructor as inputs. The other subclass, `ArmaEstimate` provided in file `arma_est.cpp`, is for the case in which the ARMA parameters must be computed from the autocorrelation matrix. Because this computation makes use of methods that will be presented for AR parameter estimation, discussion of the details will be deferred until Sec. 22.5. Once an instance of either `ArmaProcess` or `ArmaSource` has been constructed, the member function `OutputSequence` belonging to the base class `ArmaProcess` can be used to generate an ARMA sequence of specified length using the specified seed value to initialize the driving white noise generator. This output function is simply a straightforward implementation of Eq. (22.1). The member function `DumpParameters` can be used to dump the ARMA parameters to the specified output stream.

22.2 Autoregressive Model

If all of the moving average parameters b_k are zero, except $b_0 = 1$, the ARMA model specified by Eq. (22.1) reduces to

$$x[n] = -\sum_{k=1}^{p} a_k x[n-k] + w[n] \qquad (22.2)$$

where $w[n]$ is a white noise sequence with variance ρ_w. This equation models an *autoregressive* (AR) process of order p [abbreviated as AR(p)].

Yule-Walker equations

The coefficients of the AR model are related to the *autocorrelation sequence* (ACS) of the output sequence via the *Yule-Walker normal equations*, which are given in matrix form as

$$\begin{bmatrix} r_{xx}[0] & r_{xx}[-1] & \cdots & r_{xx}[-p] \\ r_{xx}[1] & r_{xx}[0] & \cdots & r_{xx}[-p+1] \\ \vdots & \vdots & \ddots & \vdots \\ r_{xx}[p] & r_{xx}[p-1] & \cdots & r_{xx}[0] \end{bmatrix} \begin{bmatrix} 1 \\ a[1] \\ \vdots \\ a[p] \end{bmatrix} = \begin{bmatrix} \rho_w \\ 0 \\ \vdots \\ 0 \end{bmatrix} \qquad (22.3)$$

If the ACS is known for lags 0 to p, Eq. (22.3) can be solved to obtain the AR parameters $a[1], a[2], \ldots a[p]$. In general, the solution of a matrix equation such as Eq. (22.3) will require a number of arithmetic operations proportional to p^3 and a memory size proportional to p^2. However, because the autocorrelation matrix $\mathbf{R}_x$ is both Toeplitz and Hermitian, the Levinson recursion (which is discussed in Sec. 22.3) can be used to solve Eq. (22.3), requiring a number of arithmetic operations proportional to p^2 and a memory size of $2p$.

Characterization of AR processes

For any asymptotically stable AR(p) process, the acf $r_{xx}(m)$ for lag m satisfies the difference equation

$$r(m) + \sum_{k=1}^{p} a_k^* r_{xx}(m-k) = 0$$

Second-order AR processes are particularly easy to characterize in closed form. This fact makes AR(2) process models attractive test sources for evaluating statistical processing algorithms, such as those found in Chaps. 24 through 27, because closed-form theoretical results can be readily obtained for comparison with the empirical results obtained from various computer experiments and examples. The variance σ_x^2 of the AR process and the variance ρ_w of the driving white noise process are related by

$$\sigma_x^2 = \left(\frac{1+a_2}{1-a_2}\right)\frac{\rho_w}{[(1+a_2)^2 - a_1^2]} \tag{22.4}$$

and

$$\rho_w = \sigma_x^2 \left(\frac{1-a_2}{1+a_2}\right)[(1+a_2)^2 - a_1^2] \tag{22.5}$$

The input correlation matrix $\mathbf{R}_x$ for a real-valued AR(2) process is given by

$$\mathbf{R}_x = \begin{bmatrix} r_{xx}(0) & r_{xx}(1) \\ r_{xx}(1) & r_{xx}(0) \end{bmatrix}$$

where $r_{xx}(0) = \sigma_x^2$
$r_{xx}(1) = \frac{-a_1}{1+a_2}\sigma_x^2$

The two eigenvalues of $\mathbf{R}_x$ are

$$\lambda_1 = \left(1 - \frac{a_1}{1+a_2}\right)\sigma_x^2$$

$$\lambda_2 = \left(1 + \frac{a_1}{1+a_2}\right)\sigma_x^2$$

The normalized eigenvectors $\mathbf{q}_1$ and $\mathbf{q}_2$ associated with the eigenvalues λ_1 and λ_2 are

$$\mathbf{q}_1 = \begin{bmatrix} 1/\sqrt{2} \\ 1/\sqrt{2} \end{bmatrix}$$

$$\mathbf{q}_2 = \begin{bmatrix} 1/\sqrt{2} \\ -1/\sqrt{2} \end{bmatrix}$$

Software notes

The class ArProcess, provided in file ar_proc.cpp, implements an AR model of a random process. There are two subclasses that inherit from ArProcess. The first of these subclasses, ArSource which is provided in file ar_src.cpp, is for the case where the vector of AR parameters $\mathbf{a} = \{a_k\}$ and driving noice variance ρ_w are known or assumed, and provided to the subclass as inputs. The other subclass, ArEstimate provided in file ar_est.cpp, is for the case in which the AR parameters are obtained by solving the Yule-Walker equations. This second subclass makes use of the class YuleWalker provided in file yulewalk.cpp to implement these equations. The class YuleWalker has two constructors. The first is for the case in which the autocorrelation matrix is known or assumed and provided in the form of an ACS corresponding to the first column of the Hermitian Toeplitz autocorrelation matrix. The second YuleWalker constructor is for the case in which the autocorrelation matrix is to be estimated from a sequence of a signal samples that is provided as an input. Because the autocorrelation matrix must be Toeplitz in order to be used in the Levinson recursion, the estimated matrix is computed using the autocorrelation method as given by Eqs. (21.48) and (21.49). Both of these constructors make use of the function LevinsonRecursion described in Sec. 22.3 to solve for the parameters a_k and ρ_w that are needed to initialize an object of class ArProcess. Once an instance of either ArProcess or ArSource has been constructed, the member function OutputSequence belonging to the base class ArProcess can be used to generate an AR sequence of specified length using the specified seed value to initialize the driving white noise generator. This output function is simply a straightforward implementation of Eq. (22.2). The member function DumpParameters can be used to dump the AR parameters to the specified output stream. A main program for accepting user inputs and exercising ArModel can be found in the file prog_22a.cpp.

22.3 Levinson Recursion

The *Levinson recursion* is a computationally efficient algorithm for finding the inverse of a Hermitian Toeplitz matrix such as the autocorrelation matrix that appears in Eq. (22.3). This algorithm was originally developed by Levinson in

1947 [3] for solving the linear prediction normal equations that will be discussed in Chap. 23. It was rediscovered by Durbin in 1960 [4] and used for fitting an AR model to a given correlation sequence.

Algorithm 22.1 Levinson recursion for solving the Yule-Walker equations.

1. Initialize

$$a_1[1] = \frac{r_{xx}[1]}{r_{xx}[0]} \tag{22.6}$$

$$\rho_1 = (1 - |a_1[1]|^2)r_{xx}[0] \tag{22.7}$$

where the r_{xx} are as in Eq. (22.3).

2. For $k = 2, 3, \ldots, p$ in succession compute

$$a_k[k] = -\frac{r_{xx}[k] + \sum_{m=1}^{k-1} a_{k-1}[m]r_{xx}[k-m]}{\rho_{k-1}} \tag{22.8}$$

$$a_k[i] = a_{k-1}[i] + a_k[k]\, a_{k-1}^*[k-i] \quad i = 1, 2, \ldots, k-1 \tag{22.9}$$

$$\rho_k = (1 - |a_k[k]|^2)\rho_{k-1} \tag{22.10}$$

3. The desired coefficients $a[i]$ are obtained as $a[i] \equiv a_p[i]$.

Software notes

The function LevinsonRecursion provided in file levin.cpp implements Algorithm 22.1. The prototype for this function is shown in Listing 22.1. The Hermitian Toeplitz autocorrelation matrix is completely specified by the elements in the first column which are passed in the complex array toeplitz. The number of a_k coefficients is specified by the integer ar_order. Note that the coefficient vector avec has ar_order+1 elements—the constant $a_0 \equiv 1$ is not counted in ar_order. Finally, the pointer ar_drv_var is used by the function to return the required variance for the driving white noise process in the corresponding AR model.

Listing 22.1 Prototype for the LevinsonRecursion function

```
int LevinsonRecursion( complex *toeplitz,
                       int ar_order,
                       complex *avec,
                       double *ar_drv_var );
```

22.4 Moving Average Model

If all of the autoregressive parameters a_k are zero, the ARMA model specified by Eq. (22.1) reduces to

$$x[n] = \sum_{k=1}^{q} b_k w[n-k] + w[n] \qquad (22.11)$$

which models a *moving average* (MA) process of order q [abbreviated as MA(q)]. As was done in (22.11), it is a common practice to scale the parameters such that $b_0 = 1$.

Estimation of MA parameters

Estimation of MA parameters is not as straightforward as the estimation of AR parameters. One widely used approach is Durbin's method which, rather than estimating MA parameters directly from the data, estimates MA parameters from the parameters of a high order AR approximation. It can be shown [1] that an infinite order AR(∞) process is equivalent to an MA(q) process when

$$a[m] + \sum_{n=1}^{q} b[n]\, a[m-n] = \delta[m] \qquad (22.12)$$

where a are the AR parameters, b are the MA parameters and $\delta[m]$ is the Kronecker delta which is defined by

$$\delta[m] = \begin{cases} 1 & \text{if } m = 0 \\ 0 & \text{if } m \neq 0 \end{cases}$$

Because any practical estimate of the AR process must have an order that is finite, it will not be possible to satisfy Eq. (22.12) exactly. For an AR(M) estimate where $M \gg q$, we can form a set of error equations of the form

$$e[m] = \hat{a}_M[m] + \sum_{n=1}^{q} b[n]\, \hat{a}_M[m-n] \qquad (22.13)$$

Given the estimate $\hat{a}_M$, the goal is to find the set of $b[i]$ that in some sense minimizes the error. One particularly convenient minimization involves forming the squared error variance

$$\rho_e = \frac{1}{M} \sum_{m} |e[m]|^2 \qquad (22.14)$$

and finding the set of $b[i]$ that minimizes ρ_e. It turns out that if Eq. (22.14) is summed over $0 < m \leq M + q$, the corresponding set of equations formed by Eq. (22.13) is structurally identical to the forward linear prediction error in the autocorrelation method of linear prediction (Sec. 23.3), and the YuleWalker class presented in Sec. 22.2 can be used to solve for the $b[i]$ that minimize ρ_e.

Algorithm 22.2 Durbin's method for estimation of MA parameters.

1. Formulate and solve (using Algorithm 22.1) the Yule-Walker equations to fit a large order AR(M), $M \gg q$ model to the available data sequence $\{x_k\}$.
2. Using the set of AR parameters $\{a_k\}$ found in step 1, form the equation

$$\begin{bmatrix} \hat{r}_{aa}[0] & \hat{r}_{aa}[-1] & \cdots & \hat{r}_{aa}[-q] \\ \hat{r}_{aa}[1] & & & \hat{r}_{aa}[-q+1] \\ \vdots & & \ddots & \vdots \\ \hat{r}_{aa}[q] & \hat{r}_{aa}[q-1] & \cdots & \hat{r}_{aa}[0] \end{bmatrix} \begin{bmatrix} 1 \\ b[1] \\ \vdots \\ b[q] \end{bmatrix} = \begin{bmatrix} \beta \\ 0 \\ \vdots \\ 0 \end{bmatrix}$$

and solve it using the Levinson recursion.

Software notes

The class MaProcess, provided in file ma_proc.cpp, implements an MA model of a random process. There are two subclasses that inherit from MaProcess. The first of these subclasses, MaSource which is provided in file ma_src.cpp, is for the case where the vector of MA parameters $\mathbf{b} = \{b_k\}$ and driving noise variance ρ_w are known or assumed, and provided to the constructor as inputs. The other subclass, MaEstimate provided in file ma_est.cpp, is for the case in which the MA parameters are obtained using Durbin's method as given in Algorithm 22.2. Once an instance of either MaSource or MaEstimate has been constructed, the member function OutputSequence belonging to the base class MaProcess, can be used to generate an MA sequence of specified length using the specified seed value to initialize the driving white noise generator. This output function is simply a straightforward implementation of Eq. (22.11). The member function DumpParameters can be used to dump the MA parameters to the specified output stream. A main program for accepting user inputs and exercising MaModel can be found in the file prog_22b.cpp.

22.5 Estimation of ARMA Parameters

Estimation of parameters for an ARMA model can at times be a difficult problem. It is usually accomplished by estimating the AR parameters, and using them to construct a filter that removes the AR portion of the observed sequence. MA parameters then are estimated for the residue left after this filtering is performed. The AR parameters are estimated using a modified form of the Yule-Walker equations.

Modified Yule-Walker equations

The AR parameters of an ARMA(p, q) model are related to the ACS of the output via the *modified Yule-Walker equations*, which are given in matrix form as

$$\begin{bmatrix} r_{xx}[q] & r_{xx}[q-1] & \cdots & r_{xx}[q-p+1] \\ r_{xx}[q+1] & r_{xx}[q] & \cdots & r_{xx}[q-p+2] \\ \vdots & \vdots & \ddots & \vdots \\ r_{xx}[q+p-1] & r_{xx}[q+p-2] & \cdots & r_{xx}[q] \end{bmatrix} \begin{bmatrix} a[1] \\ a[2] \\ \vdots \\ a[p] \end{bmatrix} = \begin{bmatrix} r_{xx}[q+1] \\ r_{xx}[q+2] \\ \vdots \\ r_{xx}[q+p] \end{bmatrix}$$

(22.15)

If the ACS is known or can be estimated for lags $q - p + 1$ to $q + p$, Eq. (22.15) can be solved to obtain the AR parameters $a[1], a[2], \ldots a[p]$. Because the autocorrelation matrix is not Hermitian, the Levinson recursion cannot be used directly to solve this equation. An extension of the Levinson recursion based on the work of Trench [5] is given in Algorithm 22.3. A detailed derivation of the extension is presented in [6]. The ACS estimator used to form the $r_{xx}[k]$ may be either the biased Eq. (21.46) or unbiased Eq. (21.45) estimator.

Algorithm 22.3 Generalized Levinson algorithm for solving the modified Yule-Walker equation.

1. Initialize

$$a_1[1] = -\frac{r_{xx}[q+1]}{r_{xx}[q]} \tag{22.16}$$

$$b_1[1] = -\frac{r_{xx}[q-1]}{r_{xx}[q]} \tag{22.17}$$

$$\rho_1 = (1 - a_1[1]b_1[1])\, r_{xx}[q] \tag{22.18}$$

2. For $k = 2, 3, \ldots, p$ in succession

   (a) compute

$$a_k[k] = -\frac{r_{xx}[q+k] + \sum_{m=1}^{k-1} a_{k-1}[m] r_{xx}[q+l-m]}{\rho_{k-1}} \tag{22.19}$$

$$a_k[i] = a_{k-1}[i] + a_k[k] b_{k-1}[k-i] \quad i = 1, 2, \ldots, k-1 \tag{22.20}$$

   (b) for $k < p$ compute

$$b_k[k] = -\frac{r_{xx}[q-k] + \sum_{m=1}^{k-1} b_{k-1}[m] r_{xx}[q-l+m]}{\rho_{k-1}} \tag{22.21}$$

$$b_k[i] = b_{k-1}[i] + b_k[k] a_{k-1}[k-i] \quad i = 1, 2, \ldots, k-1 \tag{22.22}$$

$$\rho_k = (1 - a_k[k] b_k[k]) \rho_{k-1} \tag{22.23}$$

Software notes

As discussed in Sec. 22.1, the class ArmaProcess implements an ARMA model of a random process. A main program for accepting user inputs and exercising ArmaProcess can be found in the file prog_22c.cpp. The constructor for the derived class ArmaEstimate, used when the ARMA parameters must be computed from the autocorrelation matrix, makes use of the class ModYuleWalker (provided in file mod_yuwa.cpp) to implement the modified Yule-Walker equations. The constructor for ModYuleWalker calls the function GeneralizedLevinson, (provided in file gen_lev.cpp) which implements Algorithm 22.3.

References

1. Kay, S. M. *Modern Spectral Estimation: Theory and Application*, Prentice-Hall, Englewood Cliffs, NJ, 1988.
2. Marple, S. L. *Digital Spectral Analysis with Applications*, Prentice-Hall, Englewood Cliffs, NJ, 1987.
3. Levinson, N. "The Wiener RMS (Root Mean Square) Error Criterion in Filter Design and Prediction," *J. Math. Phys.*, Vol. 25, pp. 261–278, 1947.
4. Durbin, J. "The Fitting of Time-Series Models," *Rev. Inst. Int. Statist.*, Vol. 28, pp. 233–243, 1960.
5. Trench, W. F. "An Algorithm for the Inversion of Finite Toeplitz Matrices," *J. Soc. Ind. Appl. Math.*, Vol. 12, pp. 515-522, Sept. 1964.
6. Press, W. H. *et al.*, *Numerical Recipes in C*, 2nd Ed., Cambridge University Press, New York, 1992.

Chapter 23

Linear Prediction

In *linear prediction*, the current value $x[n]$ of a random signal sequence x is estimated or *predicted* as a linear combination of P past observations of the sequence. Linear prediction is one of the most widely used types of optimal filtering. In terms of total installed instances, linear prediction's use in speech coding applications alone must place it at the top. It is also widely used in applications such as radar and sonar processing that involve tracking of moving objects.

23.1 Linear Estimation

Let $s[n]$ be a signal which is corrupted by noise $w[n]$ prior to being observed as $x[n]$ at the input to a linear filter as shown in Fig. 23.1. The filter's output $\hat{d}[n]$ is an estimate of the *desired* signal $d[n]$. The error $\varepsilon[n]$ is the difference between $d[n]$ and $\hat{d}[n]$. This configuration is referred to as the *linear estimation problem*. The desired signal $d[n]$ must be selected to satisfy requirements of the intended application, and then the filter must be designed in such a way that the error $\varepsilon[n]$ is minimized.

A number of different names are attached to the linear estimation problem depending on the form of the observed signal $x[n]$ and the desired signal $d[n]$. The usual terminology is summarized in Table 23.1. When the goal is to estimate the original signal $s[n]$ from the observed signal $x[n] = s[n] + w[n]$, the desired signal is $d[n] = s[n]$, and the estimation is referred to as *filtering*. When the goal is to estimate future values of $s[n]$ from the observed signal $x[n]$, the desired signal is $d[n] = s[n + M]$, $M > 0$, and the estimation is referred to as *prediction*. When the goal is to estimate past values of $s[n]$, the desired signal is $d[n] = s[n - M]$, $M > 0$, and the estimation is referred to as *smoothing*. All three of the cases mentioned so far include noise in the observations and no noise in the desired signal. A very important form of estimation, linear prediction has as its apparent goal the estimation of future values of the observed signal $x[n]$. Depending on the context, this case will be treated with noise

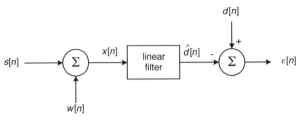

Figure 23.1 Model for linear estimation problem.

TABLE 23.1 Types of Linear Estimation

Operation	Observations	Desired signal
filtering	$x[n] = s[n] + w[n]$	$d[n] = s[n]$
prediction	$x[n] = s[n] + w[n]$	$d[n] = s[n+M]; M > 0$
smoothing	$x[n] = s[n] + w[n]$	$d[n] = s[n-M]; M > 0$
linear prediction	$x[n] = s[n]$	$d[n] = s[n+M]; M > 0$
	$x[n] = s[n] + w[n]$	$d[n] = x[n+M]; M > 0$

in both the observed signal and desired signal, i.e., $x[n] = s[n] + w[n]$ and $d[n] = x[n+M]$, $M > 0$, or with noise absent in both the observed signal and desired signal, i.e., $x[n] = s[n]$ and $d[n] = s[n+M]$, $M > 0$. Although the apparent goal is the estimation of future values of $x[n]$, in most practical applications the *estimation* is only a means of obtaining linear prediction coefficients that are widely used for bandwidth efficient coding of speech signals for recording or transmission.

23.2 Linear Predictive Filtering

A *forward linear predictor* of order P estimates the value $x[n]$ of a random data sequence x as a linear combination of the P prior observations $x[n-P]$, $x[n-P+1]$, ... $x[n-1]$

$$\hat{x}_f[n] = -\sum_{k=1}^{P} a_f[k]\, x[n-k] \tag{23.1}$$

The error of such a prediction is simply the difference between the observed value $x[n]$ and the estimated value $\hat{x}_f[n]$.

$$e_f[n] = x[n] - \hat{x}_f[n] \tag{23.2}$$

$$= x[n] + \sum_{k=1}^{P} a_f[k]\, x[n-k] \tag{23.3}$$

The *forward linear prediction coefficients* $a_f[k]$ are selected to minimize the mean-square error E_P, which is given by

$$E_P = G \sum_m (e_f[m])^2 \qquad (23.4)$$

By substituting Eq. (23.3) into Eq. (23.4) we obtain

$$E_P = G \sum_m \left(x[m] + \sum_{k=1}^{P} a_f[k]\, x[m-k] \right)^2 \qquad (23.5)$$

By defining $a_f[0] \equiv 1$, this equation can be written more compactly as

$$E_P = G \sum_m \left(\sum_{k=0}^{P} a_f[k]\, x[m-k] \right)^2 \qquad (23.6)$$

For the summation to be an average, the constant gain term G should be set equal to the reciprocal of the number of terms in the summation.

The range of the summation in Eq. (23.6) can be specified in one of two different ways that lead to different solutions for the coefficients $a_f[k]$. The two different approaches have come to be called the *autocorrelation method* and the *covariance method*. These methods are discussed further in the following sections.

23.3 Autocorrelation Method

The autocorrelation method of linear prediction is based on the assumption that exactly N observations of the data sequence x are available for estimation of the linear prediction coefficients. The available interval for $x[n]$ is assumed to be either $1 \leq n \leq N$ or $0 \leq n \leq N-1$ depending on whether the origin of the indexing scheme employed is 1 or 0. Texts such as [1] that are not tied to specific computer implementations could easily use either scheme, but indexing from 0 through $N-1$ seems to prevail. Texts such as [2] and [3] that present software implementations written in FORTRAN tend to use indexing that runs from 1 through N because this agrees with the way array indexing works in FORTRAN. Similarly, for texts that include C or C++ software implementations, indexing from 0 through $N-1$ is the most natural choice. Therefore, let us assume that the data sequence $x[n]$ is observed over the interval $0 \leq n \leq N-1$. The data sequence is assumed to be identically zero outside of this interval. Under this assumption, Eq. (23.3) will yield nonzero values for $e_f[n]$ over the interval $0 \leq n \leq N+P-1$. The coefficients $a_f[k]$ that minimize E_P summed over this range are found by solving the matrix equation

$$\mathbf{R}_P \begin{bmatrix} 1 \\ \mathbf{a}_f \end{bmatrix} = \begin{bmatrix} \sigma^2 \\ \mathbf{0}_P \end{bmatrix} \qquad (23.7)$$

where

$$\mathbf{a}_f = \begin{bmatrix} a_f[1] \\ \vdots \\ a_f[P] \end{bmatrix}, \quad \mathbf{R}_P = \begin{bmatrix} r[0,0] & r[0,1] & \cdots & r[0,P] \\ r[1,0] & r[1,1] & \cdots & r[1,P] \\ \vdots & & \ddots & \vdots \\ r[P,0] & r[P,1] & \cdots & r[P,P] \end{bmatrix}$$

and $\mathbf{0}_P$ is a column vector containing P zero-valued elements. The elements of $\mathbf{R}_P$ are obtained as

$$r[i,j] = \frac{1}{N} \sum_{n=0}^{N+P-1} x^*[n-i]x[n-j] \qquad (23.8)$$

Inspection of Eq. (23.8) reveals that $r[i,j] = r^*[j,i]$, and therefore $\mathbf{R}_P$ is Hermitian. For $0 \leq i - j \leq P$, the summation in (23.8) can be rewritten as

$$r[i,j] = \frac{1}{N} \sum_{k=0}^{N-(1+i-j)} x[k+(i-j)]\, x^*[k] \qquad (23.9)$$

revealing the fact that $r[i,j]$ depends only on the difference $(i-j)$ and that consequently $\mathbf{R}_P$ is a Toeplitz matrix. This form of the correlation matrix corresponds to Eq. (21.49) developed in Sec. 21.14, where it was noted that the correlation matrix can be expressed as

$$\mathbf{R}_P = \frac{1}{N} \mathbf{X}^H \mathbf{X} \qquad (23.10)$$

where

$$\mathbf{X} = \begin{bmatrix} \mathbf{L} \\ \mathbf{T} \\ \mathbf{U} \end{bmatrix}$$

and the lower triangular $P \times (P+1)$ matrix $\mathbf{L}$, rectangular $(N-P) \times (P+1)$ matrix $\mathbf{T}$, and upper triangular $P \times (P+1)$ matrix $\mathbf{U}$ are defined as

$$\mathbf{L} = \begin{bmatrix} x[0] & \cdots & 0 & 0 \\ \vdots & \ddots & \vdots & \vdots \\ x[P-1] & \cdots & x[0] & 0 \end{bmatrix} \qquad (23.11)$$

$$\mathbf{T} = \begin{bmatrix} x[P] & \cdots & x[0] \\ \vdots & \ddots & \vdots \\ x[N-P-1] & & x[P] \\ \vdots & \ddots & \vdots \\ x[N-1] & \cdots & x[N-P-1] \end{bmatrix} \qquad (23.12)$$

$$\mathbf{U} = \begin{bmatrix} 0 & x[N-1] & \cdots & x[N-P] \\ \vdots & \vdots & \ddots & \vdots \\ 0 & 0 & \cdots & x[N-1] \end{bmatrix} \qquad (23.13)$$

Equation (23.7) is identical to the Yule-Walker normal equation [Eq. (22.3)] presented in Sec. 22.2. Therefore, the Levinson recursion can be used to solve for the coefficients $a_f[k]$.

Software notes

The class UnquantDirectFormIir presented in Chap. 19 can be used to generate test sequences for demonstrating linear prediction techniques. In general, the implementation of an IIR filter will produce an ARMA sequence. To generate a purely AR sequence, the numerator of the filter's transfer function must be set to a constant.

Once a test signal is generated, the class ArModel presented in Chap. 22 can be used to estimate the AR parameters from this sequence. A second instance of the class UnquantDirectFormIir then can be used to generate an AR sequence from the estimated coefficients. This approach is the one taken in the program ex23_01 which is used for Example 23.1.

Example 23.1 Consider the sequence $x[n]$ defined by

$$x[n] = u[n] + \sum_{k=1}^{4} x[n-k]a[k] \qquad (23.14)$$

where $a[1] = 3.504$
$a[2] = -5.026$
$a[3] = 3.432$
$a[4] = -0.9596$

and the excitation $u[n]$ is the unit sample function

$$u[n] = \begin{cases} 1 & n = 0 \\ 0 & \text{otherwise} \end{cases}$$

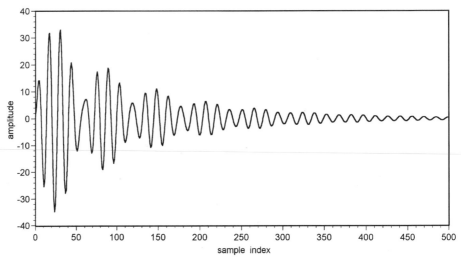

Figure 23.2 Autoregressive signal for Example 23.1.

Program ex23_01 can be used to generate a portion of this sequence and form linear prediction estimates of the coefficients $a[k]$. The sequence $x[n]$ is plotted in Fig. 23.2. If each of the correlations in $\mathbf{R}_P$ is based on 500 samples of the sequence, the estimated coefficients obtained by solving Eq. (23.7) are

$$\hat{a}[1] = 3.34773 \qquad \hat{a}[2] = -4.59546$$
$$\hat{a}[3] = 3.00484 \qquad \hat{a}[4] = -0.807598$$

(Note: The coefficient estimates produced by program ex23_01 must be multiplied by -1.) The sequence produced by using these estimates in Eq. (23.14) is plotted in Fig. 23.3. This sequence is not a particularly good match for the original sequence. The match can be improved by using better correlation estimates.

If each of the correlations in $\mathbf{R}_P$ is based on 1000 samples of the sequence, the estimated coefficients will be obtained as

$$\hat{a}[1] = 3.504 \qquad \hat{a}[2] = -5.026$$
$$\hat{a}[3] = 3.432 \qquad \hat{a}[4] = -0.959598$$

These estimates are close to the original coefficients, and the sequence produced by using them in Eq. (23.14) is virtually identical to the original sequence.

More software notes

The original coefficients for Example 23.1 were not conjured out of thin air. The approach that was actually taken is defined by Algorithm 23.1.

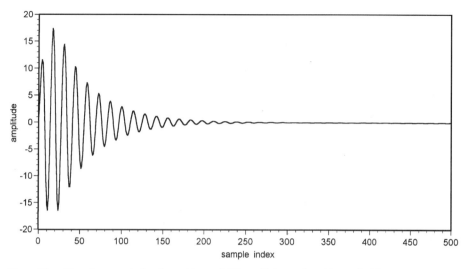

Figure 23.3 Signal generated using estimated AR coefficients from Example 23.1.

Algorithm 23.1 Generating good autoregressive sequences for testing linear prediction techniques.

1. Place conjugate pairs of poles in the complex plane at locations corresponding to damped sinusoids that would produce an interesting test signal. (For Example 23.1, the pole locations were $s = -0.01 \pm 0.4j$ and $s = -0.01 \pm 0.5j$.)
2. Perform the bilinear transformation to obtain coefficients for an IIR implementation of the filter defined by the poles selected in step 1.
3. Discard the numerator coefficients of the IIR transfer function produced in step 2, and set the numerator equal to unity.
4. Retain the denominator coefficients of the IIR transfer function. (The actual values obtained while generating Example 23.1 were $a[1] = 3.503671161$, $a[2] = -5.025949147$, $a[3] = 3.4322728$, $a[4] = -0.9596454685$. These values were rounded to obtain the values presented in the example, but the unrounded values could just as easily have been used.)

Example 23.2 The program ex23_02 can be used for the computations in this example. Use Algorithm 23.1 to generate an autoregressive sequence based on analog s-plane poles at $s = -0.1 \pm 0.1j$, $s = -0.005 \pm 0.4j$, and $s = -0.005 \pm 0.45j$. The bilinear transformation yields as denominator coefficients

$$a[1] = 5.353819612$$

$$a[2] = -12.33737815$$

$$a[3] = 15.61703573$$

$$a[4] = -11.43468826$$

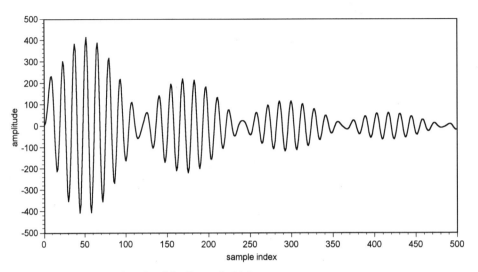

Figure 23.4 Autoregressive signal for Example 23.2.

$$a[5] = 4.58788993$$

$$a[6] = -0.7875596531$$

The sequence generated by these coefficients is plotted in Fig. 23.4.

23.4 Covariance Method

The rows in **L** as defined by Eq. (23.11) end in a number of zeros, and the rows in **U** as defined by Eq. (23.13) begin with a number of zeros. In some applications, it is objectionable to use these zeros in the calculation of $\mathbf{R}_P$. In the *covariance method*, the inclusion of zero-valued samples in the calculation of $\mathbf{R}_P$ is avoided by defining $\mathbf{R}_P$ using only the center partition of **X**

$$\mathbf{R}_P = \frac{1}{N-P}\mathbf{T}^H\mathbf{T} \qquad (23.15)$$

Where **T** is given by Eq. (23.12). The individual elements of $\mathbf{R}_P$ are obtained as

$$r[i, j] = \frac{1}{N-P}\sum_{n=P}^{N-1} x^*[n-i]\, x[n-j]$$

It can be shown [2] that defining $\mathbf{R}_P$ in this manner is equivalent to summing Eq. (23.6) over the range $P \leq n \leq N - 1$. When $\mathbf{R}_P$ is defined as in Eq. (23.15), the resulting matrix is Hermitian but not Toeplitz. Therefore, the Levinson recursion cannot be used to solve for $a_f[k]$. Instead, the coefficients are found by using the Cholesky decomposition to solve the equation

$$\mathbf{R}_P \mathbf{a} = -\mathbf{r}_P \tag{23.16}$$

where $\mathbf{R}_P = \begin{bmatrix} r_{xx}[1,1] & r_{xx}[1,2] & \cdots & r_{xx}[1,p] \\ r_{xx}[2,1] & r_{xx}[2,2] & \cdots & r_{xx}[2,p] \\ \vdots & & \ddots & \vdots \\ r_{xx}[p,1] & r_{xx}[p,2] & \cdots & r_{xx}[p,p] \end{bmatrix}$

$$\mathbf{a} = \begin{bmatrix} a[1] \\ a[2] \\ \vdots \\ a[p] \end{bmatrix} \qquad \mathbf{r}_P = \begin{bmatrix} r_{xx}[1,0] \\ r_{xx}[2,0] \\ \vdots \\ r_{xx}[p,0] \end{bmatrix}$$

Cholesky decomposition

If $\mathbf{A}$ is a symmetric matrix, it can be factored into

$$\mathbf{A} = \mathbf{L}\mathbf{D}\mathbf{L}^H \tag{23.17}$$

where $\mathbf{L} = \{l_{ij}\}$ is a lower triangular matrix and $\mathbf{D} = \{d_i\}$ is a diagonal matrix. If $\mathbf{A}$ is also positive definite, all the elements on the diagonal of $\mathbf{D}$ will be positive and $\mathbf{D}$ can be split to yield

$$\mathbf{A} = (\mathbf{L}\sqrt{\mathbf{D}})(\mathbf{L}\sqrt{\mathbf{D}})^H$$
$$= \mathbf{R}\mathbf{R}^H \tag{23.18}$$

The decomposition represented in Eq. (23.18) is the *Cholesky decomposition*, and that in Eq. (23.17) is sometimes called the *modified Cholesky decomposition*.

Algorithm 23.2 Using the modified Cholesky decomposition to solve the set of linear equations $\mathbf{A}\mathbf{x} = \mathbf{b}$ when $\mathbf{A}$ is a positive definite matrix.

1. Decompose the $\mathbf{A}$ matrix as

$$\mathbf{A} = \mathbf{L}\mathbf{D}\mathbf{L}^H \tag{23.19}$$

where $\mathbf{L} = \{l_{ij}\}$ is a lower triangular matrix and $\mathbf{D} = \{d_i\}$ is a diagonal matrix.

(a) Initialize

$$d_1 = a_{11} \tag{23.20}$$

(b) For $i = 2, 3, \ldots N$ in succession compute

$$l_{ij} = \frac{a_{i1}}{d_1} \quad \text{for } j = 1 \tag{23.21}$$

$$l_{ij} = \frac{a_{ij}}{d_j} - \sum_{k=1}^{j-1} \frac{l_{ik} d_k l_{jk}^*}{d_j} \quad \text{for } j = 2, 3, \ldots i - 1 \tag{23.22}$$

$$d_i = a_{ii} - \sum_{k=1}^{i-1} d_k |l_{ik}|^2 \tag{23.23}$$

2. With **A** decomposed as in (23.19), the set of equations to be solved becomes

$$\mathbf{LDL^H x = b} \tag{23.24}$$

Substituting

$$\mathbf{y = DL^H x} \tag{23.25}$$

into Eq. (23.24) yields

$$\mathbf{Ly = b} \tag{23.26}$$

Solve for **y** as follows:

(a) Initialize

$$y_1 = b_1 \tag{23.27}$$

(b) For $k = 2, 3, \ldots N$ in succession, compute

$$y_k = b_k - \sum_{j=1}^{k-1} l_{kj} y_j \tag{23.28}$$

3. Multiply both sides of (23.25) by $\mathbf{D}^{-1}$ to obtain

$$\mathbf{L^H x = D^{-1} y} \tag{23.29}$$

Solve for **x** as follows:

(a) Initialize

$$x_N = \frac{y_N}{d_N} \tag{23.30}$$

(b) For $k = N-1, N-2, \ldots 1$ in succession, compute

$$x_k = \frac{y_k}{d_k} - \sum_{j=k+1}^{N} l_{jk}^* x_j \tag{23.31}$$

Software notes

The function `CholeskyDecomp` provided in file `cholesky.cpp` implements Algorithm 23.2. The prototype for this function is shown in Listing 23.1. The correlation matrix is passed into this function via `ax` which points to an object of type `complex_matix`. This matrix is assumed to be Hermitian symmetric, and only the upper triangular portion needs to be specified. The lower triangular portion is ignored even if defined. The number of x_k coefficients is specified by the integer `ord`. The vector **b** is passed into the function via `bx` which is a pointer to `complex`. The final parameter, `epsilon`, is a tolerance that is used to test for an ill-conditioned **A** matrix. The functions `CovarMethCorrMtx` and `CovarMethRightHandVect` provided in file `covmeth.cpp` can be used to generate respectively the correlation matrix and R.H.S. correlation vector in the formats needed by `CholeskyDecomp`.

Listing 23.1 Prototype for the `CholeskyDecomp` function

```
int CholeskyDecomp( int order,
                    complex_matrix *ax,
                    complex *bx,
                    double epsilon );
```

Example 23.3 Use program ex23_03 to apply the covariance method of linear prediction to the sequence defined in Example 23.1. If each of the correlations in $\mathbf{R}_P$ is based on 200 samples of the sequence, the estimated coefficients obtained by solving Eq. (23.16) are exact matches for the original coefficients. This performance is noticeably better than the performance of the autocorrelation method, which did rather poorly with 500 samples per correlation. In fact, the covariance method provides good matches for this particular sequence, with as few as 10 samples per correlation. However, the addition of a small amount of noise to the sequence will degrade the performance of the covariance method to a level roughly comparable to the autocorrelation method.

23.5 Lattice Filters

In some applications, the linear prediction coefficients and the error sequence $e[n]$ are of more interest than the estimated signal $x[n]$. In these cases, it is convenient to augment the linear predictor to form the *prediction error filter* shown in Fig. 23.5. The straightforward implementation of such a filter would involve a transversal filter implementation of the linear predictor. However, it can be shown [6, 7] that the prediction error filter also can be implemented as a lattice of the sort shown in Fig. 23.6. The lattice implementation tends to be more tolerant of quantization and roundoff.

A prediction error filter of order M is completely specified by either the set of tap weights $\{a_M[k], k = 1, 2, \ldots M\}$ or by the set of reflection coefficients $\kappa_k, k = 1, 2, \ldots M$ plus input variance $r_{xx}[0]$. With regard to the relationships between these alternative specifications, there are four possible analysis scenarios:

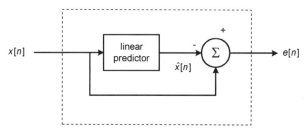

Figure 23.5 Prediction error filter.

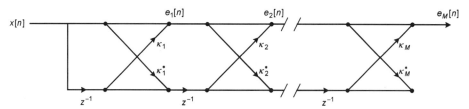

Figure 23.6 Lattice implementation of the prediction error filter.

1. The ACS of the input process is either known or able to be estimated for lags 0 through M, and the requirement is to compute the set of tap weights $a_M[1], a_M[2], \ldots a_M[M]$ for a prediction error filter of order M.

2. The ACS of the input process is either known or able to be estimated for lags 0 through M, and the requirement is to compute the set of reflection coefficients $\kappa_1, \kappa_2, \ldots \kappa_M$.

3. The reflection coefficients $\kappa_1, \kappa_2, \ldots \kappa_M$ are known and the ACS of the input process is known for lag 0. The requirement is to compute the set of tap weights $a_M[1], a_M[2], \ldots a_M[M]$ for a prediction error filter of order M.

4. The set of tap weights $a_M[1], a_M[2], \ldots a_M[M]$ for a prediction error filter of order M is known and the requirement is to compute the corresponding set of reflection coefficients $\kappa_1, \kappa_2, \ldots \kappa_M$.

The first two cases are handled by the Levinson recursion as shown in Algorithm 23.3. This algorithm is similar to Algorithm 22.1 with the coefficients $a_m[m]$ being explicitly identified as k_m. For the third case, where the reflection coefficients are known, it is not necessary to compute Eqs. (23.32) or (23.34), and Algorithm 23.3 can be simplified to Algorithm 23.4. For the fourth case, in which the tap weights of the M-th order prediction error filter are known, it is necessary to use the inverse form of the Levinson recursion to compute the tap weights of the prediction error filters of orders $M-1, M-2, \ldots 1$. The reflection coefficients are then obtained as

$$\kappa_m = a_m[m] \quad m = 1, 2, \ldots M$$

The details of this process are spelled out in Algorithm 23.5.

Linear Prediction

Algorithm 23.3 Levinson recursion for generating prediction error tap weights and the reflection coefficients for the corresponding lattice.

1. Initialize

$$\kappa_1 = a_1[1] = \frac{r_{xx}[1]}{r_{xx}[0]} \quad (23.32)$$

$$\rho_1 = \left(1 - |\kappa_1|^2\right) r_{xx}[0] \quad (23.33)$$

2. For $m = 2, 3, \ldots, M$ in succession compute

$$\kappa_m = -\frac{r_{xx}[m] + \sum_{k=1}^{m-1} a_{m-1}[k] r_{xx}[m-k]}{\rho_{m-1}} \quad (23.34)$$

$$a_m[i] = a_{m-1}[i] + \kappa_m\, a^*_{m-1}[m-k] \quad k = 1, 2, \ldots, m-1 \quad (23.35)$$

$$a_m[m] = \kappa_m \quad (23.36)$$

$$\rho_m = \left(1 - |\kappa_m|^2\right) \rho_{m-1} \quad (23.37)$$

Algorithm 23.4 Levinson recursion for computing prediction error tap weights from the reflection coefficients for the corresponding lattice.

1. acf $r_{xx}[0]$ and reflection coefficients $\kappa_1, \kappa_2, \ldots \kappa_M$ are known. Initialize

$$\rho_1 = \left(1 - |\kappa_1|^2\right) r_{xx}[0] \quad (23.38)$$

2. For $m = 2, 3, \ldots, M$ in succession compute

$$a_m[i] = a_{m-1}[i] + \kappa_m\, a^*_{m-1}[m-k] \quad k = 1, 2, \ldots, m-1 \quad (23.39)$$

$$a_m[m] = \kappa_m \quad (23.40)$$

$$\rho_m = \left(1 - |\kappa_m|^2\right) \rho_{m-1} \quad (23.41)$$

Algorithm 23.5 Inverse Levinson recursion for computing reflection coefficients from the prediction error filter's tap weights.

1. Tap weights $a_M[k], k = 0, 1, \ldots M$ are known. Initialize

$$\kappa_M = a_M[M] \quad (23.42)$$

2. For $m = M, M-1, \ldots 2$ in succession compute

$$a_{m-1}[k] = \frac{a_m[k] - \kappa_m a^*_m[m-k]}{1 - |\kappa_m|^2} \quad k = 0, 1, \ldots m \quad (23.43)$$

$$\kappa_{m-1} = a_{m-1}[m-1] \quad (23.44)$$

References

1. Rabiner, R. Lawrence and Ronald W. Schafer. *Digital Processing of Speech Signals*. Prentice-Hall, Englewood Cliffs, NJ, 1978.
2. Marple, S. Lawrence. *Digital Spectral Analysis with Applications*. Prentice-Hall, Englewood Cliffs, NJ, 1987.
3. Kay, M. Steven, *Modern Spectral Estimation: Theory & Application*. Prentice-Hall, Englewood Cliffs, NJ, 1988.
4. Wiener, Norbert. *Extrapolation, Interpolation, and Smoothing of Stationary Time Series*. MIT Press, Cambridge, MA, 1949.
5. Orfanidis, J. Sophocles, *Optimum Signal Processing: An Introduction*. Macmillan Pub. Co., New York, 1985.
6. Therrien, W. Charles, *Discrete Random Signals and Statistical Signal Processing*. Prentice-Hall, Englewood Cliffs, NJ, 1992.
7. Makhoul, John. "Linear Prediction: A Tutorial Review," *Proc. IEEE*, vol. 63, pp. 561–580, April 1975.

Chapter 24

Adaptive Filters

Adaptive filters are digital filters with coefficients that can change over time. The general idea is to assess how well the existing coefficients are performing and then adapt the coefficient values to improve performance. This approach is useful in two somewhat different application categories. The first category involves filtering requirements that are stationary but unknown. In this case an adaptive filter can be initialized with a guess and then allowed to converge to a better solution. The second category involves filtering requirements that may be loosely bounded in some way, but which vary over time. In the first category, speed of convergence often is a secondary consideration behind the steady-state error remaining in the converged filter. In the second category, steady-state error is important, but the adaptation speed must be sufficient to allow the filter coefficients to track the time-varying requirements. The trade between convergence speed and steady-state error is a fundamental issue in adaptive signal processing.

Adaptive filters consist of three basic sections, as depicted in Fig. 24.1:

1. A *filtering section* that operates on the available input signal $x[k]$ to produce an output signal $y[k]$. This output is assumed to be an estimate of some desired signal $d[k]$. The filter can be either an FIR or IIR design, but the use of FIR designs is far more common. The FIR portion of an adaptive filter is sometimes called an *adaptive linear combiner*.

2. An *error section* that computes an estimation error by comparing the filter output $y[k]$ and the desired signal $d[k]$. When the desired signal is readily available, this operation is easily accomplished by computing the difference between $y[k]$ and $d[k]$. However, in many practical applications, $d[k]$ is not readily available and must be obtained via processing of some other signal.

3. An *adaptation section* that uses the estimation error formed in (2) to compute new tap weights for the filter in (1).

Many different algorithms can be incorporated in the adaptation section. These algorithms vary greatly in their computational complexity and in their

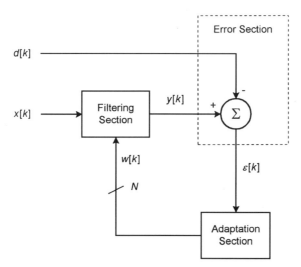

Figure 24.1 Basic structure of an adaptive filter.

performance. Much of the work in designing an adaptive filter is involved in selecting the algorithm that offers the most acceptable balance between performance and cost of implementation and operation. Selection of the order and quantization plan for the filter also are important design considerations. In general, the adaptation algorithms for FIR and IIR filters are different, and some of the simplest and most effective algorithms are for use with FIR filters.

This chapter examines some of the performance issues common to all adaptive filters and presents three specific adaptive algorithms: the method of *steepest descent* (SD), the *least-mean-square* (LMS) algorithm and the *recursive least-squares* (RLS) algorithm.

24.1 Adaptive Linear Combiner

The *adaptive linear combiner* (ALC) is the cornerstone of virtually all linear adaptive signal processing. This structure is similar to an FIR filter, with one important exception. In an ALC, the tap weights are time varying, while in an FIR filter the tap weights are fixed. Most of the work in adaptive signal processing is concerned with ways, optimally or at least beneficially, to set and vary the tap weights in an ALC. There is considerable overlap with the theory of linear estimation. The ALC forms a linear estimate of some *desired* signal; the error between this estimate and the desired signal drives the adaptation of the ALC tap weights. The better the estimate gets, the smaller the changes that are made to the tap weights.

The output $y[k]$ of the adaptive linear combiner at time k can be expressed in vector form as

$$y[k] = \mathbf{x}^T[k]\,\mathbf{w}[k] = \mathbf{w}^T[k]\,\mathbf{x}[k] \qquad (24.1)$$

where $\mathbf{x}[k]$ is the column vector of the N most recent input samples:

$$\mathbf{x}[k] = \begin{bmatrix} x[k] \\ x[k-1] \\ x[k-2] \\ \vdots \\ x[k-N+1] \end{bmatrix}$$

and $\mathbf{w}[k]$ is the column vector of tap weights at iteration k

$$\mathbf{w}[k] = \begin{bmatrix} w_0[k] \\ w_1[k] \\ \vdots \\ w_{N-1}[k] \end{bmatrix}$$

Equation (24.1) can be expressed in the form of a summation as

$$y[k] = \sum_{n=0}^{N-1} x[k-n] w_n[k]$$

24.2 Properties of the Performance Surface

Given a desired signal $d[k]$ and the adaptive system's estimate $y[k]$ of this signal at time k, the error at time k is given by

$$\varepsilon[k] = d[k] - y[k]$$

The mean-square error is obtained as

$$\xi = E\{\varepsilon^2[k]\} = E\{d^2[k]\} + \mathbf{w}^T[k]\,\mathbf{R}\,\mathbf{w}[k] - 2\mathbf{p}^T\mathbf{w}[k] \qquad (24.2)$$

where $\mathbf{R} = E[\mathbf{x}\,\mathbf{x}^T]$
$\mathbf{p} = E\{d[k]\,\mathbf{x}[k]\}$

The column vector $\mathbf{p}$ is the cross correlation between the input sequence and the desired response at time k, the column vector $\mathbf{w}$ is the *weight vector*, and $\mathbf{R}$ is the *input correlation matrix*. For a stationary input process, both $\mathbf{p}$ and $\mathbf{R}$ are fixed, so ξ at time k can be viewed as a function of the weight vector at time k

$$\xi[k] = f_\xi(\mathbf{w}[k])$$

For an adaptive filter built around an adaptive linear combiner, the *mean-square error* (MSE) as a function of the weights forms a hyperparaboloid which is a parabolic surface in $(N+1)$ dimensional space where N is the number of weights. This surface will always have a single global minimum and no local minima. In the context of adaptive signal processing, such a surface is called the *performance surface* or *error surface*.

The two-weight case provides the best introductory example because it has an error surface that is a paraboloid in 3-space and that is relatively easy to depict in a two-dimensional diagram. The error function for a single-weight case would be even easier to draw, but the resulting two-dimensional parabola is not particularly good for illustrating some of the characteristics exhibited by multiple-weight error surfaces.

Figure 24.2 depicts a portion of the error surface for an adaptive filter with two weights. The minimum of this surface or "bottom of the bowl" lies at the point $(\tilde{w}_0, \tilde{w}_1, \xi_{min})$ where $\tilde{w}_0$ and $\tilde{w}_1$ are the two components of the optimum weight vector $\tilde{\mathbf{w}}$. Notice that, in general, $\xi_{min} \neq 0$. If we cut the error surface shown in Fig. 24.2 with planes parallel to the plane defined by the w_0 and w_1 axes, we obtain concentric ellipses of constant MSE as shown in Fig. 24.3.

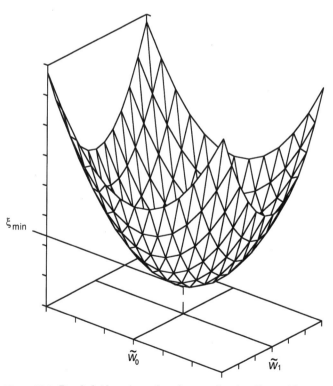

Figure 24.2 Paraboloid error surface for an adaptive filter with two weights.

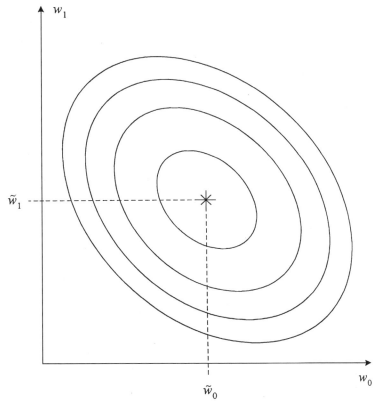

Figure 24.3 Ellipses of constant mean-square error obtained by cutting the error surface in Fig. 24.2.

These ellipses are centered on the point $(\tilde{w}_0, \tilde{w}_1)$, which corresponds to the point of minimum MSE. This coordinate system based on $w[0]$ and $w[1]$ is called the *natural coordinate system*. We can impose a new set of coordinates **v** with axes labeled v_0 and v_1, which have their origins at the point $(\tilde{w}_0, \tilde{w}_1)$. In this *translated coordinate system*, two lines labeled $\bar{v}_0$ and $\bar{v}_1$ can be drawn normal to all the ellipses, as shown in Fig. 24.4. These lines are the *principal axes* of the ellipses, also referred to as the principal axes of the error surface. For a given error surface, these lines are unique. Although an infinite number of lines can be drawn normal to a single ellipse, only lines coincident with the principal axes will be normal to all of the concentric ellipses.

Often it will be convenient to work in the *principal coordinate system*, which uses the performance surface's principal axes $\bar{v}_0, \bar{v}_1, \ldots, \bar{v}_{N-1}$ as its coordinate axes. To change between natural coordinates and translated coordinates, use the transformations

$$\mathbf{v} = \mathbf{w} - \tilde{\mathbf{w}} \tag{24.3}$$

$$\mathbf{w} = \mathbf{v} + \tilde{\mathbf{w}} \tag{24.4}$$

408 Chapter Twenty-Four

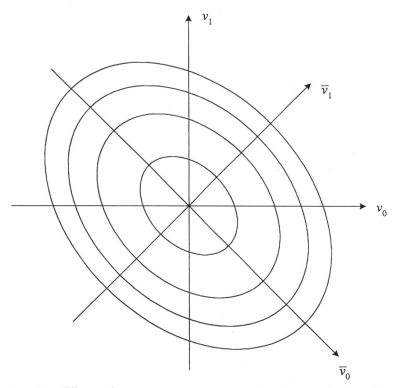

Figure 24.4 Ellipses of constant mean-square error plotted in translated coordinates.

To change between translated coordinates and principal coordinates, use the transformations

$$\bar{\mathbf{v}} = \mathbf{Q}^T \mathbf{v} \tag{24.5}$$

$$\mathbf{v} = \mathbf{Q}\bar{\mathbf{v}} \tag{24.6}$$

and to change between natural coordinates and principal coordinates, use the transformations

$$\bar{\mathbf{v}} = \mathbf{Q}^T(\mathbf{w} - \tilde{\mathbf{w}}) \tag{24.7}$$

$$\mathbf{w} = \mathbf{Q}\bar{\mathbf{v}} + \tilde{\mathbf{w}} \tag{24.8}$$

where $\mathbf{Q}$ is the eigenvector matrix of $\mathbf{R}$

$$\mathbf{Q} = [\mathbf{q}_0 \quad \mathbf{q}_1 \quad \cdots \quad \mathbf{q}_{N-1}]$$

The column vector $\mathbf{q}_n$ is the eigenvector corresponding to the eigenvalue λ_n.

The matrix $\mathbf{Q}$ is not unique in that the ordering of the $\mathbf{q}_n$ will depend on the ordering of the λ_n. If the λ_n are in order of increasing value as n goes from 0 to $N-1$, the resulting $\mathbf{Q}$ will be different than if the λ_n are in order of decreasing value. Different configurations of $\mathbf{Q}$ will result in the different transformations in Eqs. (24.5)–(24.8). This issue will be explored further in Sec. 24.3. However, there is no single, *correct* way to order the λ_n and $\mathbf{q}_n$; so long as one is consistent the choice can be arbitrary.

In the general case, contours of constant error on the hyperparaboloid error surface will be hyperellipses that are parallel to the *weight hyperplane*, the hyperplane defined by the $w_0, w_1, \ldots, w_{N-1}$ axes. When projected onto the weight hyperplane, the form of these hyperellipses is given by

$$\mathbf{w}^T[k]\,\mathbf{R}\,\mathbf{w}[k] - 2\mathbf{p}^T\mathbf{w}[k] = \text{constant}$$

The error surface can be expressed in terms of the three different coordinate systems as

$$\xi = \xi_{\min} + (\mathbf{w} - \tilde{\mathbf{w}})^T \mathbf{R}\,(\mathbf{w} - \tilde{\mathbf{w}}) \qquad (24.9)$$

$$= \xi_{\min} + \mathbf{v}^T \mathbf{R} \mathbf{v} \qquad (24.10)$$

$$= \xi_{\min} + \bar{\mathbf{v}}^T \mathbf{\Lambda} \bar{\mathbf{v}} \qquad (24.11)$$

where $\mathbf{\Lambda} = \text{diag}[\lambda_0, \lambda_1, \ldots, \lambda_{N-1}]$ is a diagonal matrix having the eigenvalues of $\mathbf{R}$ along its diagonal.

The second partial derivatives of the error surface ξ, with respect to the principal axes, can be obtained from the eigenvalues λ_n of the input correlation matrix $\mathbf{R}$

$$\frac{\partial^2 \xi}{\partial \bar{v}_n^2} = 2\lambda_n \quad n = 0, 1, \ldots, N$$

Gradient

The *gradient* $\nabla f_\xi(\mathbf{w}[k])$ is the partial derivative of the MSE with respect to the weight vector

$$\nabla f_\xi(\mathbf{w}[k]) = \frac{\partial \xi}{\partial \mathbf{w}[k]} = \begin{bmatrix} \frac{\partial \xi}{\partial w_0[k]} \\ \frac{\partial \xi}{\partial w_1[k]} \\ \vdots \\ \frac{\partial \xi}{\partial w_{N-1}[k]} \end{bmatrix} \qquad (24.12a)$$

$$= 2\mathbf{R}\mathbf{w}[k] - 2\mathbf{p} \qquad (24.12b)$$

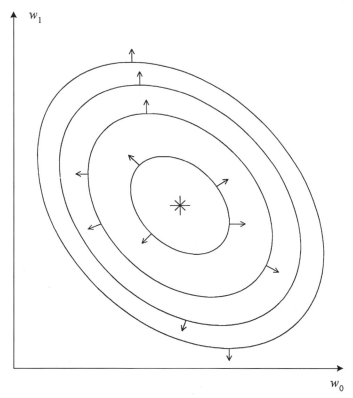

Figure 24.5 Gradients normal to the ellipses of constant mean-square error.

The gradient is a vector in N-space. Among all possible directional derivatives of f_ξ at the point defined by $\mathbf{w}[k]$, the derivative in the direction of $\nabla f_\xi(\mathbf{w}[k])$ has the largest value, and this value is $\|\nabla f_\xi(\mathbf{w}[k])\|$. The derivative in the direction opposite to that of $\nabla f_\xi(\mathbf{w}[k])$ has the smallest value, and this value is $-\|\nabla f_\xi(\mathbf{w}[k])\|$. In other words, the vector $-\nabla f_\xi(\mathbf{w}[k])$ points from $\mathbf{w}[k]$ in the direction of maximum *decrease* in ξ. This fact is the basis of the *method of steepest descent*, to be discussed in Sec. 24.4.

As shown in Fig. 24.5, the gradient at any point will always be normal to the constant error ellipse passing through that point. Although the negative gradient points in the direction of maximum decrease in ξ, it does not in general point toward the global minimum of the error surface. However, the negative gradient does point toward the global minimum when the point at which the gradient is evaluated lies on one of the principal axes of the error surface.

The gradient also can be expressed in terms of the translated coordinate system as

$$\nabla f_\xi(\mathbf{v}[k]) = \frac{\partial \xi}{\partial \mathbf{v}[k]} = 2\mathbf{R}\,\mathbf{v}[k]$$

In the development and analysis of adaptive filters, it is a common practice to use an abbreviated notation of ∇_k to denote $\nabla f_\xi(\mathbf{w}[k])$ or $\nabla f_\xi(\mathbf{v}[k])$.

Gradient estimation

In most adaptive filtering applications, the function $f_\xi(\mathbf{w})$ is not known, so it will not be possible to compute the gradient using Eq. (24.12a). If the gradient is needed, the appropriate partial derivatives must be estimated from the available data. The first two partial derivatives of $f_\xi(\mathbf{w})$ with respect to any single weight w_n can be estimated as

$$\frac{\partial \xi}{\partial w_n} \approx \frac{f_\xi(\mathbf{w} + \mathbf{\Delta}_n) - f_\xi(\mathbf{w} - \mathbf{\Delta}_n)}{2\delta} \tag{24.13}$$

where $\mathbf{\Delta}_n$ is a column vector the same length as $\mathbf{w}$, containing 0 in every row except for row n, which contains the perturbation δ.

24.3 Constructing Test Cases for Adaptive Filters

So far we have considered adaptive filters in the abstract. Before we can explore specific examples of adaptation behavior, we need to establish some convenient test cases that can be used to exercise the various algorithms. In assessing the performance of various adaptive algorithms, it will be helpful to have prior knowledge of the optimal weight values to which the algorithm ideally should converge. For the moment, let us assume that a realization of a random process will be used as input to an adaptive filter that is configured as a linear predictor. Two somewhat different situations can arise depending upon whether the random process is a *regular* random process or a *predictable* random process. In a regular random process, the realizations themselves are random in nature. In a predictable random process, the realizations are deterministic. (See Chap. 21 for more on this distinction.) As we will discover in subsequent sections, the behavior of an adaptive filter will be noticeably different depending upon whether the input signal is a realization of a regular random process or a realization of a predictable random process.

Test signals from regular random processes

The methods of Chap. 23 can be used to construct a parametric model of a random process with known parameters. Specifically, the methods of Sec. 23.2 can be used to generate an autoregressive process that is supplied as input to an adaptive filter configured as a linear predictor. A two-tap adaptive linear predictor is shown in Fig. 24.6, along with a model of an AR(2) process. The optimal values of the adaptive coefficients will be given by

$$\tilde{w}_0 = -a_1 \qquad \tilde{w}_1 = -a_2$$

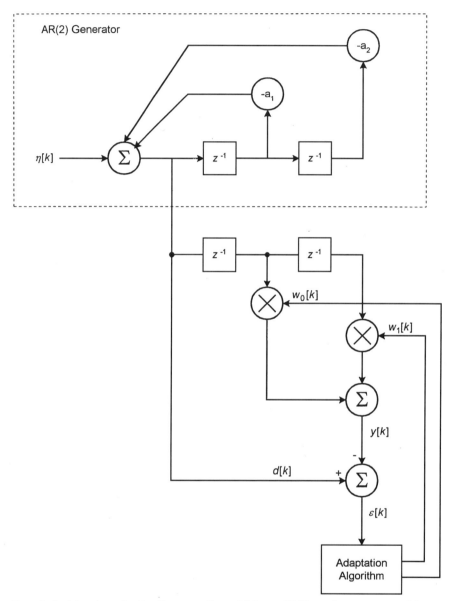

Figure 24.6 A two-tap adaptive linear predictor fed by an AR(2) random process model.

An asymptotically stable autoregressive process of order 2 is a convenient test signal for demonstrating the behavior of various adaptive filters. The correlation matrix for such an AR(2) process will have two eigenvalues, and the convergence behavior of an adaptive filter will change as the ratio between the two eigenvalues is changed. Therefore, there will often be a need to design a test signal that exhibits a specific ratio $\chi = \lambda_1/\lambda_0$. When ellipses of constant MSE are plotted for a given AR(2) process, the ratio of the ellipses' semi major axes a_m and semi minor axes b_m is given by:

$$\frac{a_m}{b_m} = \begin{cases} \sqrt{\chi} & \lambda_1 \geq \lambda_0 \\ \frac{1}{\sqrt{\chi}} & \lambda_1 < \lambda_0 \end{cases} \qquad (24.14)$$

For some test signals, χ can become so large that the diagrams of constant MSE become impractical to draw, with the various ellipses collapsing into near line segments. For illustration purposes, we will want to use ratios in the range $1 < \chi < 6$.

Algorithm 24.1 provides details for the specification and characterization of AR(2) processes for exercising adaptive filters.

Algorithm 24.1 Constructing good AR(2) examples.

This algorithm specifies and characterizes an autoregressive process of the form

$$u[k] = \eta[k] - a_1 u[k-1] - a_2 u[k-2] \qquad (24.15)$$

where η is a discrete-time zero-mean white noise process with variance σ_η^2.

1. Select a value for a_1.
2. Select a value for the desired eigenvalue ratio χ.
3. Determine a_2 using

$$a_2 = \frac{1 - \chi - a_1(\chi + 1)}{\chi - 1} \qquad (24.16)$$

4. Check the resulting process for asymptotic stationarity:

   (a) Form the characterisitic equation

   $$z^2 + a_1 z + a_2 = 0$$

   (b) Find the poles of AR process model's filter by finding the roots of the characteristic equation

   $$p = \frac{-a_1 \pm \sqrt{a_1^2 - 4a_2}}{2} \qquad (24.17)$$

   If both poles have a magnitude less than 1, then the process is asymptotically stationary. If either pole has a magnitude greater than 1, return to step 1.

5. The variance σ_η^2 of the driving white noise process and the variance σ_u^2 of the AR process are related by a function of the coefficients. Select a desired value for either σ_η^2 or σ_u^2 and then determine the other using the appropriate equation

$$\sigma_u^2 = \left(\frac{1+a_2}{1-a_2}\right) \frac{\sigma_\eta^2}{(1+a_2)^2 - a_1^2} \qquad (24.18)$$

$$\sigma_\eta^2 = \sigma_u^2 \left(a_1^2 - a_2^2 + 1 - \frac{2a_1^2}{a_2 + 1}\right) \qquad (24.19)$$

6. The autocorrelation matrix $\mathbf{R}$ is

$$\mathbf{R} = \begin{bmatrix} r_0 & r_1 \\ r_1 & r_0 \end{bmatrix}$$

where

$$r_0 = \sigma_u^2 \qquad (24.20)$$

$$r_1 = \frac{-a_1}{1+a_2}\sigma_u^2 \qquad (24.21)$$

7. Determine the eigenvalues of $\mathbf{R}$ as

$$\lambda_0 = \left(1 + \frac{a_1}{1+a_2}\right)\sigma_u^2 \qquad (24.22)$$

$$\lambda_1 = \left(1 - \frac{a_1}{1+a_2}\right)\sigma_u^2 \qquad (24.23)$$

8. When the λ_n are defined as in step 7, the normalized eigenvector matrix $\mathbf{Q}$ will always be given by

$$\mathbf{Q} \triangleq [\mathbf{q}_0 \ \mathbf{q}_1] = \frac{1}{\sqrt{2}}\begin{bmatrix} 1 & 1 \\ -1 & 1 \end{bmatrix} \qquad (24.24)$$

for any AR(2) process.

9. Equations for the contours of constant MSE can be obtained from Eqs. (24.9) through (24.11), which for the 2-weight case reduce to

$$\xi = \xi_{\min} + r_0(w_0 - \tilde{w}_0)^2 + r_0(w_1 - \tilde{w}_1)^2$$
$$+ 2r_1(w_0 - \tilde{w}_0)(w_1 - \tilde{w}_1) \qquad (24.25)$$

$$= \xi_{\min} + r_0 v_0^2 + r_0 v_1^2 + 2r_1 v_0 v_1 \qquad (24.26)$$

$$= \xi_{\min} + \bar{v}_0^2 \lambda_0 + \bar{v}_1^2 \lambda_1 \qquad (24.27)$$

where $\xi_{\min} = \sigma_\eta^2$.

As given previously in Eq. (24.5), the weight vector is transformed from translated coordinates into principal coordinates using

$$\bar{\mathbf{v}} = \mathbf{Q}^T \mathbf{v}$$

When $\mathbf{Q}$ is defined as in Eq. (24.24), the resulting transformation from v to $\bar{v}$ is a simple rotation by $-45°$, as shown in Fig. 24.7. In Haykin [2], the definitions of λ_0 and λ_1 are reversed and the matrix $\mathbf{Q}$ becomes

$$\mathbf{Q} = \begin{bmatrix} 1 & 1 \\ 1 & -1 \end{bmatrix}$$

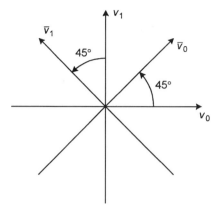

Figure 24.7 Principal coordinates $\bar{\mathbf{v}}$ obtained as a simple rotation of the translated coordinates **v**.

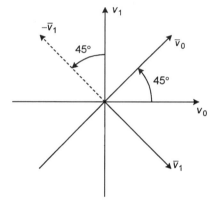

Figure 24.8 Principal coordinates $\bar{\mathbf{v}}$ obtained as a rotation of the translated coordinates **v** followed by a negation of the rotated v_1 axis.

Using this matrix in Eq. (24.5) does not result in a transformation that is a simple rotation. This transformation corresponds to a rotation by $+45°$, followed by a negation of the v_1, as depicted in Fig. 24.8.

Example 24.1 Construct and characterize an AR(2) process having $a_1 = -0.9$, $\chi = 2$, and $\sigma_u^2 = 1.0$.

solution Using Eq. (24.16), determine a_2 as

$$a_2 = \frac{1 - 2 - (-0.9)(2+1)}{2 - 1} = 0.35$$

The roots of the characteristic equation $z^2 - 0.9x + 1.7 = 0$ are obtained using Eq. (24.17):

$$p = \frac{0.9 \pm \sqrt{0.81 - 6.8}}{2} = 0.45 \pm 1.22372j$$

$$|p| = 1.7$$

The magnitude of the poles is greater than 1, so the process is not asymptotically stationary. It is not possible to construct a stationary AR(2) process with the given values of a_1 and χ.

Example 24.2 Construct and characterize an AR(2) process having $a_1 = -0.6$, $\chi = 2$, and $\sigma_u^2 = 1.0$.

solution Using Eq. (24.16), determine a_2 as

$$a_2 = \frac{1 - 2 - (-0.6)(2+1)}{2-1} = 0.8$$

The roots of the characteristic equation $z^2 - 0.6x + 0.8 = 0$ are obtained using Eq. (24.17):

$$p = \frac{0.6 \pm \sqrt{0.36 - 3.2}}{2} = 0.3 \pm 0.8426j$$

$$|p| = 0.89441$$

The magnitude of the poles is less than 1, so the process is asymptotically stationary. Using Eq. (24.19) we find

$$\sigma_\eta^2 = 0.32 \sigma_u^2 = 0.32$$

Using Eqs. (24.22) and (24.23), the eigenvalues are found to be

$$\lambda_0 = 0.6666 \qquad \lambda_1 = 1.3333$$

Finally, the constant-MSE ellipses are defined in terms of the principal coordinates by the equation

$$\xi = \xi_{\min} + \bar{v}_0^2 \lambda_0 + \bar{v}_1^2 \lambda_1$$

$$= 0.32 + 0.6666\,\bar{v}_0^2 + 1.3333\,\bar{v}_1^2$$

When the initial weight vector $\mathbf{w}[0] = \mathbf{0}$, we have

$$\bar{\mathbf{v}}[0] = \mathbf{Q}^T(\mathbf{w}[0] - \tilde{\mathbf{w}}) = \begin{bmatrix} 1 & -1 \\ 1 & 1 \end{bmatrix} \begin{bmatrix} -0.6 \\ 0.8 \end{bmatrix} = \begin{bmatrix} -0.98995 \\ 0.14142 \end{bmatrix}$$

and the initial MSE is

$$\xi[0] = 0.32 + 0.6666(0.98995)^2 + 1.3333(0.14142)^2$$

$$= 0.99993$$

The constant-MSE ellipse for $\xi = 0.99993$ has a semimajor axis $a[0]$ and semiminor axis $b[0]$ given by

$$a[0] = \sqrt{\frac{0.99993 - 0.32}{0.6666}} = 1.00995$$

$$b[0] = \sqrt{\frac{0.99993 - 0.32}{1.3333}} = 0.71412$$

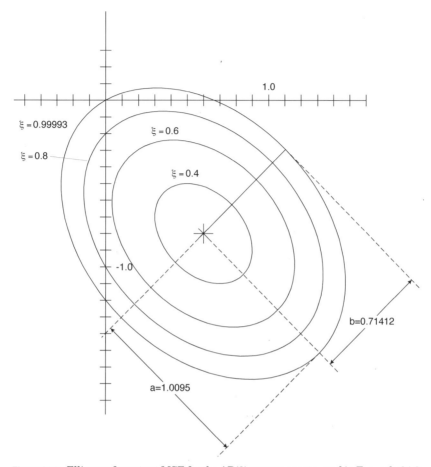

Figure 24.9 Ellipses of constant MSE for the AR(2) process constructed in Example 24.2.

The ratio of $a[0]/b[0]$ equals $\sqrt{2}$, which agrees with Eq. (24.14). A plot containing several constant MSE ellipses for this process is shown in Fig. 24.9.

Test signals from predictable random processes

Using trigonometric identities it can be shown that

$$\cos\left(\frac{2\pi k}{N}\right) = w_0 \sin\left(\frac{2\pi k}{N}\right) + w_1 \sin\left(\frac{2\pi(k-1)}{N}\right) \qquad (24.28)$$

where

$$w_0 = \cot\frac{2\pi}{N} \qquad w_1 = -\csc\frac{2\pi}{N} \qquad N > 2$$

Based on Eq. (24.28), we can construct the two-tap adaptive filter shown in

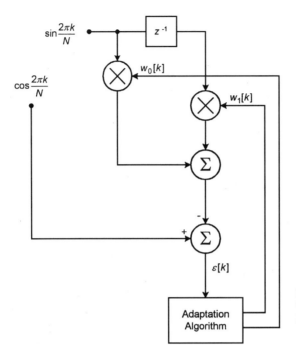

Figure 24.10 A two-weight ALC used to estimate a sinusoidal process.

Fig. 24.10. The correlation matrix $\mathbf{R}$ for the input to this filter is given by

$$\mathbf{R} = \begin{bmatrix} r_0 & r_1 \\ r_1 & r_0 \end{bmatrix}$$

$$r_0 = \frac{1}{2}$$

$$r_1 = \frac{1}{2}\cos\frac{2\pi}{N}$$

The eigenvalues λ_n are found by solving

$$\det[\mathbf{R} - \lambda\mathbf{I}] = 0$$

$$\left(\frac{1}{2} - \lambda\right)^2 - \left(\frac{1}{2}\cos\frac{2\pi}{N}\right)^2 = 0$$

$$\lambda^2 - \lambda + \frac{1}{4}\sin^2\left(\frac{2\pi}{N}\right) = 0$$

$$\lambda = \frac{1}{2} \pm \frac{1}{2}\cos\frac{2\pi}{N}$$

As in the case for regular random processes, there often will be a need to design a test signal that exhibits a specific ratio $\chi = \lambda_1/\lambda_0$. It turns out that the ratio varies depending upon the value of N:

$$\chi \triangleq \frac{\lambda_1}{\lambda_0} = \frac{1 + \cos(2\pi/N)}{1 - \cos(2\pi/N)}$$

$$\left(1 - \cos\frac{2\pi}{N}\right)\chi = 1 + \cos\frac{2\pi}{N}$$

$$\chi - 1 = (1 + \chi)\cos\frac{2\pi}{N}$$

$$\cos^{-1}\frac{\chi - 1}{\chi + 1} = \frac{2\pi}{N}$$

$$N \approx \frac{2\pi}{\cos^{-1}\frac{\chi-1}{\chi+1}}$$

Algorithm 24.2 provides details for the specification and characterization of a sinusoidal process for exercising adaptive filters.

Algorithm 24.2 Constructing a second-order autoregressive model of a sinusoid.

This algorithm characterizes the following autoregressive model of a sinusoidal process

$$\cos\frac{2\pi k}{N} = w_0 \sin\frac{2\pi k}{N} + w_1 \sin\frac{2\pi(k-1)}{N}$$

1. Select a value for the desired eigenvalue ratio χ.
2. Determine an integer value for N using

$$N \approx \frac{2\pi}{\cos^{-1}\left(\frac{\chi-1}{\chi+1}\right)} \quad (24.29)$$

3. Determine the optimal weight values $\tilde{w}_0$ and $\tilde{w}_1$ as

$$\tilde{w}_0 = \cot\frac{2\pi}{N} \quad (24.30)$$

$$\tilde{w}_1 = -\csc\frac{2\pi}{N} \quad (24.31)$$

4. The correlation matrix $\mathbf{R}$ is

$$\mathbf{R} = \begin{bmatrix} r_0 & r_1 \\ r_1 & r_0 \end{bmatrix}$$

where

$$r_0 = \frac{1}{2} \tag{24.32}$$

$$r_1 = \frac{1}{2}\cos\frac{2\pi}{N} \tag{24.33}$$

5. Determine the eigenvalues of **R** as

$$\lambda_0 = \frac{1}{2}\left(1 - \cos\frac{2\pi}{N}\right) \tag{24.34}$$

$$\lambda_1 = \frac{1}{2}\left(1 + \cos\frac{2\pi}{N}\right) \tag{24.35}$$

6. When the λ_n are defined as in step 5, the normalized eigenvector matrix **Q** will be given by

$$\mathbf{Q} = \frac{1}{\sqrt{2}}\begin{bmatrix} 1 & 1 \\ -1 & 1 \end{bmatrix} \tag{24.36}$$

7. Equation (24.28) is exact; therefore, $\xi_{\min} = 0$ and the contours of constant MSE can be obtained from Eqs. (24.9) through (24.11) which reduce to

$$\xi = r_0(w_0 - \tilde{w}_0)^2 + r_0(w_1 - \tilde{w}_1)^2 + 2r_1(w_0 - \tilde{w}_0)(w_1 - \tilde{w}_1) \tag{24.37}$$

$$= r_0 v_0^2 + r_0 v_1^2 + 2r_1 v_0 v_1 \tag{24.38}$$

$$= \bar{v}_0^2 \lambda_0 + \bar{v}_1^2 \lambda_1 \tag{24.39}$$

Example 24.3 Use Algorithm 24.2 to construct and characterize a sinusoidal signal with $\chi \approx 3$ for testing a two-tap adaptive filter.

solution Using Eq. (24.29), we find the desired value for N as

$$N = \frac{2\pi}{\cos^{-1}\left(\frac{3-1}{3+1}\right)} = 6$$

The optimal weight values are

$$\tilde{w}_0 = \cot\frac{\pi}{3} = 0.57735$$

$$\tilde{w}_1 = -\csc\frac{\pi}{3} = -1.154701$$

The correlation values r_0 and r_1 are

$$r_0 = \frac{1}{2} \qquad r_1 = \frac{1}{2}\cos\frac{\pi}{3} = \frac{1}{4}$$

The eigenvalues of $\mathbf{R}$ are obtained as

$$\lambda_0 = \frac{1}{2}\left(1 - \cos\frac{\pi}{3}\right) = 0.25$$

$$\lambda_1 = \frac{1}{2}\left(1 + \cos\frac{\pi}{3}\right) = 0.75$$

The constant-MSE ellipses are defined in terms of the principal coordinates by the equation

$$\xi = \bar{v}_0^2 \lambda_0 + \bar{v}_1^2 \lambda_1$$
$$= 0.25\, \bar{v}_0^2 + 0.75\, \bar{v}_1^2$$

When the initial weight vector $\mathbf{w}[0] = \mathbf{0}$, the corresponding values for $\bar{\mathbf{v}}[0]$ are obtained as

$$\bar{\mathbf{v}}[0] = \mathbf{Q}^{\mathrm{T}}(\mathbf{w}[0] - \tilde{\mathbf{w}}) = \frac{1}{\sqrt{2}}\begin{bmatrix} 1 & -1 \\ 1 & 1 \end{bmatrix}\begin{bmatrix} -0.57735 \\ 1.154701 \end{bmatrix} = \begin{bmatrix} -1.224745 \\ 0.408248 \end{bmatrix}$$

and the initial MSE is

$$\xi[0] = 0.25(1.224745)^2 + 0.75(0.408248)^2$$
$$= 0.5$$

The constant-MSE ellipse for $\xi = \xi[0] = 0.5$ has a semimajor axis $a[0]$ and semiminor axis $b[0]$ given by

$$a[0] = \sqrt{\frac{\xi[0]}{\lambda_0}} = 1.4142$$

$$b[0] = \sqrt{\frac{\xi[0]}{\lambda_1}} = 0.816497$$

A plot containing several constant-MSE ellipses for this process is provided in Fig. 24.11.

Test signals from mixed random processes

Any random process can be decomposed into the sum of a regular random process and a predictable random process. Sometimes the term *mixed random process* is used to emphasize the mixed nature of a particular random process that is not purely random or not purely predictable. Consider a process that is essentially a predictable random process (such as a sinusoid) with a relatively small amount of added noise. The two-tap adaptive filter system given in Fig. 24.10 can be modified to include added noise as shown in Fig. 24.12.

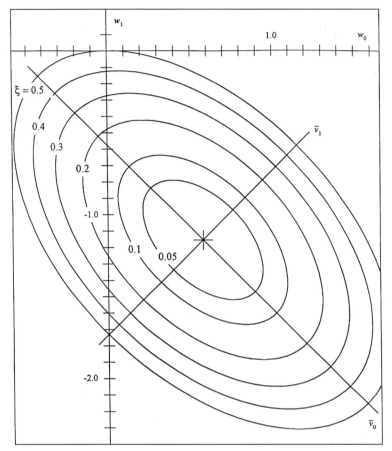

Figure 24.11 Ellipses of constant MSE for the sinusoidal process constructed in Example 24.3.

The addition of the noise results in a few changes to the characterization of the process as indicated in Algorithm 24.3.

Algorithm 24.3 Constructing a second-order autoregressive model of a sinusoid plus noise.

This algorithm characterizes the mixed random process corresponding to the signal labeled $x[k]$ in Fig. 24.12.

1. Select a value for the desired eigenvalue ratio χ.
2. Determine an integer value for N using

$$N \approx \frac{2\pi}{\cos^{-1}\left(\frac{(\chi-1)(1+\phi)}{\chi+1}\right)} \qquad (24.40)$$

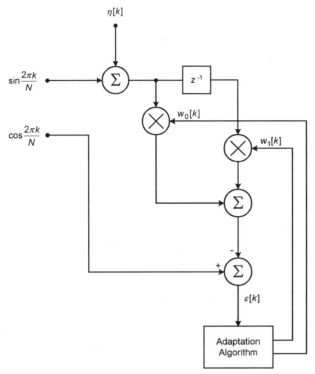

Figure 24.12 A two-weight ALC used to estimate a sinusoidal process from noisy samples.

3. Determine the optimal weight values $\tilde{w}_0$ and $\tilde{w}_1$ as

$$\tilde{w}_0 = \frac{\cos\frac{2\pi}{N}\sin\frac{2\pi}{N}}{(1+2\phi)^2 - \cos^2\left(\frac{2\pi}{N}\right)} \tag{24.41}$$

$$\tilde{w}_1 = \frac{-(1+2\phi)\sin\frac{2\pi}{N}}{(1+2\phi)^2 - \cos^2\left(\frac{2\pi}{N}\right)} \tag{24.42}$$

4. Assuming that the random noise samples are statistically independent, the correlation matrix **R** of the input to the filter is given by

$$\mathbf{R} = \begin{bmatrix} r_0 & r_1 \\ r_1 & r_0 \end{bmatrix}$$

$$r_0 = \frac{1}{2} + \phi \tag{24.43}$$

$$r_1 = \frac{1}{2}\cos\frac{2\pi}{N} \tag{24.44}$$

where ϕ is the power of the random noise

$$\phi = E\{\eta^2[k]\}$$

5. Determine the eigenvalues of $\mathbf{R}$ as

$$\lambda_0 = \frac{1}{2}\left(1 + 2\phi - \cos\frac{2\pi}{N}\right) \tag{24.45}$$

$$\lambda_1 = \frac{1}{2}\left(1 + 2\phi + \cos\frac{2\pi}{N}\right) \tag{24.46}$$

6. When the λ_n are defined as in step 5, the normalized eigenvector matrix $\mathbf{Q}$ will be given by

$$\mathbf{Q} = \frac{1}{\sqrt{2}}\begin{bmatrix} 1 & 1 \\ -1 & 1 \end{bmatrix}$$

7. The contours of constant MSE can be obtained from the equation

$$\xi = \left(\frac{1}{2} + \phi\right)(w_0^2 + w_1^2) + w_0 w_1 \cos\left(\frac{2\pi}{N}\right) + w_1 \sin\left(\frac{2\pi}{N}\right) + \frac{1}{2} \tag{24.47}$$

Find $\xi_{\min}$ by substituting $w = \tilde{w}$ and the given values for ϕ and N in Eq. (24.47).

8. The ellipses of constant MSE now can be characterized in a more convenient form by substituting $\xi_{\min}$ from step 7 and λ_0, λ_1 from step 5 into

$$\xi = \xi_{\min} + \bar{v}_0^2 \lambda_0 + \bar{v}_1^2 \lambda_1 \tag{24.48}$$

Example 24.4 Use Algorithm 24.3 to construct and characterize a sinusoidal signal plus noise for testing a two-tap adaptive filter. The noise variance is $\phi = 0.05$ and the desired eigenvalue ratio is $\chi = 3$.

solution Using Eq. (24.40), we find the desired value for N as

$$N \approx \frac{2\pi}{\cos^{-1}\left(\frac{(2)(1.05)}{4}\right)} = 6.1715$$

Since N must be an integer, we can choose either $N = 6$ or $N = 7$. When $N = 6$, the actual eigenvalue ratio is $\chi = 2.818$, and when $N = 7$ the actual ratio is $\chi = 3.923$. We should select $N = 6$ because it yields the ratio closest to the desired value of $\chi = 3$. Using Eqs. (24.41) and (24.41), we find the optimal weights

$$\tilde{\mathbf{w}} = \begin{bmatrix} 0.451055 \\ -0.992321 \end{bmatrix}$$

The correlation values r_0 and r_1 are

$$r_0 = 0.55 \quad r_1 = 0.25$$

The eigenvalues of **R** are obtained as

$$\lambda_0 = \frac{1}{2}\left(1.1 - \cos\frac{\pi}{3}\right) = 0.3$$

$$\lambda_1 = \frac{1}{2}\left(1.1 + \cos\frac{\pi}{3}\right) = 0.8$$

The minimum MSE $\xi_{\min} = 0.070313$ is found by substituting $\tilde{\mathbf{w}}$ into Eq. (24.47). Thus the constant-MSE ellipses are defined in terms of the principal coordinates by

$$\xi = 0.070313 + 0.3\bar{v}_0^2 + 0.8\bar{v}_1^2$$

When the initial weight vector $\mathbf{w}[0] = \mathbf{0}$, the corresponding values for $\bar{\mathbf{v}}[0]$ are obtained as

$$\begin{aligned}
\bar{\mathbf{v}}[0] &= \mathbf{Q}^{\mathrm{T}}(\mathbf{w}[0] - \tilde{\mathbf{w}}) \\
&= \frac{1}{\sqrt{2}}\begin{bmatrix} 1 & -1 \\ 1 & 1 \end{bmatrix}\begin{bmatrix} -0.451055 \\ 0.992321 \end{bmatrix} \\
&= \begin{bmatrix} -1.020621 \\ 0.382733 \end{bmatrix}
\end{aligned}$$

and the initial MSE is

$$\begin{aligned}
\xi[0] &= 0.070313 + 0.3(1.020621)^2 + 0.8(0.382733)^2 \\
&= 0.5
\end{aligned}$$

A plot containing several constant-MSE ellipses for this process is provided in Fig. 24.13.

24.4 Method of Steepest Descent

At any point on an adaptive filter's error surface, the negative gradient points in the direction of *steepest descent*, that is, the direction of greatest decrease in ξ. Starting at the point determined by the weight vector $\mathbf{w}[k]$, the method of *steepest descent* follows the negative gradient for a short distance (as regulated by the stepsize parameter μ) to a new point determined by the new weight vector $\mathbf{w}[k+1]$

$$\mathbf{w}[k+1] = \mathbf{w}[k] + \mu(-\nabla_k) \tag{24.49}$$

The method then follows the negative gradient (evaluated at $\mathbf{w}[k+1]$) for a short distance to determine $\mathbf{w}[k+2]$. This process is repeated until the weight vector reaches the optimum $\tilde{\mathbf{w}}$. In general, the weights will follow a smooth curved trajectory from $\mathbf{w}[0]$ to $\tilde{\mathbf{w}}$. In the special case where the initial weight

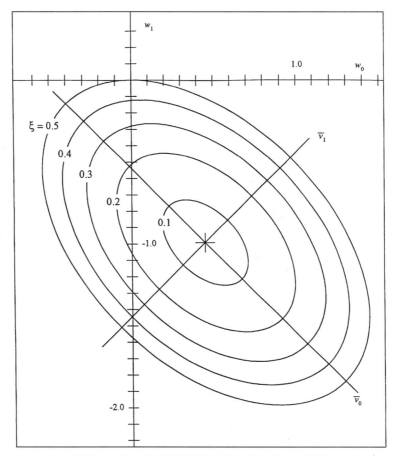

Figure 24.13 Ellipses of constant MSE for the noisy sinusoidal process constructed in Example 24.4.

vector lies on one of the principal axes of the performance surface (and the exact gradients are used) the trajectory will be a straight line from $\mathbf{w}[0]$ to $\tilde{\mathbf{w}}$.

The need for direct evaluation of the gradient can be eliminated by substituting (24.12b) for ∇_k to yield

$$\mathbf{w}[k+1] = \mathbf{w}[k] + 2\mu(\mathbf{p} - \mathbf{R}\mathbf{w}[k]) \qquad (24.50)$$

This equation requires knowledge of $\mathbf{p}$ and $\mathbf{R}$. In some applications, it might be feasible to estimate $\mathbf{p}$ and $\mathbf{R}$ from the available data. This is occasionally done in non-real-time analysis of time series data, but it is almost never done in practical real-time adaptive filters. The true utility of Eq. (24.50) lies in the insights into adaptive filter operation and performance measures that can be obtained from postulated scenarios when $\mathbf{p}$ and $\mathbf{R}$ are known exactly. Some of these performance measures are developed in Sec. 24.5.

In practical real-time application of the steepest descent method, the gradient used in Eq. (24.49) can be estimated by using Eq. (24.13) to estimate the partial derivatives in (24.12a). The values of $f_\xi(\mathbf{w} \pm \mathbf{\Delta}_n)$ are estimated by setting the weight vector to $\mathbf{w} \pm \mathbf{\Delta}_n$ and measuring $\varepsilon[k]$. Ergodicity is assumed and a short-time average of $\varepsilon^2[k]$ values is used in place of the expected value $E\{\varepsilon^2[k]\}$ in Eq. (24.2).

Stability

It can be shown (e.g. [1]) that the steepest descent algorithm is stable and convergent if and only if μ satisfies

$$0 < \mu < \frac{1}{\lambda_{\max}}$$

where $\lambda_{\max}$ is the largest eigenvalue of the input correlation matrix $\mathbf{R}$.

Example 24.5 Using the exact gradient, apply the method of steepest descent to the AR(2) process defined in Example 24.2. Use three different starting points: ($w_0 = 0.0, w_1 = 0.0$), ($w_0 = 0.6666, w_1 = 0.0$), and the point that lies at the lower right end of principal axis $\tilde{\mathbf{v}}_0$.

solution Obtain an expression for ξ in terms of the natural coordinate weight vector $\mathbf{w}$. Starting with Eq. (24.9)

$$\xi = \xi_{\min} + (\mathbf{w} - \tilde{\mathbf{w}})^T \mathbf{R} (\mathbf{w} - \tilde{\mathbf{w}})$$

we substitute

$$\xi_{\min} = 0.32$$

$$\tilde{\mathbf{w}} = \begin{bmatrix} 0.6 \\ -0.8 \end{bmatrix}$$

$$r_0 = \sigma_u^2 = 1.0$$

$$r_1 = \frac{-a_1}{1+a_2}\sigma_u^2 = \frac{1}{3}$$

to obtain

$$\xi = w_0^2 + w_1^2 - \frac{2}{3}w_0 + \frac{6}{5}w_1 + \frac{2}{3}w_0 w_1 + 1$$

Differentiating with respect to w_0 and w_1 yields the gradient

$$\nabla = \begin{bmatrix} 2w_0 + \frac{2}{3}w_1 - \frac{2}{3} \\ 2w_1 + \frac{2}{3}w_0 + \frac{6}{5} \end{bmatrix} \quad (24.51)$$

File ex24_05.cpp contains a simple program that uses Eq. (24.51) to implement the method of steepest descent for the given AR(2) process. Two of the specified starting points can be used directly as inputs to this program. Some manipulations are necessary to find the natural coordinates of the third starting point. The major principal axis and the ellipse for $\xi = 0.99993$ intersect at the point defined by ($\bar{v}_0 = 1.00995, \bar{v}_1 = 0.0$). The corresponding natural coordinates are obtained by using Eq. (24.8)

$$\mathbf{w} = \mathbf{Q}\bar{\mathbf{v}} + \tilde{\mathbf{w}}$$

$$= \frac{1}{\sqrt{2}} \begin{bmatrix} 1 & 1 \\ -1 & 1 \end{bmatrix} \begin{bmatrix} 1.00995 \\ 0.0 \end{bmatrix} + \begin{bmatrix} 0.6 \\ -0.8 \end{bmatrix}$$

$$= \begin{bmatrix} 1.31414 \\ -1.51414 \end{bmatrix}$$

Figure 24.14 shows the weight-convergence trajectories for the three specified starting points.

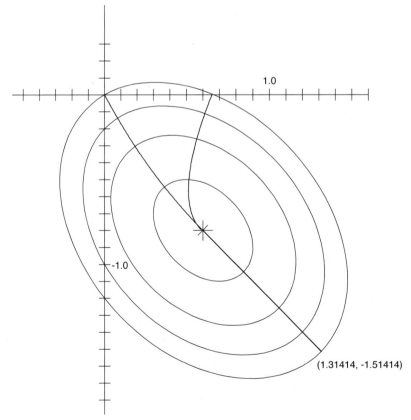

Figure 24.14 Ellipses of constant MSE for the steepest descent method with two weights. The curved trajectory from $\mathbf{w}[0]$ to $\tilde{\mathbf{w}}$ is normal to each ellipse.

24.5 Performance Measures

Learning curve

A plot of MSE versus iteration number is called the *learning curve* of an adaptive filter. For some adaptation methods, such as the method of steepest descent, it is possible to express the MSE at iteration k in closed form. For other methods, such as the LMS algorithm discussed in Sec. 24.6, the learning curve must be determined empirically. Even if the learning curve itself must be determined empirically, some characteristics of the curve, such as time constants, may still be determined analytically.

For the case of the steepest-descent method, the MSE at iteration k is given by

$$\xi[k] = \xi_{\min} + \sum_{n=0}^{N-1} (\bar{v}_n[0])^2 \lambda_n (1 - 2\mu\lambda_n)^{2k} \quad (24.52)$$

where the λ_n are the eigenvalues of $\mathbf{R}$, and the $\bar{v}_n[0]$ are the initial tap weights expressed in terms of the principal coordinate system.

Example 24.6 Plot the learning curve for the steepest-descent method as applied to the specified AR(2) process in Example 24.5. Start at $(\bar{v}_0 = 1.00995, \bar{v}_1 = 0.0)$, and use $\mu = 0.01$.

solution File ex24_06.cpp contains a simple program that uses Eq. (24.52) to generate the learning curve for the given AR(2) process. The resulting curve is shown in Fig. 24.15.

Misadjustment and excess MSE

The optimum weight vector $\tilde{\mathbf{w}}$ that minimizes the MSE can be found analytically by setting the gradient equal to zero and solving for $\mathbf{w}$:

$$\nabla f_\xi(\mathbf{w}) = 2\mathbf{R}\mathbf{w} - 2\mathbf{p} = 0 \quad (24.53)$$

$$\tilde{\mathbf{w}} = \mathbf{R}^{-1}\mathbf{p}$$

This result is simply the Weiner-Hopf equation. The minimum MSE achieved when the weight vector is optimized is given by

$$\xi_{\min} = E\{d^2[k]\} - \mathbf{p}^T\mathbf{R}^{-1}\mathbf{p} \quad (24.54)$$

$$= E\{d^2[k]\} - \mathbf{p}^T\tilde{\mathbf{w}}$$

(A detailed derivation of this result can be found in [1].) For a given combination of $\mathbf{R}$, $\mathbf{p}$, and $E\{d^2[k]\}$, the MSE cannot be reduced below this value of $\xi_{\min}$. For

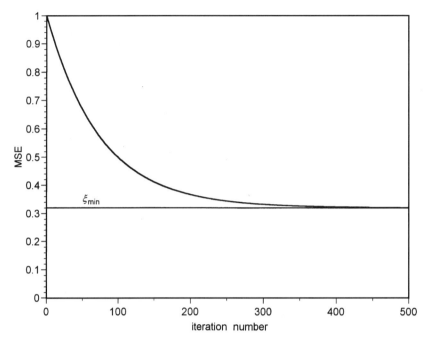

Figure 24.15 Learning curve for Example 24.6.

an ideal, noise-free adaptation algorithm, the weight vector will converge to $\tilde{\mathbf{w}}$ and the MSE can be made equal to $\xi_{\min}$. However, for practical algorithms that include some noise, there will be additional error due to small deviations of the weight vector around the optimal value of $\tilde{\mathbf{w}}$. The *excess mean-square error* is defined as

$$\text{excess MSE} = E[\xi[k] - \xi_{\min}]$$

$$= E[(\mathbf{w}[k] - \tilde{\mathbf{w}})^T \mathbf{R}(\mathbf{w}[k] - \tilde{\mathbf{w}})] \quad (24.55)$$

$$= E[\mathbf{v}^T[k]\, \mathbf{R}\, \mathbf{v}[k]] \quad (24.56)$$

$$= E[\bar{\mathbf{v}}^T[k]\, \mathbf{\Lambda}\, \bar{\mathbf{v}}[k]] \quad (24.57)$$

where $\mathbf{\Lambda} = \text{diag}[\lambda_0, \lambda_1, \ldots, \lambda_{N-1}]$ with λ_n being the eigenvalues of the input correlation matrix $\mathbf{R}$. Equations (24.55), (24.56), and (24.57) respectively are in terms of natural coordinates, translated coordinates, and principal coordinates.

A normalized measure of excess error is the *misadjustment M*, defined as

$$M \triangleq \frac{(\text{excess MSE})}{\xi_{\min}} \quad (24.58)$$

24.6 The LMS Algorithm

The least-mean-square or LMS algorithm, first introduced in 1960 by Widrow and Hoff [4], is the most widely used of all the adaptive filtering algorithms that have been developed. This algorithm is relatively simple to implement, yet it provides good performance in a wide range of applications.

The weight update equation for the method of steepest descent discussed in Sec. 24.4 is

$$\mathbf{w}[k+1] = \mathbf{w}[k] - \mu \nabla_k \tag{24.59}$$

where the gradient defined in Eq. (24.12a) can be expanded as:

$$\nabla_k = \frac{\partial \xi[k]}{\partial \mathbf{w}[k]} = \begin{bmatrix} \frac{\partial \xi[k]}{\partial w_0[k]} \\ \frac{\partial \xi[k]}{\partial w_1[k]} \\ \vdots \\ \frac{\partial \xi[k]}{\partial w_{N-1}[k]} \end{bmatrix} = \begin{bmatrix} \frac{\partial}{\partial w_0[k]} E\{\varepsilon^2[k]\} \\ \frac{\partial}{\partial w_1[k]} E\{\varepsilon^2[k]\} \\ \vdots \\ \frac{\partial}{\partial w_{N-1}[k]} E\{\varepsilon^2[k]\} \end{bmatrix}$$

If we neglect the expectation and take $\varepsilon^2[k]$ as an estimate of $\xi[k]$, we obtain a gradient estimate of the form

$$\hat{\nabla}_k = \frac{\partial \varepsilon^2[k]}{\partial \mathbf{w}[k]} = \begin{bmatrix} \frac{\partial \varepsilon^2[k]}{\partial w_0[k]} \\ \frac{\partial \varepsilon^2[k]}{\partial w_1[k]} \\ \vdots \\ \frac{\partial \varepsilon^2[k]}{\partial w_{N-1}[k]} \end{bmatrix} \tag{24.60}$$

By the chain rule

$$\frac{\partial \varepsilon^2}{\partial w} = \varepsilon \frac{\partial \varepsilon}{\partial w} + \varepsilon \frac{\partial \varepsilon}{\partial w} = 2\varepsilon \frac{\partial \varepsilon}{\partial w}$$

Since
$$\varepsilon[k] = d[k] - \sum_{m=0}^{N-1} x[k-m] w_m[k]$$

the chain rule can be used to obtain the derivative of $\varepsilon[k]$ with respect to weight $w_n[k]$ as

$$\frac{\partial \varepsilon[k]}{\partial w_n[k]} = -x[k-n]$$

Therefore Eq. (24.60) can be rewritten as

$$\hat{\nabla}_k = 2\varepsilon[k] \begin{bmatrix} \frac{\partial \varepsilon[k]}{\partial w_0[k]} \\ \frac{\partial \varepsilon[k]}{\partial w_1[k]} \\ \vdots \\ \frac{\partial \varepsilon[k]}{\partial w_{N-1}[k]} \end{bmatrix} = -2\varepsilon[k]\,\mathbf{x}[k]$$

Substituting this estimate for the gradient in Eq. (24.59) yields

$$\mathbf{w}[k+1] = \mathbf{w}[k] + 2\mu\,\varepsilon[k]\,\mathbf{x}[k] \qquad (24.61)$$

Equation (24.61) defines the LMS algorithm.

Example 24.7 Apply the LMS method with $\mu = 0.05$ to the AR(2) process defined in Example 24.2. Plot (1) the learning curve averaged over 550 trials, (2) the MSE versus iteration number for a single trial, and (3) the convergence trajectory in the weight plane for a single trial.

solution File ex24_07.cpp contains a program that queries for the required parameters and then creates an instance of the class LmsFilter that inherits from the class AdaptiveFir. The LMS tap updates are performed by the LmsFilter::UpdateTaps method provided in Listing 24.1. The input signal is generated using an instance of class ArSource which was discussed in Chap. 23. The learning curve averaged over 550 trials is shown in Fig. 24.16. The MSE quickly drops to a value near $\xi_{\min}$ and then exhibits relatively small (in the range $\xi_{\min}$ to $\xi_{\min} + 0.1$) fluctuations. At first this looks like a fairly good adaptation until we look at the single-trial MSE plot in Fig. 24.17 and convergence trajectory in Fig. 24.18. These plots show that without the averaging of the learning curve, the MSE can fluctuate wildly and the weight trajectory is quite erratic, covering a large area around the optimal weight setting. These fluctuations can be reduced by using a smaller value of μ.

Example 24.8 Repeat Example 24.7 with $\mu = 0.005$.

solution The program contained in file ex24_07.cpp also can be used for this example. The learning curve averaged over 550 trials is shown in Fig. 24.19. Notice that the MSE does not get close to $\xi_{\min}$ until around iteration 200. Compare this to Fig. 24.16 in which the MSE gets close to $\xi_{\min}$ in less than 50 iterations. The MSE for a single trial, shown in Fig. 24.20, is still quite erratic; but the convergence trajectory in Fig. 24.21 exhibits a greatly reduced amount of *bottom of the bowl* noise around the optimum weight point.

Example 24.9 Apply the LMS method with $\mu = 0.05$ to the noisy sinusoidal process defined in Example 24.4.

solution File ex24_09.cpp contains a program that queries for the required parameters and then constructs an instance of class LmsFilter. The sinusoidal part of the noisy signal is generated directly in the main program. The learning curve is shown

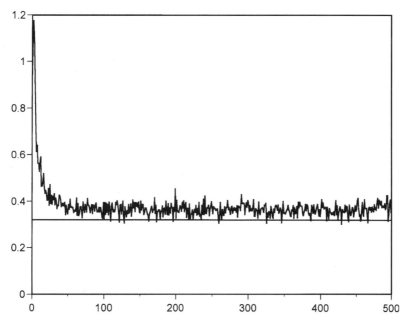

Figure 24.16 Learning curve for Example 24.7.

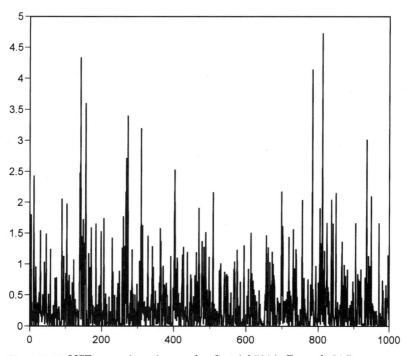

Figure 24.17 MSE versus iteration number for trial 501 in Example 24.7.

434 Chapter Twenty-Four

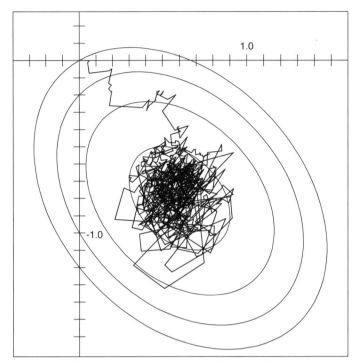

Figure 24.18 Weight-convergence trajectory for Example 24.7.

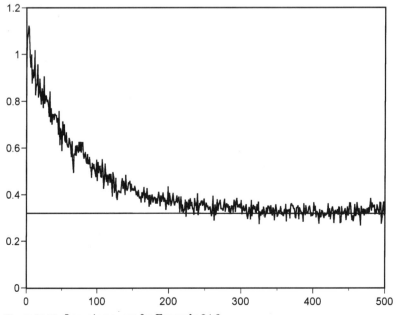

Figure 24.19 Learning curve for Example 24.8.

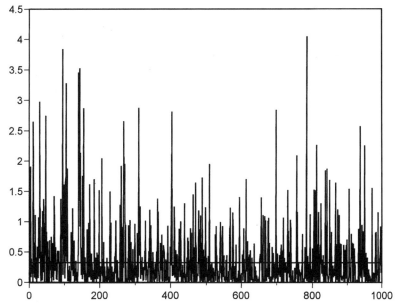

Figure 24.20 MSE versus iteration number for trial 501 in Example 24.8.

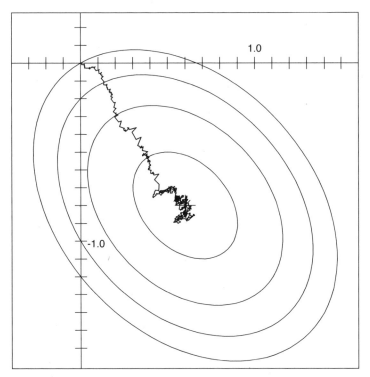

Figure 24.21 Weight-convergence trajectory for Example 24.8.

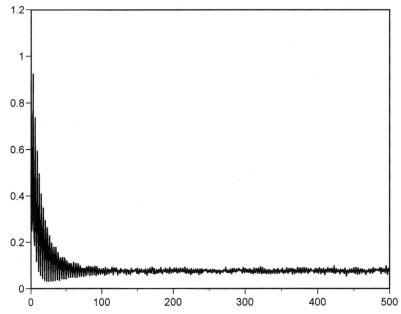

Figure 24.22 Learning curve for Example 24.9.

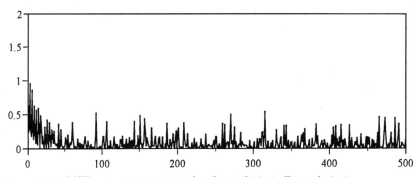

Figure 24.23 MSE versus iteration number for trial 501 in Example 24.9.

in Fig. 24.22. The time constant of this curve appears to be about the same as the time constant for Example 24.7, but the random fluctuations are much smaller. As we might expect this is indicative of smaller fluctuations in the single-trial MSE curve and the convergence trajectory, which are shown in Fig. 24.23 and Fig. 24.24, respectively.

24.7 Recursive Least-Squares Algorithm

The *recursive least-squares* (RLS) algorithm provides performance that is usually superior relative to the LMS algorithm, at a moderate increase in computational complexity. Details of the RLS algorithm are provided in Algorithm 24.4.

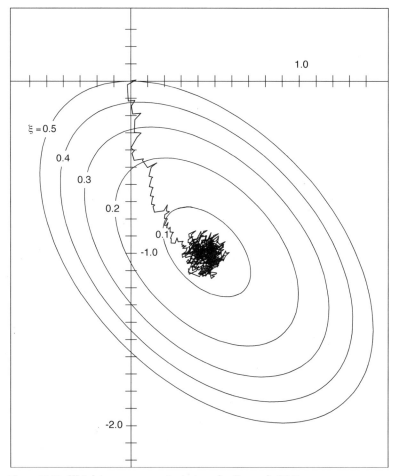

Figure 24.24 Weight-convergence trajectory for Example 24.9.

Algorithm 24.4 Recursive least-squares adaptive filtering.

Definitions:

P is an $N \times N$ matrix
I is an $N \times N$ identity matrix
u is an $N \times 1$ column vector
x is an $N \times 1$ column vector of input samples
w is an $N \times 1$ column vector of tap weights
δ is a small positive constant (typically $\delta = 0.01$)
$\lambda = 1 - \epsilon$ where ϵ is a small positive constant

1. Initialize

$$\mathbf{P}[0] = \delta \mathbf{I} \qquad \mathbf{w}[0] = \mathbf{0}$$

2. For each iteration $k = 1, 2, \ldots$ compute

$$\varepsilon[k] = d[k] - \mathbf{w}^H[k-1]\,\mathbf{x}[k]$$

$$\mathbf{u} = \frac{\mathbf{P}[k-1]\,\mathbf{x}[k]}{\lambda + \mathbf{x}^H[k]\,\mathbf{P}[k-1]\,\mathbf{x}[k]}$$

$$\mathbf{w}[k] = \mathbf{w}[k-1] + \varepsilon^*[k]\,\mathbf{u}$$

$$\mathbf{P}[k] = \mathbf{P}[k-1] - \mathbf{u}\,\mathbf{x}^H[k]\,\mathbf{P}[k-1]$$

Example 24.10 Apply the RLS method with $\delta = 0.005$ and $\lambda = 1.0$ to the AR(2) process defined in Example 24.2.

solution File ex24_10.cpp contains a program that queries for the required parameters and then creates an instance of the class RlsFilter that inherits from the class AdaptiveFir. The RLS tap updates are performed by the RlsFilter::UpdateTaps method provided in Listing 24.2. The learning curve shown in Fig. 24.25 gets close to $\xi_{\min}$ within 10 iterations. This is significantly faster than the LMS case shown in Fig. 24.16, and the *bottom of the bowl* noise shown in Fig. 24.26 is much smaller than the corresponding noise for the LMS case shown in Fig. 24.18.

As demonstrated in this example, the RLS algorithm offers superior performance in comparison to the LMS algorithm. However, this performance comes at a price. The LMS algorithm is simple, with low computational requirements. The RLS algorithm involves several matrix multiplications, and has computational requirements many times larger than the LMS algorithm.

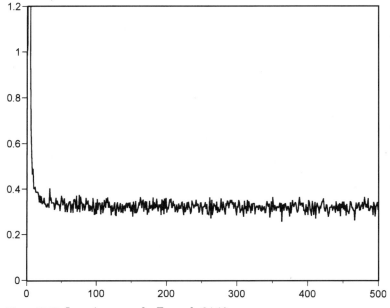

Figure 24.25 Learning curve for Example 24.10.

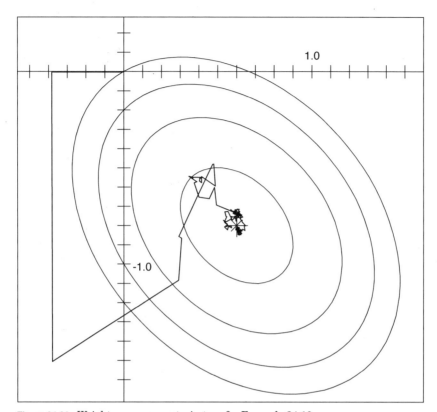

Figure 24.26 Weight-convergence trajectory for Example 24.10.

References

1. Widrow, B. and S. D. Stearns. *Adaptive Signal Processing*, Prentice-Hall, Englewood Cliffs, NJ, 1985.
2. Haykin, S. *Adaptive Filter Theory*, 3ed., Prentice-Hall, Upper Saddle River, NJ, 1996.
3. Kalouptsidis, N. and S. Theodoridis, eds. *Adaptive System Identification and Signal Processing Algorithms*, Prentice-Hall International (UK) Ltd., Hertfordshire, UK, 1993.
4. Widrow, B. and M. E. Hoff Jr. "Adaptive Switching Circuits," *IRE WESCON Conv. Rec.*, pt. 4, pp. 96–104, 1960.

Chapter 25

Classical Spectral Estimation

As discussed in Chap. 22, the *power spectral density* (PSD) and *autocorrelation function* (ACF) of a wide-sense stationary random process comprise a Fourier transform pair. Therefore, one approach to estimating the PSD is to first estimate the ACF and then compute the Fourier transform of this estimate. PSD estimates based on this approach are called *correlograms*. An alternative type of PSD estimates called *periodograms* are based on first computing the Fourier transform of the data and then performing some averaging to reduce the variability of the estimate.

25.1 Introduction to Periodograms

One intuitive approach for estimating the power spectrum of a random process is to ignore the fact that the time signal is random and simply compute the DFT over some convenient interval of time, then compute the squared magnitude for each sample in the DFT's output. This intuitive approach can be placed on a somewhat more formal footing, as follows. If a random process is ergodic, it can be shown [3] that its PSD can be defined in terms of a time average as

$$S_{xx}(f) = \lim_{M \to \infty} E\left\{ \frac{1}{(2M+1)T} \left| T \sum_{n=-M}^{M} x[n] \exp(-j2\pi f n T) \right|^2 \right\} \quad (25.1)$$

If we eliminate the limiting operation by assuming a finite data set of N samples for $x[n]$ and eliminate the expectation operation by restricting our attention to a single sample function, Eq. (25.1) becomes

$$\tilde{S}_{xx}(f) = \frac{1}{NT} \left| T \sum_{n=0}^{N-1} x[n] \exp(-j2\pi f n T) \right|^2$$

$$= \frac{T}{N} \left| \sum_{n=0}^{N-1} x[n] \exp(-j2\pi f n T) \right|^2 \quad (25.2)$$

This estimate of the PSD is often called the *sample spectrum*, due to the fact that it is computed over a finite interval of one particular *sample function* of the random process. The sample spectrum is also called the *unmodified periodogram*. Because the values of the random signal keep changing, the sample spectrum will vary greatly, depending on the specific time interval that is selected for input to the DFT. (In statistical terms, the sample spectrum is not a *consistent estimator* of the power spectral density.) The way to mitigate this variability of the sample spectrum is to perform some sort of averaging. The resulting averaged estimator is called a *modified periodogram*. There are several different types of modified periodograms, each of which implements the averaging in a different way. Several different averaging schemes will be examined in subsequent sections.

Example 25.1 For comparing the various types of periodogram, the class SinesInAwgn (in file sinawgn.cpp) creates a time sequence which is the sum of a pseudorandom part and a deteministic part comprising a number of sinusoidal components. For the pseudorandom part, the GaussRandom generator (discussed in Sec. 3.6) is used to generate zero-mean normal deviates with unity variance. These deviates are then multiplied by a user-specified value of standard deviation. To this random part is added the deterministic part

$$\sum_{m=0}^{M-1} A_m \sin(2\pi f_m n T)$$

where the amplitudes A_m, frequencies f_m, and sampling interval T are specified by the user.

SinesInAwgn can be exercised using the program ex25_01. A sample sequence produced by this program is shown in Fig. 25.1. This sequence contains 256 samples at

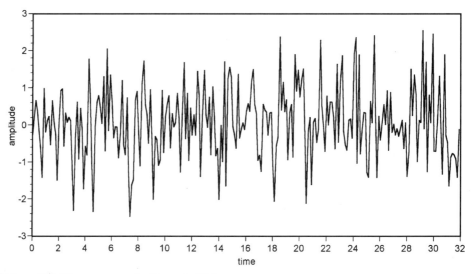

Figure 25.1 Time sequence for Example 25.1.

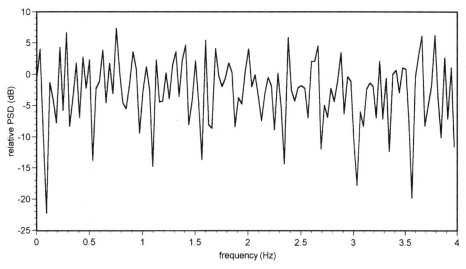

Figure 25.2 Sample spectrum for the time sequence of Fig. 25.1.

a sampling interval of 0.125 seconds. The noise variance is $\sigma^2 = 1.0$, and the single sinusoid has an amplitude of 0.2 and a frequency of 0.5. Relative to the noise, the amplitude of the sinusoidal component is so small that it is not readily apparent in the time sequence as shown in the figure. The sample spectrum corresponding to this particular signal sequence is shown in Fig. 25.2. Notice that there is no dominant spectral line at $f = 0.5$ corresponding to the sinusoidal component. The spectrum is quite erratic, with numerous deep notches.

The results of Example 25.1 are typical of the performance that can be expected when using the sample spectrum to estimate the PSD of a random process. The sample spectrum used *as is* almost never provides adequate performance for most spectrum estimation applications. As shown in subsequent sections, the major value of the sample spectrum is as a building block in more sophisticated estimation algorithms.

25.2 Daniell Periodogram

In the Daniell periodogram, averaging of the sample spectrum is accomplished by averaging over a number of adjacent spectral frequencies. The m-th value of the periodogram would be obtained by averaging $2P + 1$ sample spectrum values for frequencies $m - P$ through $m + P$, i.e.,

$$\hat{S}_D[m] = \frac{1}{2P+1} \sum_{k=m-P}^{m+P} \tilde{S}_{xx}[k] = \frac{T}{N(2P+1)} \sum_{k=m-P}^{m+P} |X[k]|^2 \quad (25.3)$$

where $X[k]$ is the DFT of the data sequence $x[n]$. The averaging in Eq. (25.3)

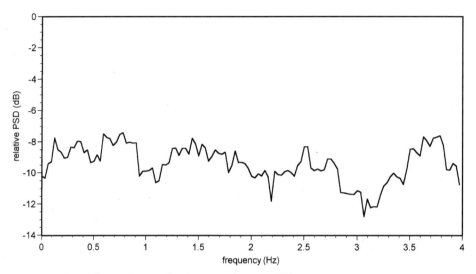

Figure 25.3 Daniell periodogram for the time sequence of Fig. 25.1.

is relatively straightforward, but the first P samples of the periodogram are special cases, because they include the values $X[-P], X[-P+1], \ldots X[-1]$. The required values can be obtained by exploiting the periodicity of inherent in frequency sequences produced by the DFT, i.e.,

$$X[-P] = X[N-P], \qquad X[-P+1] = X[N-P+1]$$

etc.

The DaniellPeriodogram function for performing this averaging is provided in file dan_pdgm.cpp. As illustrated by the next example, the Daniell periodogram sometimes provides only limited improvement over the unmodified periodogram.

Example 25.2 Program ex25_02, can be used to compute the Daniell periodogram for the pseudorandom time sequence defined in Example 25.1. The result for the case of $P = 5$ is shown in Fig. 25.3. A spectral line at $f = 0.5$ corresponding to the sinusoidal component of the signal cannot be discerned.

Example 25.2 demonstrates the relatively poor performance of the Daniell periodogram. For many applications, better estimation techniques must be used. However, these better techniques almost always require larger amounts of data on which to base the estimates. When only a limited amount of data is available, the Daniell periodogram may be the only viable choice and can often yield adequate results when the spectrum is smooth, well-behaved and lacking in narrow features. Such a spectrum is exhibited by signal produced by *continuous phase frequency shift keying*.

Example 25.3 Consider the case of *M-ary continuous phase frequency shift keying* (CPFSK). If the modulation signal consists of rectangular pulses of width T_{sym} and the peak frequency deviation is f_d, it can be shown [1, 2] that the power spectral density of the modulated signal is given by

$$S(f) = \frac{T_{sym}}{4M} \left[\sum_{n=1}^{M} A_n^2(f) + \frac{2}{M} \sum_{n=1}^{M} \sum_{m=1}^{M} A_n(f) A_m(f) B_{n+m}(f) \right] \quad (25.4)$$

where $A_n(f) = \frac{\sin\{\pi T_{sym}[f - f_d(2n-1-M)]\}}{\pi T_{sym}[f - f_d(2n-1-M)]}$

$B_n(f) = \frac{\cos(2\pi f T_{sym} - \alpha_n) - \beta \cos \alpha_n}{1 + \beta^2 - 2\beta \cos(2\pi f T_{sym})}$

$\alpha_n = 2\pi f_d T_{sym}(n - 1 - M)$

$\beta = \frac{2}{M} \sum_{n=1}^{M/2} \cos[2\pi f_d (2n-1) T_{sym}]$

These equations are implemented in the global function CpfskSpectrum which is provided in file fsk_spec.cpp. Program ex25_02 has been modified to create program ex25_03 which uses CpfskSource() instead of SineInAwgn() as a signal source and thereby generates a segment of CPFSK signal and estimates its PSD using the Daniell periodogram. This program also calls CpfskSpectrum to generate samples of the theoretical PSD at the same frequencies at which the periodogram is computed. The theoretical PSD and the corresponding Daniell periodogram ($P = 5$) are compared in Fig. 25.4 for the case of $T_{sym} = 1$, $T_{samp} = 0.125$, $M = 2$, and $f_d = 0.25$. For this particular case, the estimated spectrum compares reasonably well with the theoretical spectrum. For the case of $f_d = 0.35$ depicted in Fig. 25.5, the comparison is not quite as good.

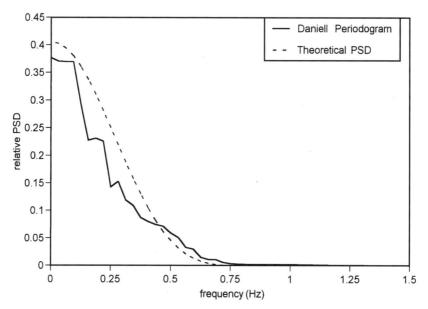

Figure 25.4 Daniell periodogram and theoretical PSD for a binary CPFSK signal having $T = 1$ and $f_d = 0.25$.

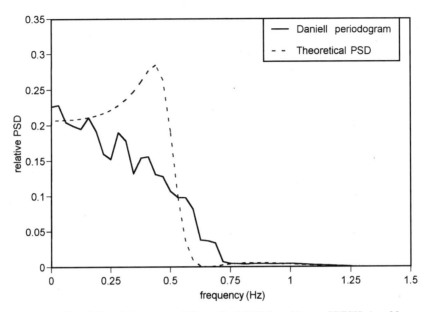

Figure 25.5 Daniell periodogram and theoretical PSD for a binary CPFSK signal having $T = 1$ and $f_d = 0.35$.

25.3 Bartlett Periodogram

In a Bartlett periodogram, a pseudoensemble of sample sequences is created by dividing one long N-sample sequence into P nonoverlapping segments of D samples each. The pth segment is related to the original N-sample sequence by

$$x_p[n] = x[pD + n]$$

The sample spectrum is for each of these D-sample segments is defined as

$$\tilde{S}_p(f) = \frac{1}{DT} \left| T \sum_{n=0}^{D-1} x_p[n] \exp(-j2\pi f nT) \right|^2 \quad (25.5)$$

In Eq. (25.5), the sample spectrum is defined as a function of continuous frequency. If we make use of the DFT to evaluate the sample spectrum at a uniformly spaced set of discrete frequencies, the resulting discrete frequency sample spectrum is given by

$$\tilde{S}_p[m] = \frac{T}{D} \left| \sum_{n=0}^{D-1} x_p[n] \exp(-j2\pi mn/D) \right|^2$$

$$= \frac{T}{D} |X_p[m]|^2 \quad m = 0, 1, \ldots D - 1 \quad (25.6)$$

where $X_p[m]$ is the DFT of the pth data segment $x_p[n]$. Finally, the P different sample spectra are arithmetically averaged at each discrete frequency to produce the Bartlett periodogram

$$\hat{S}_B[m] = \frac{1}{P} \sum_{p=0}^{P-1} \tilde{S}_p[m] \tag{25.7}$$

$$= \frac{T}{DP} \sum_{p=0}^{P} |X_p[m]|^2 \quad m = 0, 1, \ldots D-1 \tag{25.8}$$

Software notes

The most straightforward approach for implementing the Bartlett periodogram in software would be to construct a function that would accept as input a long segment of the time sequence to be analyzed. This long segment would then be divided into shorter segments inside of the function. This approach would require a data buffer with at least PD locations to hold the long input sequence plus two data buffers with at least D locations each, one to hold each sample spectrum as it is produced and one to hold the sum of the sample spectra that must be accumulated to perform the necessary averaging. If sufficient memory is available, this is not a problem. However, if P or D is very large, or if the processing must be performed on a processor with minimal memory (such as an embedded processor in a piece of consumer audio gear) another implementation approach is called for. Two alternative strategies can be characterized as a *supply-driven* strategy and a *demand-driven* strategy, based on how the periodogram function gets input data from the calling program. Both approaches would require three data buffers with at least D locations each. In a supply-driven approach, the calling program would be responsible for segmenting the input data and feeding it to the periodogram function [say BartlettPeriodogram()] one segment at a time. This would entail a specialized call each time the calling program determines that BartlettPeriodogram() needs to reinitialize itself to begin accumulating a brand new average. This specialized call would be followed by $P-1$ regular calls to *supply* the remaining data segments. In the demand-driven approach the calling program would call BartlettPeriodogram() once only for each complete periodogram. The Bartlett Periodogram() function itself would determine when it is ready for the next data segment, and would *demand* this data by calling some external function [say GetNextSegment()], that would return a pointer to a buffer containing the next segment of input data. The demand-driven approach has the advantage of hiding the details of the segmentation logic inside of the periodogram function. This feature will become increasingly attractive as we investigate more sophisticated periodogram algorithms that involve more complicated segmentation logic. Therefore, the function BartlettPeriodogram() has been implemented using a demand-driven design. Furthermore, it has been designed to use a pointer for calling GetNextSegment(). Depending upon the value of the pointer provided by the calling program, the function BartlettPeriodogram() can wind up calling one of several different versions of GetNextSegment(). One version might obtain the data segment from a disk file, another version might obtain the data segment from a tape, a third version might read data from an A/D port, and a fourth version might use software to generate synthetic data for testing and demonstration purposes (as is done in ex25_04). The ultimate source of the data would be transparent to the BartlettPeriodogram() function—the interface to each version of GetNextSegment() would remain the same. Furthermore, as used in this chapter, the function GetNextSegment() is a virtual member function of the base class SignalSource. Several derived classes, such as SinesInAwgn and CpfskSource, redefine GetNextSegment() to satisfy the particular requirements of the specific signal types they provide. The complete implementation of BartlettPeriodogram() is provided in file bartpdgm.cpp.

Example 25.4 The program ex25_04 can be used to compute the Bartlett periodogram for a longer version of the time sequence described in Example 25.1. The resulting periodogram for averaging over 16 segments is shown in Fig. 25.6. Compared to Fig. 25.2, the estimate of the wideband noise spectrum has become noticeably less erratic, and a definite peak is discernable at $f = 0.5$ Hz. Still further improvement can be seen for the case of averaging over 512 segments, which is shown in Fig. 25.7. However, this improvement does not come cheaply: averaging over 512 segments requires 131,072 input samples compared to only 4096 input samples for averaging over 16 segments.

Example 25.5 Compute the Bartlett periodogram for the case of a binary CPFSK signal with $T_{\text{sym}} = 1$, $T_{samp} = 0.125$, and $f_d = 0.35$. The program ex25_05 can be used to perform these calculations. The theoretical PSD and the corresponding Bartlett

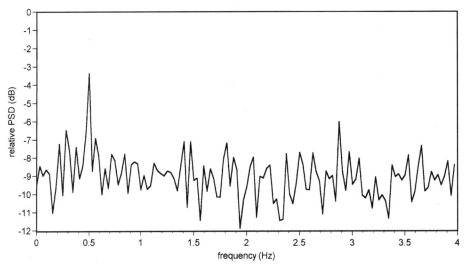

Figure 25.6 Bartlett periodogram for the time sequence of Example 25.1, with averaging over 16 segments.

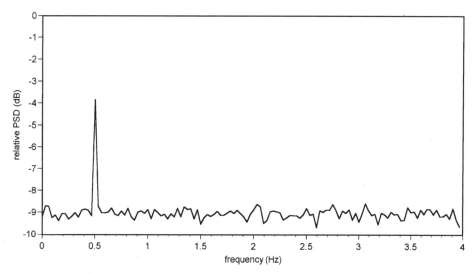

Figure 25.7 Bartlett periodogram for the time sequence of Example 25.1, with averaging over 512 segments.

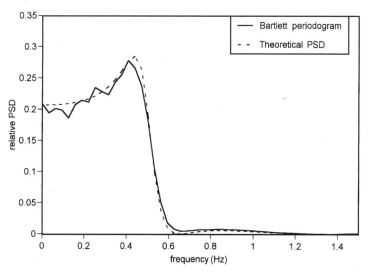

Figure 25.8 Bartlett periodogram and theoretical PSD for a binary CPFSK signal having $T = 1$ and $f_d = 0.35$.

periodogram are compared in Fig. 25.8. This result is considerably better than the Daniell periodogram that was computed in Example 25.3 for the same CPFSK signal.

Section 25.5 describes a segmenting scheme that allows segments to overlap, thereby providing the benefits of averaging over many segments even when only moderate numbers of input samples are available.

25.4 Windowing and Other Issues

Periodograms are based on DFT's and suffer from most of the same limitations that were discussed for the DFT in Chap. 12. As shown there, leakage can be reduced by applying a tapered *data window* to the time sequence data prior to calculation of the DFT. Application of the window function will cause the total energy in each signal segment to change. Consequently, the results of each DFT must be scaled by $1/U$ where

$$U = \frac{1}{D} \sum_{n=0}^{D-1} |w[n]|^2$$

Thus, if a window $w[n]$ is applied to each segment in a Bartlett periodogram, Eq. (25.6) becomes

$$\hat{S}_B[m] = \frac{T}{DPU} \sum_{p=0}^{P-1} |Y_p[m]|^2$$

where $Y_p[m]$ is the DFT of the p-th windowed data segment $w[n]x_p[n]$.

Example 25.6 Program ex25_06 can be used for the computations in this example. Consider the case of two sinusoids embedded in white Gaussian noise. One sinusoid has a frequency of 2.48 and an amplitude of 100; the other has a frequency of 2.3 and an amplitude of 5. The noise has zero mean and unity variance. If a Bartlett periodogram with 512 segments of 256 samples each is used to estimate the PSD of this signal, the result will be as shown in Fig. 25.9. The sidelobe content of the stronger signal almost completely masks the presence of the weaker signal. If a von Hann window is applied to each data segment prior to computation of the DFT, the resulting periodogram will be as shown in Fig. 25.10. Two peaks corresponding to the two sinusoids are now visible. As shown in Figs. 25.11 and 25.12, similar (but inferior) results can also be obtained using either a Hamming window or a triangular window.

In Example 25.6, both sinusoids existed over the entire duration of each data segment, so the usual sidelobe *scalloping* is not evident. Lobe width in the frequency domain is inversely proportional to pulse width in the time domain, and when the *pulse* of sinewave spans the entire DFT input record, the frequency domain lobes degenerate to the width of a single sample. To see sidelobe scalloping in the frequency domain result, it is necessary to pad the input record with zeros. The fft() function has been modified to take a fourth argument. The new prototype is

```
void fft( complex *time_seq,
          complex *freq_seq,]
          int num_samps,
          int fft_len );
```

The third argument is now interpreted as the number of data samples in the array time_seq. The fourth argument is the length of the FFT that is to be

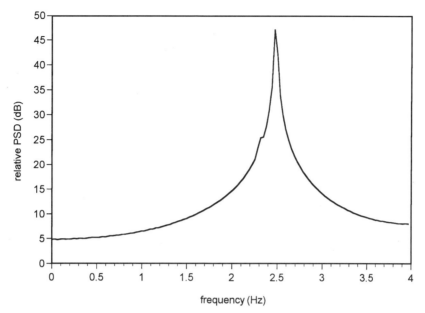

Figure 25.9 Bartlett periodogram for Example 25.6.

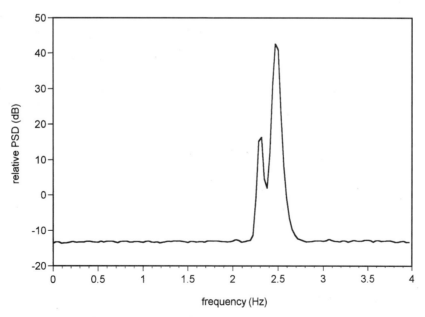

Figure 25.10 Bartlett periodogram for Example 25.6 when a von Hann window is used to reduce sidelobe content of the sinusoids' spectra.

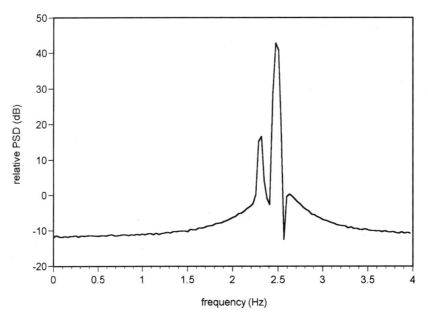

Figure 25.11 Bartlett periodogram for Example 25.6 when a Hamming window is used to reduce sidelobe content of the sinusoids' spectra.

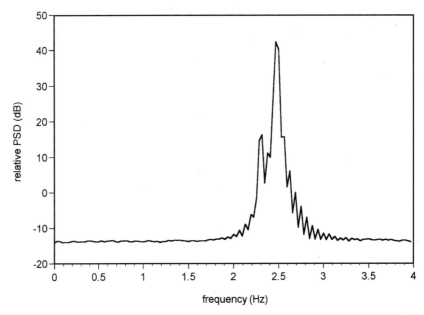

Figure 25.12 Bartlett periodogram for Example 25.6 when a triangular window is used to reduce sidelobe content of the sinusoids' spectra.

performed and also the number of samples to be placed in the array `freq_seq`. The function will pad the input time sequence with `fft_len-num_samps` zeros before performing the indicated FFT.

Example 25.7 The program `ex25_07` uses the new four-argument `fft()` function to recompute the periodograms of Example 25.6 with some zero padding of the input segments. Each input segment contains 256 samples and the FFT size is 2048 samples. The results are shown in Figs. 25.13 through 25.16. Only the frequency interval from 1.5 to 3.5 is plotted to better display the details of the sidelobe structure.

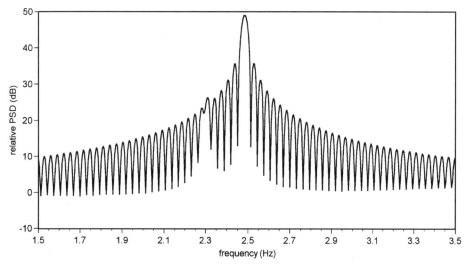

Figure 25.13 Bartlett periodogram for Example 25.7.

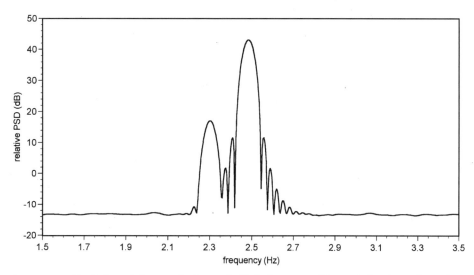

Figure 25.14 Bartlett periodogram for Example 25.7 when a von Hann window is used to reduce sidelobe content of the sinusoids' spectra.

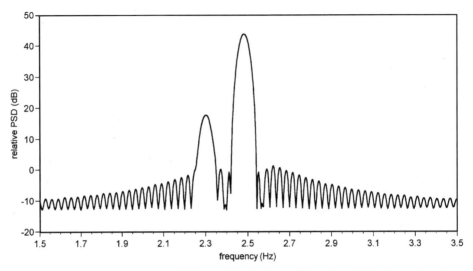

Figure 25.15 Bartlett periodogram for Example 25.7 when a Hamming window is used to reduce sidelobe content of the sinusoids' spectra.

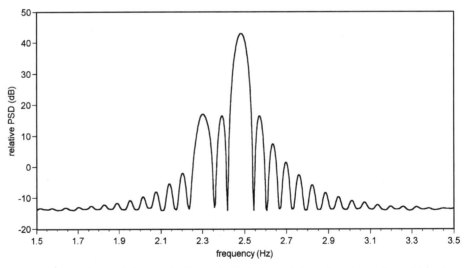

Figure 25.16 Bartlett periodogram for Example 25.7 when a triangular window is used to reduce sidelobe content of the sinusoids' spectra.

25.5 Welch Periodogram

In a Welch periodogram, a pseudoensemble of sample sequences is created by dividing one long N-sample sequence into P overlapping segments of D samples each, with a shift of $S < D$ samples between consecutive segments. A data window $w[n]$ is then applied to each segment. The windowed pth segment is related to the original N-sample sequence by

$$y_p[n] = w[n]x[n+pS] \tag{25.9}$$

The discrete frequency sample spectrum of the windowed pth segment is given by

$$\tilde{S}_p[m] = \frac{T}{UD}|Y_p[m]|^2 \quad m = 0, 1, \ldots D-1 \tag{25.10}$$

where $Y_p[m]$ is the DFT of the pth windowed data segment $y_p[n]$ and

$$U = \sum_{n=0}^{D-1} w^2[n] \tag{25.11}$$

The P different sample spectra are arithmetically averaged at each discrete frequency to produce the Welch periodogram

$$\hat{S}_W[m] = \frac{T}{P}\sum_{p=0}^{P-1} \tilde{S}_p[m]$$

$$= \frac{T}{UDP}\sum_{p=0}^{P}|Y_p[m]|^2 \quad m = 0, 1, \ldots D-1 \tag{25.12}$$

It can be shown [5] that $\hat{S}_W[m]$ can be calculated as

$$\hat{S}_W[m] = \frac{T}{2UDP}\sum_{p=1}^{P/2}(|B_p[m]|^2 + |B_p[N-m]|^2) \tag{25.13}$$

where $B_p[m] = Y_{2p-1}[m] + jY_{2p}[m]$
$\qquad\qquad = \text{DFT}\{y_{2p-1}[n] + jy_{2p}[n]\}$
$\qquad\qquad = \text{DFT}\{w[n](x_{2p-1}[n] + jx_{2p}[n])\}$

Instead of the P DFT's of D samples each required by Eq. (25.12), only $P/2$ DFT's of D samples each are required by Eq. (25.13).

25.6 Correlograms

As noted in Chap. 21, the PSD and ACF of a wide-sense stationary random process comprise a Fourier transform pair. This relationship can be exploited to provide an alternative to the periodogram for estimating the PSD of a random process. Specifically

$$\hat{S}_C(f) = T\sum_{m=-L}^{L} \hat{r}_{xx}[m]\exp(-j2\pi fmT)$$

where $\hat{r}_{xx}[m]$ is one of the sample correlation functions given in Sec. 21.9. To keep the variance of the estimate from becoming too large, the maximum lag index L is usually restricted to some small fraction of the total number of data samples N. A maximum of $L \leq N/10$ often is observed as a rule of thumb.

References

1. Rorabaugh, C. Britton. *Communications Formulas & Algorithms.* McGraw-Hill, New York, 1990.
2. Proakis, G. John. *Digital Communications.* McGraw-Hill, New York, 1983.
3. Marple, S. Lawrence. *Digital Spectral Analysis with Applications.* Prentice-Hall, Englewood Cliffs, NJ, 1987.
4. Kay, M. Steven. *Modern Spectral Estimation: Theory & Application.* Prentice-Hall, Englewood Cliffs, NJ, 1988.
5. Welch, D. Peter. "The Use of Fast Fourier Transform for the Estimation of Power Spectra: A Method Based on Time Averaging over Short Modified Periodograms," *IEEE Trans. Audio Electroacoust.*, vol. AU-15, pp. 70–73, June 1967.

Chapter

26

Modern Spectral Estimation

As discussed in Sec. 22.2, the PSD of an AR(p) process can be obtained as

$$P_{\text{AR}}(f) = \frac{T\rho_w}{|A(f)|^2} \qquad (26.1)$$

where $A(f) = \sum_{k=0}^{p} a_k \exp(-j2\pi f k)$ (26.2)
 T = sampling interval
 ρ_w = variance of the white noise driving the AR process

Likewise, as discussed in Sec. 22.1, the PSD of an ARMA(p,q) process can be obtained as

$$P_{\text{ARMA}}(f) = T\rho_w \left|\frac{B(f)}{A(f)}\right|^2 \qquad (26.3)$$

where $A(f)$ is given by Eq. (26.2) and

$$B(f) = \sum_{k=0}^{q} b_k \exp(-j2\pi f k) \qquad (26.4)$$

A number of methods already discussed in earlier chapters can be used to form estimates $\hat{a}_k$ and $\hat{b}_k$ of the AR and MA coefficients. These estimates can be substituted into Eqs. (26.2) and (26.4) to generate an estimated PSD using either Eq. (26.1) or (26.3).

26.1 Yule-Walker Method

In the Yule-Walker method the AR coefficients in Eq. (26.2) are estimated using the Yule-Walker normal equations (23.3) with the biased estimate of the

ACF used to populate the correlation matrix **R**. The resulting equations can be solved for the AR coefficients by means of the Levinson recursion presented in Sec. 22.3. This method can be used with relatively short data records but the spectral resolution can sometimes be rather poor, as demonstrated in Example 26.2.

Example 26.1 Use the Yule-Walker method to estimate the PSD of an AR(2) process with $a_1 = -0.6$, $a_2 = 0.8$, and $\rho_w = 0.32$. Generate estimates for data records of length $N = 100$ and $N = 1000$. Compare these estimates to the true PSD generated, using the known values of a_1, a_2, and ρ_w in Eq. (26.1). Assume a sampling interval of $T = 1$.

solution The program ex26_01 can be used to generate the desired estimates. Using a seed of 11123313, the estimated AR coefficients are obtained as:

For $N = 100$: $\quad \hat{a}_1 = -0.615163 \quad \hat{a}_2 = 0.72939$

For $N = 1000$: $\quad \hat{a}_1 = -0.629441 \quad \hat{a}_2 = 0.824018$

The estimated PSDs are compared to the true PSD in Fig. 26.1.

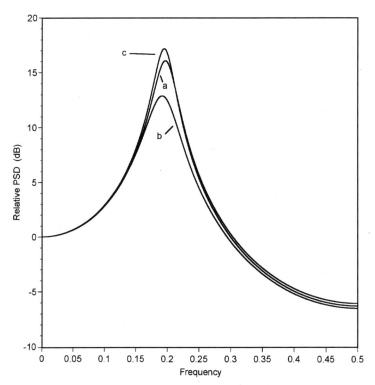

Figure 26.1 Power spectral densities for Example 26.1: (a) true PSD, (b) estimated PSD for $N = 100$, (c) estimated PSD for $N = 1000$.

Example 26.2 Use the Yule-Walker method with $N = 100$ to estimate the PSD of the AR(4) process defined by

$$a_1 = -0.99413 \qquad a_2 = 2.23221$$
$$a_3 = -0.994129 \qquad a_4 = 0.999998$$
$$\rho_w = 0.05$$

Compare this estimate to the true PSD generated using the known values of a_1 through a_4.

solution The program ex26_01 can also be used for this example. The estimated AR coefficients are

$$\hat{a}_1 = -0.958852 \qquad \hat{a}_2 = 1.62376$$
$$\hat{a}_3 = -0.71748 \qquad \hat{a}_4 = 0.433365$$

The two PSDs are plotted in Fig. 26.2. Notice that the estimated spectrum is unable to resolve the two spectral peaks that appear in the true spectrum.

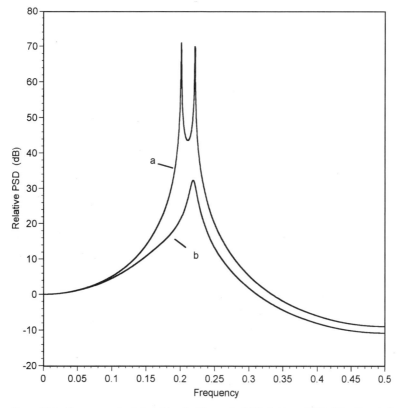

Figure 26.2 Power spectral densities for Example 26.2: (a) true PSD, (b) estimated PSD for $N = 100$.

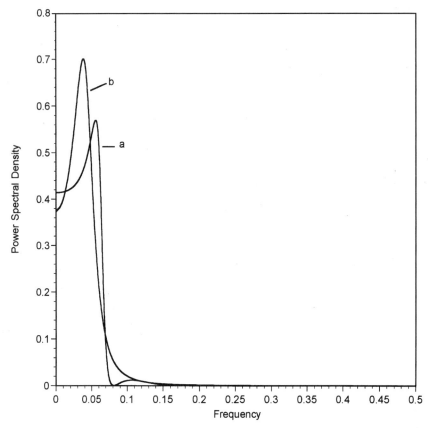

Figure 26.3 Power spectral densities for Example 26.3: (a) true PSD, (b) estimated PSD for $N = 100$.

Example 26.3 Use the Yule-Walker method for an AR(2) process with $N = 100$ to estimate the PSD of a binary CPFSK signal with $T_{\text{sym}} = 1$, $T_{\text{samp}} = 0.125$, $f_c = 0.0$, and $f_d = 0.35$ (i.e., $h = 0.7$). Compare this estimate to the theoretical spectrum of the CPFSK signal given by Eq. (25.4).

solution The program ex26_03 can be used to generate both the desired estimate and the corresponding true PSD. The two PSDs are plotted in Fig. 26.3. The true PSD actually contains nulls which can only be seen when the PSDs are plotted in decibels, as they are in Fig. 26.4. Notice that the AR estimate of the spectrum does not contain even a hint of these nulls.

26.2 Burg Method

The Burg method estimates reflection coefficients by minimizing estimates of the prediction error power. The Levinson recursion then can be used to obtain the AR coefficients from the reflection coefficients. The Burg method, along with the applicable portion of the Levinson recursion, are summarized in Algorithm 26.1.

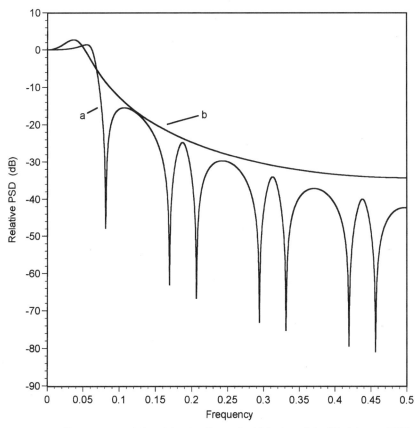

Figure 26.4 Power spectral densities for Example 26.3 plotted in dB: (a) true PSD, (b) estimated PSD for $N = 100$.

Algorithm 26.1 Burg method for AR spectral estimation.

1. Initialize

$$\hat{r}_{xx}[0] = \frac{1}{N} \sum_{n=0}^{N-1} |x[n]|^2$$

$$\hat{\rho}_0 = \hat{r}_{xx}[0]$$

$$f_0[n] = x[n] \quad n = 1, 2, \ldots, N-1$$

$$b_0[n] = x[n] \quad n = 0, 1, \ldots, N-2$$

2. For $k = 1, 2, \ldots, p$ compute

$$\hat{\kappa}_k = \frac{-2 \sum_{n=k}^{N-1} f_{k-1}[n] b_{k-1}^*[n-1]}{\sum_{n=k}^{N-1} \left(|f_{k-1}[n]|^2 + |b_{k-1}[n-1]|^2 \right)}$$

$$\hat{\rho}_k = \left(1 - |\kappa_k|^2\right) \hat{\rho}_{k-1}$$

$$\hat{a}_k[i] = \begin{cases} \hat{a}_{k-1}[i] + \hat{\kappa}_k \hat{a}_{k-1}^*[k-i] & i = 1, 2, \ldots, k-1 \\ \hat{\kappa}_k & i = k \end{cases}$$

$$f_k[n] = f_{k-1}[n] + \hat{\kappa}_k b_{k-1}[n-1] \quad n = k+1, k+2, \ldots, N-1$$

$$b_k[n] = b_{k-1}[n-1] + \hat{\kappa}_k^* f_{k-1}[n] \quad n = k, k+1, \ldots, N-2$$

Example 26.4 Repeat Example 26.2 using the Burg method in place of the Yule-Walker method.

solution The program ex26_04 can be used to generate the desired estimates. The estimated AR coefficients are

$$\hat{a}_1 = -0.990951 \quad \hat{a}_2 = 2.18985$$
$$\hat{a}_3 = -0.97841 \quad \hat{a}_4 = 0.971388$$

Only the PSD estimate generated by the Burg method is shown in Fig. 26.5, because it would be hard to distinguish from the theoretical spectrum if both were plotted. Comparison of this figure with Fig. 26.2 reveals that the Burg method is superior to the Yule-Walker method when it comes to resolving closely spaced spectral peaks.

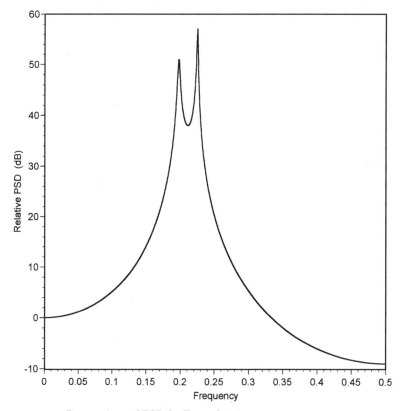

Figure 26.5 Burg-estimated PSD for Example 26.4.

26.3 RLS Method

The RLS adaptive filtering technique discussed in Sec. 24.7 can be used to estimate the coefficients for an AR process as was illustrated in Example 24.19. Such estimates can be used in Eq. (26.1) to generate an estimate of the PSD. If there are sufficient data and computational capacity available, the PSD estimates obtained vis RLS estimation of the AR coefficients can be as good as or even superior to PSD estimates obtained using any of the other techniques discussed in this chapter.

Example 26.5 Use the RLS adaptive filtering technique (with $\delta = 0.01$ and $\lambda = 0.95$) to estimate the AR coefficients for the AR(4) process defined by

$$a_1 = -1.78218 \quad a_2 = 2.68388$$
$$a_3 = -1.75488 \quad a_4 = 0.969738$$
$$\rho_w = 0.05$$

The true PSD of this process is shown in Fig. 26.6. Use the estimated AR coefficients after iteration 100 to compute an estimate of the PSD.

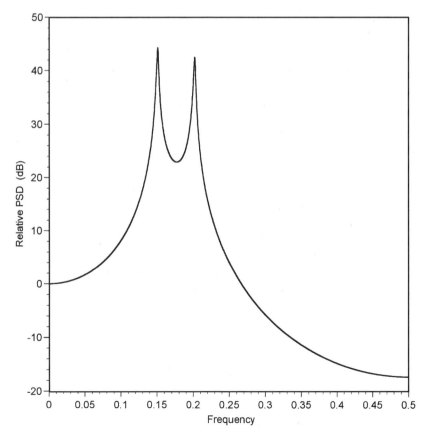

Figure 26.6 True PSD for AR(4) process in Example 26.5.

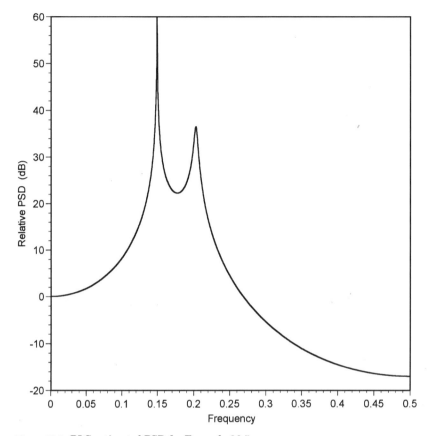

Figure 26.7 RLS-estimated PSD for Example 26.5.

solution The program ex26_05 can be used to generate the specified test signal and estimate the AR coefficients using the RLS method. The estimated coefficients at iteration 100 are

$$\hat{a}_1 = -1.76296 \quad \hat{a}_2 = 2.65508$$
$$\hat{a}_3 = -1.72733 \quad \hat{a}_4 = 0.969896$$

The resulting PSD estimate is shown in Fig. 26.7.

26.4 LMS Method

The LMS adaptive filtering technique discussed in Sec. 24.6 can be used to estimate the coefficients for an AR process as was illustrated in Example 24.7. However, the convergence of these estimates is not nearly as rapid as it is for the coefficients estimated using the RLS method.

Example 26.6 Use the LMS adaptive filtering technique with $\mu = 0.001$ to estimate the AR parameters for the AR(4) process defined in Example 26.5. Use the estimated

TABLE 26.1 Estimated AR Coefficients for Example 26.6

	True	$N = 5000$	$N = 10{,}000$	$N = 20{,}000$
a_1	−1.78218	−1.20542	−1.57999	−1.74623
a_2	2.68388	2.1.70087	2.29764	2.59961
a_3	−1.75488	−0.776344	−1.37482	−1.67995
a_4	0.969738	0.378576	0.775831	0.915822

coefficients after iterations 5000, 10,000, and 20,000 to compute estimates of the PSD for this process.

solution The program ex26_06 can be used to generate the specified test signal and estimate the AR coefficients using the LMS method. The estimated coefficients are compared to the specified coefficients in Table 26.1. The PSDs estimated after iterations 5000 and 10,000 are shown in Fig. 26.8. The PSD estimated after iteration 20,000 is shown in Fig. 26.9.

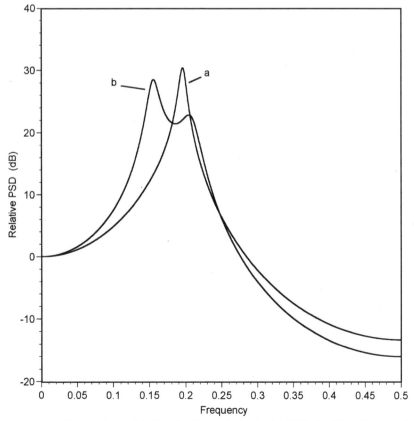

Figure 26.8 LMS-estimated PSDs for Example 26.6: (a) PSD for $N = 5000$ and (b) PSD for $N = 10{,}000$.

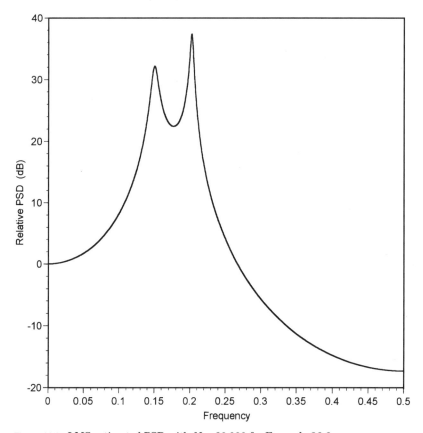

Figure 26.9 LMS-estimated PSD with $N = 20,000$ for Example 26.6.

26.5 Spectral Estimation of Noisy AR Processes

Autoregressive spectral estimation techniques such as the Yule-Walker method and Burg method are somewhat sensitive to the presence of additive noise in the observations of the process. This sensitivity is not surprising because the additive noise is a moving-average process rather than an autoregressive one. There are several possible approaches for dealing with the presence of noise in samples of an AR process:

1. Using an AR model having an order that is significantly larger than the order of the noise-free AR process to be estimated. This approach can sometimes be effective, but spurious peaks may be introduced into the spectrum if the order is increased too much.

2. Using ARMA estimation instead of AR estimation. Estimation of ARMA coefficients often is very difficult, and the presence of additive noise does not make it any easier. However, sometimes this is a viable approach (see Example 26.8).

3. Filtering the data to remove or reduce the observation noise. This approach is most useful when the signal of interest and the observation noise occupy

different parts of the frequency spectrum, allowing the noise to be removed by a simple frequency-selective filter. This situation can arise in applications where the sampling rate is high relative to the desired signal's bandwidth, making the system bandwidth much wider than the signal bandwidth. Typically, the noise spectrum will be relatively flat across the system bandwidth, and that portion that lies beyond the signal bandwidth can be easily removed via filtering. However, this approach is viable only if the spectral occupancy of the desired signal is known.

4. Modification of the AR parameter estimation techniques to compensate for the presence of the observation noise. The simplest manifestation of this approach is a simple correction applied to the diagonal elements of the correlation matrix **R** to remove the contribution to these elements that is due to the observation noise. If the observation noise is assumed to be uncorrelated from sample to sample, the presence of the noise will not have any impact on the off-diagonal elements of **R**. The trick is deciding just how much of a correction to apply to the diagonal elements—too much can be as bad as not enough.

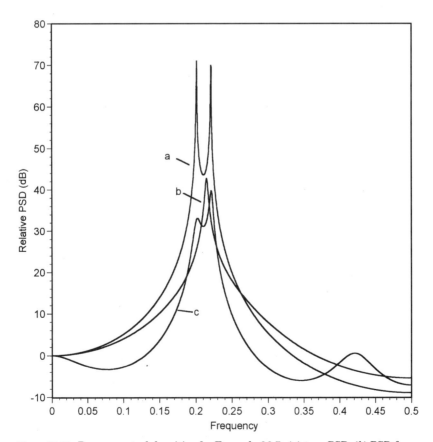

Figure 26.10 Power spectral densities for Example 26.7: (*a*) true PSD, (*b*) PSD from AR(4) estimation, and (*c*) PSD from AR(8) estimation.

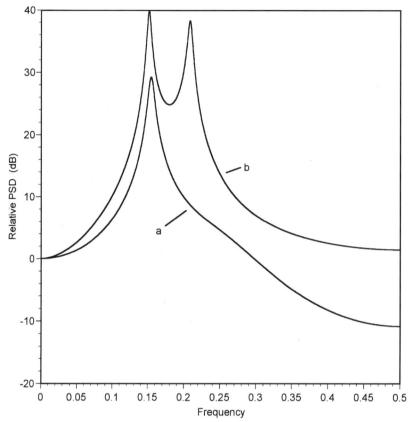

Figure 26.11 Power spectral densities for Example 26.8: (a) PSD for Burg-estimated AR(4) model and (b) PSD for ARMA(4,4) model estimated using modified Yule-Walker method.

Example 26.7 Use the Burg method with $N = 100$ to estimate the PSD of the process comprising the AR(4) process from Example 26.2 plus AWGN having a variance of 0.1. Generate PSD estimates based on AR(4) and AR(8) process models.

solution The program ex26_07 can be used to generate the desired estimates. The estimated PSDs are compared to the true PSD in Fig. 26.10.

Example 26.8 Use the modified Yule-Walker method from Sec. 22.5 to fit an ARMA (4, 4) model to the process comprising the AR(4) process from Example 26.5, plus AWGN having a variance of 0.1. Generate the PSD corresponding to this model and compare it to the PSD obtained from an AR(4) model fit to the same test signal using the Burg method. The true PSD of this process is shown in Fig. 26.6. Use $N = 100$ when forming the correlation estimate. Use an AR order of 20 in Durbin's method.

solution The program ex26_04 can be used to generate the AR(4)-based PSD, and ex26_08 can be used to estimate the coefficients for the ARMA model. The resulting PSD, calculated using Eq. (26.3), is compared to the Burg PSD in Fig. 26.11.

Chapter 27
Speech Processing

Speech signals exhibit certain characteristics which allow specialized processing methods to be employed. Some of these methods are useful for limited signal classes in addition to speech but, in general, these techniques exploit known characteristics of speech signals and are therefore not applicable to arbitrary signals. This chapter examines the salient characteristics of speech and explores several processing techniques that have been developed expressly for use on speech signals.

27.1 Speech Signals

There are several speech processing and speech coding techniques that are based on a simplified model of how speech is produced in the human vocal tract. This model is based on the assumption that speech can be represented as the output of a linear, *slowly* time-varying system. The properties of the linear system vary slowly enough that, over a short segment of speech, they can be assumed time-invariant. Thus, short segments of speech are modeled as the output of a linear time-invariant system. This system is excited by either a quasi periodic impulse train to model a segment of *voiced* speech, or by a random noise signal to model a segment of *unvoiced* speech. Speech is produced by forcing air from the lungs and then modulating the flow of this air as it moves from the lungs, through the vocal tract, to the lips. The pertinent features of the vocal tract are diagrammed in Fig. 27.1. Speech sounds are classified into the three distinct classes of *voiced* sounds, *unvoiced* sounds, and *plosive* sounds. Each of these three classes of sound is produced by a different mechanism at work in the vocal tract.

Voiced sounds are produced by forcing air through the glottis (the opening between the vocal cords) when tension of the vocal cords is adjusted such that they oscillate and thereby modulate the air flow into quasi periodic pulses. These pulses excite the resonances in the remainder of the vocal tract. Different sounds are produced as muscles work to change the shape of the vocal tract, and

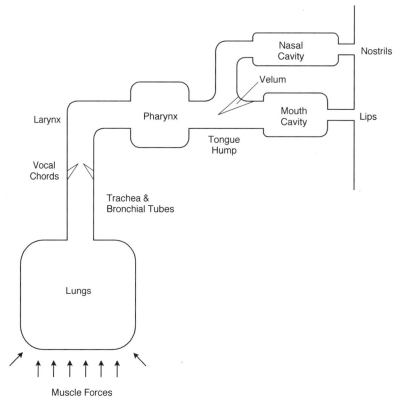

Figure 27.1 Essential features of the human vocal tract.

thereby change its resonant frequencies, or *formant frequencies*. The rate of the pulses is called the *fundamental frequency* or *pitch*. In signal processing, these two terms often are used interchangeably, but in the study of human hearing, the term *pitch* refers to the *perceived* fundamental frequency rather than to the *actual* fundamental frequency. The fundamental frequency typically falls in the range of 50 to 250 Hz for men and 120 to 500 Hz for women. Examples of voiced sounds are the *a* in *baby*, the *ee* in *beet*, and the *i* in *lie*. A short segment of a male voice pronouncing the *i* in *lie* is shown in Fig. 27.2. The quasi periodic nature of a voiced sound is evident in this waveform. The segment shown covers about 46.4 msec and represents only a portion of the complete *i* sound.

Unvoiced or *fricative* sounds are generated when air is forced through a constriction at a velocity high enough to induce turbulence. This turbulent flow produces broad-spectrum noise that excites the vocal tract. Examples of unvoiced sounds are the *f* and *sh* in *fish* and the *s* in *silly*. A short segment of a male voice pronouncing the *sh* in *fish* is shown in Fig. 27.3. There is no periodic structure apparent in this waveform.

Plosive sounds are created by completely closing off air flow, allowing pressure to build up behind the closure, and then abruptly releasing the pressure.

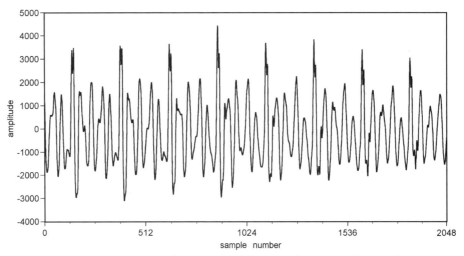

Figure 27.2 A short section of speech consisting of the *i* sound pronounced by a male.

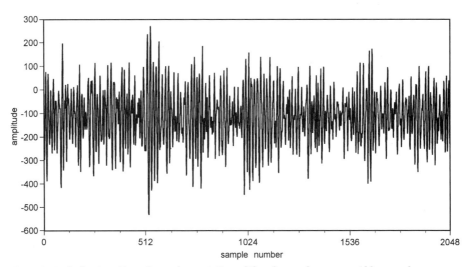

Figure 27.3 A short section of speech consisting of the *sh* sound pronounced by a male.

Examples of plosive sounds are the *p* in *punch* and the *b* in *butcher*. The closure and sudden release are usually made at the lips; therefore ventriloquists tend to have trouble pronouncing plosives correctly without a noticeable lip movement.

27.2 Cepstral Analysis

Cepstral alanysis of speech signals was motivated originally [1] by a desire to separate (for voiced sounds) the signal content due to glottal excitation from

TABLE 27.1 Paraphrased Terminology for Cepstral Analysis

Traditional term	Cepstral term
spectrum	cepstrum
filtering	liftering
frequency	quefrency
harmonic	rahmonic
period	repiod
analysis	alanysis
magnitude	gamnitude
polar	lopar
phase	saphe

the signal content due to the response of the vocal tract. The paraphrasing of *spectral analysis* to *cepstral alanysis* is due to [2], where it was introduced to reduce confusion caused by the fact that the authors found themselves "operating on the frequency side in ways customary on the time side and vice versa." This paper was concerned specifically with the detection of echos in seismograph recordings, but it has become somewhat notorious in the DSP community because of its vocabulary of paraphrased terminology (listed here in Table 27.1) and its unusual title, "The Quefrency Alanysis of Time Series for Echos: Cepstrum, Pseudo-Autocovariance, Cross-Cepstrum and Saphe Cracking." The DSP community is somewhat divided on whether *cepstrum* is pronounced as *sep-strum* or *kep-strum*. *Cepstrum* is a paraphrase of *spectrum* and the *c* in spectrum is pronounced as a *k*; hence, *kep-strum*. However, in virtually every English word beginning with *ce*, the *c* is pronounced as an *s*. One notable exception is *Celtic*, which is pronounced as either *Sell-tick* or *Kell-tick*. Most of the recent literature (e.g. [3]) retains the use of cepstrum and cepstral but rejects alanysis in favor of the usual analysis.

The glottal excitation is typically rich in harmonic content, and this content is modulated by the filtering response of the vocal tract. Cepstral analysis provides a way to separate the excitation from the filter response and thereby facilitate the estimation of the excitation's fundamental frequency and the filter's resonant frequencies (formants). Assume that a sequence of voiced speech samples is the result of convolving the glottal excitation sequence $e[n]$ with the vocal tract's discrete impulse response $\theta[n]$,

$$s[n] = e[n] * \theta[n] \tag{27.1}$$

In the frequency domain, the convolution relationship in Eq. (27.1) becomes a multiplication relationship

$$S(\omega) = E(\omega)\Theta(\omega) \tag{27.2}$$

Even in the frequency domain, there is no easy way to separate the excitation from the filter response. However, recognizing that $\log AB = \log A + \log B$, the multiplication relationship in Eq. (27.2) can be transformed into an additive relationship by taking the logarithm

$$\log S(\omega) = \log E(\omega) + \log \Theta(\omega) \tag{27.3}$$

In general, $S(\omega)$, $W(\omega)$, and $\Theta(\omega)$ will be complex-valued. The use of complex logarithms can be avoided by taking the magnitude prior to taking the logarithm

$$\log |S(\omega)| = \log |E(\omega)| + \log |\Theta(\omega)| \tag{27.4}$$

Motivated by Eq. (27.4), the *real cepstrum* of a signal $s[n]$ is defined as

$$c[n] = \mathcal{F}_{\text{DTFT}}^{-1}\{\log |\mathcal{F}_{\text{DTFT}}\{s[n]\}|\} \tag{27.5}$$

$$= \frac{1}{2\pi} \int_{-\pi}^{\pi} \log |S(\omega)| e^{j\omega n} \, d\omega \tag{27.6}$$

where

$$S(\omega) = \sum_{n=-\infty}^{\infty} s[n] e^{-j\omega n} \tag{27.7}$$

Motivated by Eq. (27.3), the *complex cepstrum* of a signal $s[n]$ is defined as

$$\gamma[n] = \mathcal{F}_{\text{DTFT}}^{-1}\{\log \mathcal{F}_{\text{DTFT}}\{s[n]\}\}$$

$$= \frac{1}{2\pi} \int_{-\pi}^{\pi} \log[S(\omega)] e^{j\omega n} \, d\omega$$

Computation of the real cepstrum

Practical application of cepstral analysis depends on machine computation of the cepstrum. Clearly, direct evaluation of Eq. (27.7) over an infinite sequence is impossible. In real world applications, the cepstrum will be computed for a finite segment of speech with that segment usually selected by a finite-duration window. Assuming that $s[n]$ is defined over a large range of n and that the window $w[n]$ is defined over the N values of n from $n = 0$ through $n = N - 1$, we can define a *frame* of speech at time m to be the product of the long-term sequence $s[n]$ with the shifted window $w[m - n]$:

$$f[n; m] \stackrel{\Delta}{=} s[n] w[m - n] \tag{27.8}$$

In cepstral analysis, $w[n]$ is often defined as a Hamming window. If we restrict our attention to the frame $f[n;m]$, the *short-term discrete-time Fourier transform* (stDTFT) is defined as

$$\tilde{S}(\omega;m) = \sum_{n=m-N+1}^{m} f[n;m]e^{-j\omega n}$$

Substituting $\tilde{S}(\omega;m)$ for $S(\omega)$ in Eq. (27.6), we obtain

$$c[n] = \mathcal{F}_{\text{DTFT}}^{-1}\{\log|\mathcal{F}_{\text{stDTFT}}\{f[n;m]\}|\} \qquad (27.9)$$

$$= \frac{1}{2\pi}\int_{-\pi}^{\pi} \log|\tilde{S}(\omega;m)|e^{j\omega n}\, d\omega \qquad (27.10)$$

$$= \frac{1}{2\pi}\int_{-\pi}^{\pi} \log\left|\sum_{l=m-N+1}^{m} f[n;m]e^{-j\omega l}\right| e^{j\omega n}\, d\omega \qquad (27.11)$$

Equation (27.11) is still not in a form that can be easily computed. The stDTFT operates on a finite-duration discrete-time frame $f[n;m]$ to produce a continuous-frequency spectrum $\tilde{S}(\omega;m)$. The log magnitude of this spectrum is then operated on by the IDTFT to produce the discrete-quefrency cepstrum $c[n]$. Neither evaluation of the stDTFT over a continuum of frequencies nor computation of the IDTFT is particularly amenable to machine calculation. One way to deal with this situation is to compute an approximation to Eq. (27.11) by sampling $\tilde{S}(\omega;m)$ and computing an IDFT based on these samples. If $\tilde{S}(\omega;m)$ is sampled at $\omega = 2\pi k/N$, $k = 0, 1, \ldots, N-1$, the stDTFT can be replaced by a *short-term* DFT (stDFT):

$$S[k;m] = \sum_{n=m-N+1}^{m} f[n;m]e^{-jk(2\pi/N)n} \quad k = 0, 1, \ldots, N-1$$

where

$$S[k;m] = \tilde{S}\left(\frac{2\pi k}{N};m\right)$$

The approximate cepstrum becomes

$$\tilde{c}[n] = \frac{1}{N}\sum_{k=0}^{N-1} \log\left|\sum_{l=m-N+1}^{m} f[l;m]\, e^{-j(2\pi/N)kl}\right| e^{-j(2\pi/N)kn} \qquad (27.12)$$

The *true* cepstrum $c[n]$ is defined for $n \in [0, \infty]$. On the other hand, the approximate cepstrum $\tilde{c}[n]$ is uniquely defined only for $n \in [0, N-1]$, with $\tilde{c}[n] \stackrel{\triangle}{=} 0$ for

n outside of this range. Based on the relationship between the IDTFT and the IDFT, it can be established that

$$\tilde{c}[n] = \sum_{p=-\infty}^{\infty} c[n+pN] \quad n = 0, 1, \ldots, N-1 \tag{27.13}$$

Equation (27.13) indicates that nonzero values of $c[n]$ for $n \geq N$ will be aliased into the values of $\tilde{c}[n]$ for $n \in [0, N-1]$. To minimize this aliasing, it is customary to *zero-pad* the speech frame after windowing, and thereby increases the effective frame length.

Example 27.1 Compute the real cepstrum for the speech segment shown in Fig. 27.2.

solution The data samples plotted in Fig. 27.2 are provided in file ex27_01.dat. These samples were collected at a rate of 44,100 samp/sec. The program ex27_01 can be used to generate the desired cepstrum. The first 1024 samples from the input file are decimated by a factor of 4, a Hamming window is applied to the surviving 256 samples, and then 3840 zeros are appended to make an effective frame length of 4096. A DFT is performed to generate the magnitude spectrum shown in Fig. 27.4. Taking the common logarithm of this spectrum results in the spectrum shown in Fig. 27.5. Finally, an IDFT is performed to produce the cepstrum shown in Fig. 27.6. The interesting low-quefrency portion of this cepstrum is enlarged in Fig. 27.7. The spike at $q = 0.0055328$ represents the pitch period. This corresponds to a fundamental frequency of 180.74 Hz.

Formant estimation

Cepstral analysis provides a way to estimate the vocal tract resonances from a frame of speech. Figure 27.7, generated for Example 27.1 shows a cepstral peak at $q = 0.0055328$ corresponding to the pitch period. The cepstrum features

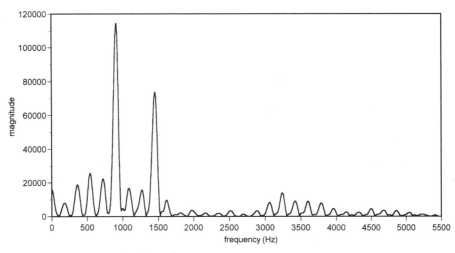

Figure 27.4 DFT output for Example 27.1.

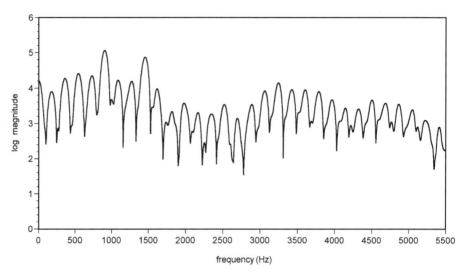

Figure 27.5 Log-magnitude spectrum for Example 27.1.

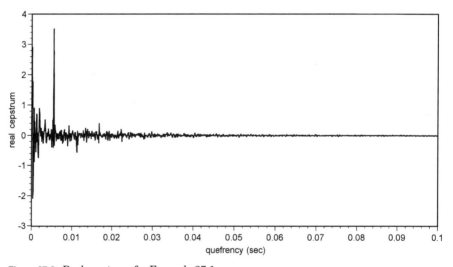

Figure 27.6 Real cepstrum for Example 27.1.

from $q = 0$ to about $q = 0.005$ correspond to the response of the vocal tract. These features can be isolated from the cepstrum by means of *short-pass liftering*. Simply put, use a quefrency window to keep the low-quefrency cepstral features while rejecting the pitch-period spike and everything beyond. Once this windowing is accomplished, simply apply a DFT to the result to produce an estimate of the vocal tract's frequency response. The formants will show up as peaks in this spectrum.

Example 27.2 The program ex27_02 can be used for the computations in this example. If the cepstrum shown in Figs. 27.6 and 27.7 is windowed to remove all content

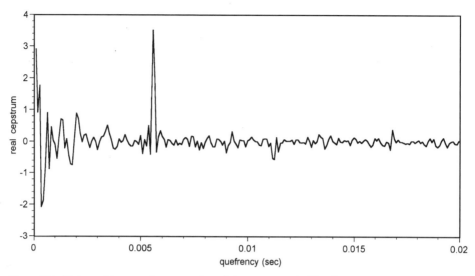

Figure 27.7 Enlarged low-quefrency portion of the cepstrum in Fig. 27.6.

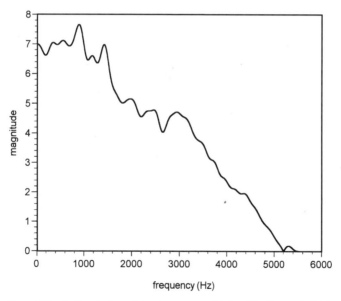

Figure 27.8 Estimated vocal tract response obtained for Example 27.2.

for $q > 0.005$, the resulting vocal tract response obtained via the DFT is shown in Fig. 27.8.

27.3 Nonlinear Quantization of Speech Signals

In applications where sampled speech is transmitted over limited bandwidth communications channels or recorded on digital media, it is advantageous to limit the number of bits per sample. The particular characteristics of speech and how it is perceived by the human ear allow for some significant economies in the design of speech processing systems that must provide intelligible speech at limited bit rates.

When a signal is uniformly quantized, one quantizer count represents the same increment of amplitude across the quantizer's entire range. For example, if a bipolar speech signal ranging between -1 volt and $+1$ volt is uniformly quantized using 8 bits, each quantizer count corresponds to $(1+1)/2^8 = 0.0078125$ volts. Small changes in a strong speech signal are nearly unnoticeable to a listener, but small changes in a weak signal are very noticeable. Therefore, it would make sense to employ a nonuniform quantization scheme that allocates more counts per volt for weak signals and fewer counts per volt for strong signals. Two such schemes, *A-law companding* and *μ-law companding*, are widely used in digital telephony applications. In addition to being justified by hearing perception-based arguments, these nonlinear schemes also exhibit better signal-to-quantization-noise ratios than does linear quantization. The word *compand* comes from a splicing of *compress* and *expand*: linear speech is *compressed* into a nonlinear format for transmission or recording, and then *expanded* back to a linear format during reception or playback.

A-law compounding

The A-law compression scheme is based on a linear characteristic for small amplitudes and a logarithmic characteristic for larger amplitudes

$$c(x) = \begin{cases} \dfrac{Ax}{1+\ln A} & 0 \le |x| \le \dfrac{x_{\max}}{A} \\ x_{\max} \dfrac{1+\ln\left(\dfrac{A|x|}{x_{\max}}\right)}{1+\ln A} \operatorname{sgn}(x) & \dfrac{x_{\max}}{A} < |x| \le x_{\max} \end{cases} \quad (27.14)$$

where x is the linear amplitude and $c(x)$ is the compressed amplitude. The linear portion of Eq. (27.14) is needed for small amplitudes because $\ln(A|x|/x_{\max})$ becomes undefined for $x = x_{\max}/A$ and negative for $|x| < x_{\max}/A$.

A-law compression is most often used with a value of $A = 87.56$. In practice, however, Eq. (27.14) is not used directly. A piecewise linear approximation designed for efficient digital implementation is used instead. The characteristic defined by Eq. (27.14) is approximated by linear pieces using the scheme depicted in Fig. 27.9. It is difficult to depict all of the linear pieces clearly in a single drawing because the abscissa spans 11 octaves. One piece of the linear approximation passes through the origin extending from $x = -x_{\max}/64$ to $x = x_{\max}/64$. There are 6 strictly positive pieces and 6 strictly negative pieces for a total of 13 pieces in all. The 6 positive pieces end at $x = (x_{\max})(2^{-k})$, $k = 5, 4, \ldots, 0$. The details of this scheme are easier to understand if we normalize the linear values such that $x_{\max} = 2048$. A linear value x_{orig} expressed in original units is

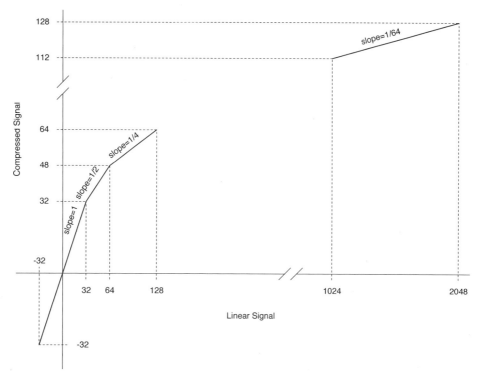

Figure 27.9 Piecewise linear compression characteristic for A87.56 PCM.

converted to the corresponding normalized value x using:

$$x = \frac{2048 x_{\text{orig}}}{x_{\max}}$$

Under this normalization scheme, the central piece of the approximation extends from $x = -32$ to $x = 32$, and the six positive pieces end at $x = 2^k$, $k = 5, 6, \ldots, 11$.

Table 27.2 provides the details of the 8-bit A87.56 compression scheme as it is usually implemented. One bit in the compressed format is reserved for the sign. Three bits are used to represent the *segment code*, thereby allowing for up to 8 positive segments and 8 negative segments. The 13 linear pieces from Fig. 27.9 are mapped into 16 segments by dividing the central piece for $|x| < 32$ into four segments: $-32 < x < -16$, $-16 < x < 0$, $0 < x < 16$, and $16 < x < 32$. The remaining 4 bits of the 8-bit compressed format are used to represent the *level code*. Each segment is divided into 16 equally spaced levels. Assuming $x_{\max} = 2048$ normalization, the *step size* for each level is 1 for segments 0 and 1. The step size for segments 2 through 7 is 2^{k-1} where k is the segment number.

TABLE 27.2 Encoding Table for A87.56 PCM

Input magnitude range	Step size Δ	Segment S	Level L
0–1			0
1–2			1
⋮	1	0	⋮
15–16			15
16–17			0
⋮	1	1	⋮
31–32			15
32–34			0
⋮	2	2	⋮
62–64			15
64–68			0
⋮	4	3	⋮
124–128			15
128–136			0
⋮	8	4	⋮
248–256			15
256–272			0
⋮	16	5	⋮
496–512			15
512–544			0
⋮	32	6	⋮
992–1024			15
1024–1088			0
⋮	64	7	⋮
1984–2048			15

Example 27.3 Assuming $x_{\max} = 1.0$, determine the 12-bit signed-magnitude linear codes and 8-bit A87.56 codes for the following values: $0.1, 0.35, -0.88, 0.9,$ and -0.005.

solution Use of Table 27.2 is made easier if the values of interest are rescaled to correspond to $x_{\max} = 2048$ as in column 2 of Table 27.3. The value in column 2 is compared to the segment ranges in Table 27.2 to determine the corresponding

TABLE 27.3 Signal Values for Example 27.3

	Linear		A87.56		
x	$2048x$	12-bit code	Segment	Level	8-bit code
0.10	204.8	0x0CC	4	9	0x49
0.35	716.8	0x2CC	6	6	0x66
−0.88	−1802.24	0xF0A	−7	12	0xFC
0.95	1945.6	0x799	7	14	0x7E
−0.005	−10.24	0x80A	−0	10	0x8A

A87.56 segment S. The level L can be obtained as

$$L = \left\lfloor \frac{2048x - B_S}{\Delta_S} \right\rfloor$$

where Δ_S is the step size for segment S and B_S is the low end of the input range for segment S. For $x = 0.1$ which falls in segment 4, we have

$$L = \left\lfloor \frac{204.8 - 128}{8} \right\rfloor = \lfloor 9.6 \rfloor = 9$$

The segments, levels and 8-bit compressed codes for each of the given values are listed in Table 27.3.

A-law expansion

Decoding the 8-bit A87.56 values into linear values makes use of the data provided in Table 27.4. It turns out that the decoding operation can be expressed in terms of a simple formula for positive segments:

$$y_P = \begin{cases} L + 0.5 & S = 0 \\ 2^{S-1}(L + 16.5) & S \geq 1 \end{cases} \quad (27.15)$$

A straightforward modification yields the formula for negative segments:

$$y_N = \begin{cases} -(L + 0.5) & S = -0 \\ -2^{|S|-1}(L + 16.5) & S \leq -1 \end{cases} \quad (27.16)$$

The resulting y values will be normalized for $x_{\max} = 2048$. They can be converted back to original units using:

$$y_{\text{orig}} = y \frac{x_{\max}}{2048}$$

TABLE 27.4 Decoding Table for A87.56 PCM

Segment S	Level L	Decoded magnitude
0	0	0.5
	1	1.5
	⋮	⋮
	15	15.5
1	0	16.5
	⋮	⋮
	15	31.5
2	0	33
	⋮	⋮
	15	63
3	0	66
	⋮	⋮
	15	126
4	0	132
	⋮	⋮
	15	252
5	0	264
	⋮	⋮
	15	504
6	0	528
	⋮	⋮
	15	1008
7	0	1056
	⋮	⋮
	15	2016

Example 27.4 Expand the 8-bit A87.56 values from Example 27.3 into linear values.

solution The 8-bit A87.56 code value 0x49 breaks down into $S = 4$ and $L = 9$. The corresponding signal value is obtained using Eq. (27.15):

$$y = 2^{(4-1)}(9 + 16.5) = 204$$

TABLE 27.5 Signal Values for Example 27.4

8-bit code	A87.56		Linear	
	Segment	Level	Normalized value	Denormalized value
0x49	4	9	204	0.099609
0x66	6	6	720	0.351563
0xFC	−7	12	−1824	−0.890625
0x7E	7	14	1952	0.953125
0x8A	−0	10	10.5	0.005127

Assuming $x_{\max} = 1.0$, the corresponding denormalized value is obtained as

$$y_{\text{orig}} = \frac{204}{2048} = 0.099609$$

The expanded values for the given A87.56 values are listed in Table 27.5.

μ-law companding

The μ-law compression scheme is based on a linear characteristic for small amplitudes and a logarithmic characteristic for larger amplitudes

$$c(x) = x_{\max} \frac{\ln\left(1 + \frac{\mu|x|}{x_{\max}}\right)}{\ln(1+\mu)} \operatorname{sgn}(x) \qquad (27.17)$$

where x is the linear amplitude and $c(x)$ is the compressed amplitude.

North American PCM standard μ-law compression uses a value of $\mu = 255$. In practice, however, Eq. (27.17) is not used directly. A piecewise linear approximation designed for efficient digital implementation is used instead. The characteristic defined by Eq. (27.17) is approximated by 15 linear pieces using the scheme depicted in Fig. 27.10. The linear values are normalized such that $x_{\max} = 4079.5$. A linear value x_{orig} expressed in original units is converted to the corresponding normalized value x using:

$$x = \frac{4079.5}{x_{\max}} x_{\text{orig}}$$

Under this normalization scheme, the central piece of the approximation extends from $x = -15.5$ to $x = 15.5$. There are seven strictly positive pieces and seven strictly negative pieces for a total of 15 pieces in all. The seven positive pieces end at normalized values of $x = (2^k - 16.5)$, $k = 6, 7, \ldots, 12$.

Table 27.6 provides the details of the 8-bit $\mu 255$ compression scheme as it is usually implemented. One bit in the compressed format is reserved for the sign. Three bits are used to represent the *segment code*, thereby allowing for up to 8

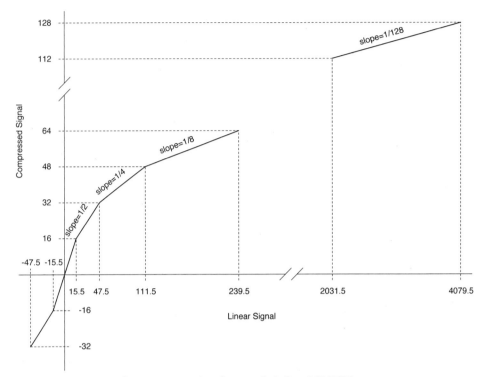

Figure 27.10 Piecewise linear compression characteristic for $\mu 255$ PCM.

positive segments and 8 negative segments. The 15 linear pieces from Fig. 27.10 are mapped into 16 segments by dividing the central piece for $|x| < 15.5$ into two segments: $-15.5 < x < 0$ and $0 < x < 15.5$. The remaining four bits of the 8-bit compressed format are used to represent the *level code*. Each segment is divided into 16 equally spaced levels, with two exceptions. Assuming $x_{\max} = 4079.5$ normalization, the *step size* for each level is 2^k where k is the segment number. The exceptions are the 0 levels for segments -0 and $+0$ which have a step size of 0.5 rather than 1.

Example 27.5 Assuming $x_{\max} = 1.0$, determine the 12-bit signed-magnitude linear codes and 8-bit A87.56 codes for the following values: $0.1, 0.35, -0.88, 0.9$, and -0.005.

solution Use of Table 27.6 is made easier if the values of interest are rescaled to correspond to $x_{\max} = 4079.5$ as in column 2 of Table 27.7. The value in column 2 is compared to the segment ranges in Table 27.6 to determine the corresponding $\mu 255$ segment S. The level L can be obtained as

$$L = \left\lfloor \frac{4079.5 \, |x| - |B_S|}{\Delta_S} \right\rfloor$$

where Δ_S is the step size for segment S and B_S is the low (i.e. smallest magnitude) end

TABLE 27.6 Encoding Table for $\mu 255$ PCM

Input magnitude range	Step size Δ	Segment S	Level L
0–0.5			0
0.5–1.5			1
⋮	1	0	⋮
14.5–15.5			15
15.5–17.5			0
⋮	2	1	⋮
45.5–47.5			15
47.5–51.5			0
⋮	4	2	⋮
107.5–111.5			15
111.5–119.5			0
⋮	8	3	⋮
231.5–239.5			15
239.5–255.5			0
⋮	16	4	⋮
479.5–495.5			15
495.5–527.5			0
⋮	32	5	⋮
975.5–1007.5			15
1007.5–1071.5			0
⋮	64	6	⋮
1967.5–2031.5			15
2031.5–2159.5			0
⋮	128	7	⋮
3951.5–4079.5			15

TABLE 27.7 Signal values for Example 27.5

	Linear		μ255		
x	$204x$	13-bit code	Segment	Level	8-bit code
0.10	407.95	0x0197	4	10	0x4A
0.35	1427.825	0x0593	6	6	0x66
−0.88	−3599.96	0x1E05	−7	12	0xFC
0.95	3875.525	0x0F23	7	14	0x7E
−0.0005	−20.3975	0x1014	−1	2	0x92

of the input range for segment S. For $x = -0.88$ which falls in segment -7, we have

$$L = \left\lfloor \frac{3589.96 - 2031.5}{128} \right\rfloor = \lfloor 12.175 \rfloor = 12$$

The segments, levels and 8-bit compressed codes for each of the given values are listed in Table 27.7

μ-law expansion

Decoding the 8-bit $\mu 255$ values into linear values makes use of the data provided in Table 27.8. It turns out that the decoding operation can be expressed in terms of a simple formula for positive segments:

$$y_F = \begin{cases} L & S = 0 \\ 2^S(L + 16.5) - 16.5 & S \geq 1 \end{cases} \qquad (27.18)$$

A straightforward modification yields the formula for negative segments:

$$y_N = \begin{cases} -L & S = -0 \\ -[2^{|S|}(L + 16.5) - 16.5 & S \leq -1 \end{cases} \qquad (27.19)$$

The resulting y values will be normalized for $x_{\max} = 4079.5$. They can be converted back to original units using:

$$y_{\text{orig}} = y \frac{x_{\max}}{4079.5}$$

Example 27.6 Expand the 8-bit $\mu 255$ values from Example 27.5 into linear values.

solution The 8-bit $\mu 255$ code value 0x4A breaks down into $S = 4$ and $L = 10$. The corresponding signal value is obtained using Eq. (27.18):

$$y = 2^4(10 + 16.5) - 16.5 = 407.5$$

TABLE 27.8 Decoding Table for $\mu 255$ PCM

Segment S	Level L	Decoded magnitude
0	0	0
	1	1
	⋮	⋮
	15	15
1	0	16.5
	⋮	⋮
	15	46.5
2	0	49.5
	⋮	⋮
	15	109.5
3	0	115.5
	⋮	⋮
	15	235.5
4	0	247.5
	⋮	⋮
	15	487.5
5	0	511.5
	⋮	⋮
	15	991.5
6	0	1039.5
	⋮	⋮
	15	1999.5
7	0	2095.5
	⋮	⋮
	15	4015.5

Assuming $x_{\max} = 1.0$, the corresponding denormalized value is obtained as

$$y_{\text{orig}} = \frac{407.5}{4079.5} = 0.09989$$

The expanded values for the given $\mu 255$ values are listed in Table 27.9.

TABLE 27.9 Signal Values for Example 27.6

8-bit code	μ255		Linear	
	Segment	Level	Normalized value	Denormalized value
0x4A	4	10	407.5	0.09989
0x66	6	6	1423.5	0.34894
0xFC	−7	12	−3631.5	−0.890183
0x7E	7	14	3887.5	0.952935
0x92	1	2	20.5	0.005025

References

1. Noll, A. Michael. "Cepstrum Pitch Determination," *J. Acoust. Soc. Am.*, Vol. 41, pp. 293–309, February 1967.
2. Bogert, B. P., M. J. R. Healy, and J. W. Tukey. "The Quefrency Alanysis of Time Series for Echos: Cepstrum, Pseudo-Autocovariance, Cross-Cepstrum and Saphe Cracking," *Proceedings of the Symposium on Time Series Analysis* (M. Rosenblatt, Ed.), pp. 209–243, Wiley, New York, 1963.
3. Deller, J. R. Jr., J. G. Proakis, and J. H. Hansen. *Discrete-Time Processing of Speech Signals*, Macmillan Pub. Co., New York, 1993.

Appendix A

Mathematical Tools

A.1 Exponentials and Logarithms

Exponentials

There is an irrational number, usually denoted as e, that is of great important in virtually all fields of science and engineering. This number is defined by

$$e \triangleq \lim_{x \to +\infty} \left(1 + \frac{1}{x}\right)^x \cong 2.71828\ldots \qquad (A.1)$$

Unfortunately, this constant remains unnamed, and writers are forced to settle for calling it *the number e* or perhaps *the base of natural logarithms*. The letter e was first used to denote the irrational in Eq. (A.1) by Leonhard Euler (1707–1783), so it would seem reasonable to refer to the number under discussion as *Euler's constant*. Such is not the case, however, as the term *Euler's constant* is attached to the constant γ defined by

$$\gamma = \lim_{N \to \infty} \left(\sum_{n=1}^{N} \frac{1}{n} - \log_e N\right) \cong 0.577215664 \qquad (A.2)$$

The number e is most often encountered in situations where it is raised to some real or complex power. The notation $\exp(x)$ often is used in place of e^x, since the former can be written more clearly and typeset more easily than the latter—especially in cases where the exponent is a complicated expression rather than a single variable. The value for e raised to a complex power z can be expanded in an infinite series

$$\exp(z) = \sum_{n=0}^{\infty} \frac{z^n}{n!} \qquad (A.3)$$

The series in Eq. (A.3) converges for all complex z having finite magnitude.

Logarithms

The *common logarithm*, or *base-10 logarithm*, of a number x is equal to the power to which 10 must be raised in order to equal x:

$$y = \log_{10} x \Leftrightarrow x = 10^y \qquad (A.4)$$

The *natural logarithm*, or *base-e logarithm*, of a number x is equal to the power to which e must be raised in order to equal x:

$$y = \log_e x \Leftrightarrow x = \exp(y) \equiv e^y \qquad (A.5)$$

Natural logarithms are also called *Napierian logarithms* in honor of John Napier (1550–1617), a Scottish amateur mathematician who in 1614 published the first account of logarithms in *Mirifici logarithmorum canonis descripto* ("A Description of the Marvelous Rule of Logarithms") (see [1]). The concept of logarithms can be extended to any positive base b, with the base-b logarithm of a number x equaling the power to which the base must be raised in order to equal x:

$$y = \log_b x \Leftrightarrow x = b^y \qquad (A.6)$$

The notation *log* without a base explicitly indicated usually denotes a common logarithm, although sometimes this notation is used to denote natural logarithms (especially in some of the older literature). More often, the notation *ln* is used to denote a natural logarithm. Logarithms exhibit a number of properties that are listed in Table A.1. Entry 1 is sometimes offered as the definition of natural logarithms. The multiplication property in entry 3 is the theoretical basis on which the design of the slide rule is based.

TABLE A.1 Properties of Logarithms

1.	$\ln x = \int_1^x \frac{1}{y} dy \quad x > 0$
2.	$\frac{d}{dx}(\ln x) = \frac{1}{x} \quad x > 0$
3.	$\log_b(xy) = \log_b x + \log_b y$
4.	$\log_b\left(\frac{1}{x}\right) = -\log_b x$
5.	$\log_b c(y^x) = x \log_b y$
6.	$\log_c c = (\log_b x)(\log_c b) = \frac{\log_b x}{\log_b c}$
7.	$\ln(1+z) = \sum_{n=1}^{\infty} (-1)^{n-1} \frac{z^n}{n} \quad |z| < 1$

Decibels

Consider a system that has an output power of P_{out} and an output voltage of V_{out} given an input power of P_{in} and an input voltage of V_{in}. The gain G, in decibels (dB), of the system is given by

$$G_{dB} = 10 \log_{10}\left(\frac{P_{out}}{P_{in}}\right) = 10 \log_{10}\left(\frac{V_{out}^2/Z_{out}}{V_{in}^2/Z_{in}}\right) \tag{A.7}$$

If the input and output impedances are equal, Eq. (A.7) reduces to

$$G_{dB} = 10 \log_{10}\left(\frac{V_{out}^2}{V_{in}^2}\right) = 20 \log_{10}\left(\frac{V_{out}}{V_{in}}\right) \tag{A.8}$$

Example A.1 An amplifier has a gain of 17.0 dB. For a 3-mW input, what will the output power be? Substituting the given data into Eq. (A.7) yields

$$17.0 \text{ dB} = 10 \log_{10}\left(\frac{P_{out}}{3 \times 10^{-3}}\right)$$

Solving for P_{out} then produces

$$P_{out} = (3 \times 10^{-3})10^{(17/10)} = 1.5 \times 10^{-1} = 150 \text{ mW}$$

Example A.2 What is the range in decibels of the values that can be represented by an 8-bit unsigned integer?

solution The smallest value is 1, and the largest value is $2^8 - 1 = 255$. Thus

$$20 \log_{10}\left(\frac{255}{1}\right) = 48.13 \text{ dB}$$

The abbreviation dBm is used to designate power levels relative to 1 milliwatt (mW). For example:

$$30 \text{ dBm} = 10 \log_{10}\left(\frac{P}{10^{-3}}\right)$$

$$P = (10^{-3})(10^3) = 10^0 = 1.0 \text{ W}$$

A.2 Complex Numbers

A complex number z has the form $a + bj$, where a and b are real and $j = \sqrt{-1}$. The *real part* of z is a, and the *imaginary part* of z is b. Mathematicians use i to denote $\sqrt{-1}$, but electrical engineers use j to avoid confusion with the traditional use of i for denoting current. For convenience, $a + bj$ is sometimes represented by the ordered pair (a, b). The *modulus* or *absolute value* of z is denoted as $|z|$ and is defined by

$$|z| = |a + bj| = \sqrt{a^2 + b^2} \tag{A.9}$$

The *complex conjugate* of z is denoted as z^* and is defined by

$$(z = a + bj) \Leftrightarrow (z^* = a - bj) \tag{A.10}$$

Conjugation distributes over addition, multiplication, and division:

$$(z_1 + z_2)^* = z_1^* + z_2^* \tag{A.11}$$

$$(z_1 z_2)^* = z_1^* z_2^* \tag{A.12}$$

$$\left(\frac{z_1}{z_2}\right)^* = \frac{z_1^*}{z_2^*} \tag{A.13}$$

Operations on complex numbers in rectangular form

Consider two complex numbers:

$$z_1 = a + bj \qquad z_2 = c + dj$$

The four basic arithmetic operations are then defined as

$$z_1 + z_2 = (a + c) + j(b + d) \tag{A.14}$$

$$z_1 - z_2 = (a - c) + j(b - d) \tag{A.15}$$

$$z_1 z_2 = (ac - bd) + j(ad + bc) \tag{A.16}$$

$$\frac{z_1}{z_2} = \frac{ac + bd}{c^2 + d^2} + j\frac{bc - ad}{c^2 + d^2} \tag{A.17}$$

Polar form of complex numbers

A complex number of the form $a + bj$ can be represented by a point in a coordinate plane as shown in Fig. A.1. Such a representation is called an *Argand diagram* [2] in honor of Jean Robert Argand (1768–1822), who published a description of this graphical representation of complex numbers in 1806 [1]. The point representing $a + bj$ can also be located using an angle θ and radius r as shown. From the definitions of sine and cosine given in Eqs. (A.25) and (A.26) of Sec. A.3, it follows that

$$a = r\cos\theta \qquad b = r\sin\theta$$

Therefore,
$$z = r\cos\theta + jr\sin\theta = r(\cos\theta + j\sin\theta) \tag{A.18}$$

The quantity $(\cos\theta + j\sin\theta)$ is sometimes denoted as "cis θ". Making use of Eq. (A.58) from Sec. A.3, we can rewrite Eq. (A.18) as

$$z = r\,\text{cis}\,\theta = r\exp(j\theta) \tag{A.19}$$

The form in Eq. (A.19) is called the *polar form* of the complex number z.

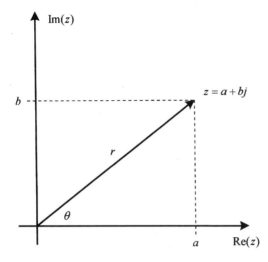

Figure A.1 Argand diagram representation of a complex number.

Operations on complex numbers in polar form

Consider three complex numbers:

$$z = r(\cos\theta + j\sin\theta) = r\exp(j\theta)$$
$$z_1 = r_1(\cos\theta_1 + j\sin\theta_1) = r_1\exp(j\theta_1)$$
$$z_2 = r_2(\cos\theta_2 + j\sin\theta_2) = r_2\exp(j\theta_2)$$

Several operations can be conveniently performed directly upon complex numbers that are in polar form, as follows.

Multiplication

$$\begin{aligned}z_1 z_2 &= r_1 r_2[\cos(\theta_1 + \theta_2) + j\sin(\theta_1 + \theta_2)]\\ &= r_1 r_2 \exp[j(\theta_1 + \theta_2)]\end{aligned} \quad (A.20)$$

Division

$$\begin{aligned}\frac{z_1}{z_2} &= \frac{r_1}{r_2}[\cos(\theta_1 - \theta_2) + j\sin(\theta_1 - \theta_2)]\\ &= \frac{r_1}{r_2}\exp[j(\theta_1 - \theta_2)]\end{aligned} \quad (A.21)$$

Powers

$$\begin{aligned}z^n &= r^n[\cos(n\theta) + j\sin(n\theta)]\\ &= r^n \exp(jn\theta)\end{aligned} \quad (A.22)$$

Roots

$$\sqrt[n]{z} = z^{1/n} = r^{1/n}\left[\cos\left(\frac{\theta + 2k\pi}{n}\right) + j\sin\left(\frac{\theta + 2k\pi}{n}\right)\right]$$

$$= r^{1/n}\exp\left[\frac{j(\theta + 2k\pi)}{n}\right] \quad k = 0, 1, 2, \ldots \quad (A.23)$$

Equation (A.22) is known as *De Moivre's theorem*. In 1730, an equation similar to Eq. (A.23) was published by Abraham De Moivre (1667–1754) in his *Miscellanea analytica* [1]. In Eq. (A.23), for a fixed n as k increases, the sinusoidal functions will take on only n distinct values. Thus, there are n different nth roots of any complex number.

Logarithms of complex numbers

For the complex number $z = r\exp(j\theta)$, the natural logarithm of z is given by

$$\begin{aligned}\ln z &= \ln[r\exp(j\theta)] \\ &= \ln\{r\exp[j(\theta + 2k\pi)]\} \\ &= (\ln r) + j(\theta + 2k\pi) \quad k = 0, 1, 2, \ldots \quad (A.24)\end{aligned}$$

The *principal value* is obtained when $k = 0$.

A.3 Trigonometry

For x, y, r, and θ as shown in Fig. A.2, the six trigonometric functions of the angle θ are defined as

$$\text{Sine:} \quad \sin\theta = \frac{y}{r} \quad (A.25)$$

$$\text{Cosine:} \quad \cos\theta = \frac{x}{r} \quad (A.26)$$

$$\text{Tangent:} \quad \tan\theta = \frac{y}{x} \quad (A.27)$$

$$\text{Cosecant:} \quad \csc\theta = \frac{r}{y} \quad (A.28)$$

$$\text{Secant:} \quad \sec\theta = \frac{r}{x} \quad (A.29)$$

$$\text{Cotangent:} \quad \cot\theta = \frac{x}{y} \quad (A.30)$$

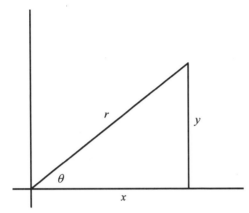

Figure A.2 An angle in the cartesian plane.

Phase shifting of sinusoids

A number of useful equivalences can be obtained by adding particular phase angles to the arguments of sine and cosine functions:

$$\cos(\omega t) = \sin\left(\omega t + \frac{\pi}{2}\right) \tag{A.31}$$

$$\cos(\omega t) = \cos(\omega t + 2n\pi) \quad n = \text{any integer} \tag{A.32}$$

$$\sin(\omega t) = \sin(\omega t + 2n\pi) \quad n = \text{any integer} \tag{A.33}$$

$$\sin(\omega t) = \cos\left(\omega t - \frac{\pi}{2}\right) \tag{A.34}$$

$$\cos(\omega t) = \cos[\omega t + (2n+1)\pi] \quad n = \text{any integer} \tag{A.35}$$

$$\sin(\omega t) = -\sin[\omega t + (2n+1)\pi] \quad n = \text{any integer} \tag{A.36}$$

Trigonometric identities

The following trigonometric identities often prove useful in the design and analysis of signal processing systems.

$$\tan x = \frac{\sin x}{\cos x} \tag{A.37}$$

$$\sin(-x) = -\sin x \tag{A.38}$$

$$\cos(-x) = \cos x \tag{A.39}$$

$$\tan(-x) = -\tan x \tag{A.40}$$

$$\cos^2 x + \sin^2 x = 1 \tag{A.41}$$

$$\cos^2 x = \frac{1}{2}[1 + \cos(2x)] \tag{A.42}$$

$$\sin(x \pm y) = (\sin x)(\cos y) \pm (\cos x)(\sin y) \tag{A.43}$$

$$\cos(x \pm y) = (\cos x)(\cos y) \mp (\sin x)(\sin y) \tag{A.44}$$

$$\tan(x + y) = \frac{(\tan x) + (\tan y)}{1 - (\tan x)(\tan y)} \tag{A.45}$$

$$\sin(2x) = 2(\sin x)(\cos x) \tag{A.46}$$

$$\cos(2x) = \cos^2 x - \sin^2 x \tag{A.47}$$

$$\tan(2x) = \frac{2(\tan x)}{1 - \tan^2 x} \tag{A.48}$$

$$(\sin x)(\sin y) = \frac{1}{2}[-\cos(x + y) + \cos(x - y)] \tag{A.49}$$

$$(\cos x)(\cos y) = \frac{1}{2}[\cos(x + y) + \cos(x - y)] \tag{A.50}$$

$$(\sin x)(\cos y) = \frac{1}{2}[\sin(x + y) + \sin(x - y)] \tag{A.51}$$

$$(\sin x) + (\sin y) = 2 \sin \frac{x + y}{2} \cos \frac{x - y}{2} \tag{A.52}$$

$$(\sin x) - (\sin y) = 2 \sin \frac{x - y}{2} \cos \frac{x + y}{2} \tag{A.53}$$

$$(\cos x) + (\cos y) = 2 \cos \frac{x + y}{2} \cos \frac{x - y}{2} \tag{A.54}$$

$$(\cos x) - (\cos y) = -2 \sin \frac{x + y}{2} \sin \frac{x - y}{2} \tag{A.55}$$

$$A \cos(\omega t + \psi) + B \cos(\omega t + \phi) = C \cos(\omega t + \theta) \tag{A.56}$$

where $C = [A^2 + B^2 - 2AB \cos(\phi - \psi)]^{1/2}$
$\theta = \tan^{-1}\left(\frac{A \sin \psi + B \sin \phi}{A \cos \psi + B \cos \phi}\right)$

$$A \cos(\omega t + \psi) + B \sin(\omega t + \phi) = C \cos(\omega t + \theta) \tag{A.57}$$

where $C = [A^2 + B^2 - 2AB \sin(\phi - \psi)]^{1/2}$
$\theta = \tan^{-1}\left(\frac{A \sin \psi - B \cos \phi}{A \cos \psi + B \sin \phi}\right)$

Euler's identities

The following four equations, called *Euler's identities*, relate sinusoids and complex exponentials.

$$e^{jx} = \cos x + j \sin x \tag{A.58}$$

$$e^{-jx} = \cos x - j \sin x \tag{A.59}$$

$$\cos x = \frac{e^{jx} + e^{-jx}}{2} \tag{A.60}$$

$$\sin x = \frac{e^{jx} - e^{-jx}}{2j} \tag{A.61}$$

Series and product expansions

Listed following are infinite series expansions for the various trigonometric functions [3].

$$\sin x = \sum_{n=0}^{\infty} \frac{(-1)^n x^{2n+1}}{(2n+1)!} \tag{A.62}$$

$$\cos x = \sum_{n=0}^{\infty} \frac{(-1)^n x^{2n}}{(2n)!} \tag{A.63}$$

$$\tan x = \sum_{n=1}^{\infty} \frac{(-1)^{n-1} 2^{2n}(2^{2n} - 1) B_{2n} x^{2n-1}}{(2n)!} \quad |x| < \frac{\pi}{2} \tag{A.64}$$

$$\cot x = \sum_{n=0}^{\infty} \frac{(-1)^n 2^{2n} B_{2n} x^{2n} - 1}{(2n)!} \quad |x| < \pi \tag{A.65}$$

$$\sec x = \sum_{n=0}^{\infty} \frac{(-1)^n E_{2n} x^{2n}}{(2n)!} \quad |x| < \frac{\pi}{2} \tag{A.66}$$

$$\csc x = \sum_{n=0}^{\infty} \frac{(-1)^{n-1} 2(2^{2n-1} - 1) B_{2n} x^{2n-1}}{(2n)!} \quad |x| < \pi \tag{A.67}$$

Values for the Bernoulli number B_n and Euler number E_n are listed in Tables A.2 and A.3, respectively. In some instances, the infinite product expansions for sine and cosine may be more convenient than the series expansions.

$$\sin x = x \prod_{n=1}^{\infty} \left(1 - \frac{x^2}{n^2 \pi^2}\right) \tag{A.68}$$

TABLE A.2 Bernoulli Numbers

$B_n N/D$ for $n = 0, 1, 2, 4, \ldots$

$B_n = 0$ for $n = 3, 5, 7, \ldots$

n	N	D
0	1	1
1	−1	2
2	1	6
4	−1	30
6	1	42
8	−1	30
10	5	66
12	−691	2730
14	7	6
16	−3617	510
18	43867	798
20	−174611	330

TABLE A.3 Euler Numbers

$E_n = 0$ for $n = 1, 3, 5, 7, \ldots$

n	E_n
0	1
2	−1
4	5
6	−61
8	1385
10	−50521
12	2,702,765
14	−199,360,981
16	19,391,512,145
18	−2,404,879,675,441
20	370,371,188,237,525

$$\cos x = \prod_{n=1}^{\infty}\left[1 - \frac{4x^2}{(2n-1)^2\pi^2}\right] \tag{A.69}$$

Orthonormality of sine and cosine

Two functions $\phi_1(t)$ and $\phi_2(t)$ are said to form an *orthogonal set* over the interval $[0, T]$ if

$$\int_0^T \phi_1(t)\phi_2(t)\, dt = 0 \tag{A.70}$$

Two functions $\phi_1(t)$ and $\phi_2(t)$ are said to form an *orthonormal set* over the interval $[0, T]$ if, in addition to satisfying Eq. (A.70), each function has unit energy over the interval

$$\int_0^T [\phi_1(t)]^2\, dt = \int_0^T [\phi_2(t)]^2\, dt = 1 \tag{A.71}$$

Consider the two signals given by

$$\phi_1(t) = A\sin(\omega_0 t) \tag{A.72}$$

$$\phi_2(t) = A\cos(\omega_0 t) \tag{A.73}$$

The signals ϕ_1 and ϕ_2 will form an orthonormal set over the interval $[0, T]$ if $\omega_0 T$ is an integer multiple of π. The set will be orthonormal as well as orthogonal if $A_2 = 2/T$. The signals ϕ_1 and ϕ_2 will form an approximately orthonormal set over the interval $[0, T]$ if $\omega_0 T \gg 1$ and $A^2 = 2/T$. The orthonormality of sine and cosine can be derived as follows.

Substitution of Eqs. (A.72) and (A.73) into Eq. (A.70) yields

$$\int_0^T \phi_1(t)\phi_2(t)\, dt = A^2 \int_0^T \sin\omega_0 t \cos\omega_0 t\, dt$$

$$= \frac{A^2}{2} \int_0^T [\sin(\omega_0 t + \omega_0 t) + \sin(\omega_0 t - \omega_0 t)]\, dt$$

$$= \frac{A^2}{2} \int_0^T \sin 2\omega_0 t\, dt = \frac{A^2}{2}\left(\frac{\cos 2\omega_0 t}{2\omega_0}\right)\Big|_{t=0}^{T}$$

$$= \frac{A^2}{4\omega_0 T}(1 - \cos 2\omega_0 T) \tag{A.74}$$

Thus, if $\omega_0 T$ is an integer multiple of π, then $\cos(2\omega_0 T) = 1$ and ϕ_1 and ϕ_2 will be orthogonal. If $\omega_0 T \gg 1$, then Eq. (A.74) will be very small and reasonably approximated by zero; thus ϕ_1 and ϕ_2 can be considered as approximately orthogonal. The energy of $\phi_1(t)$ on the interval $[0, T]$ is given by

$$E_1 = \int_0^T [\phi_1(t)]^2 \, dt = A^2 \int_0^T \sin^2 \omega_0 t \, dt$$

$$= A^2 \left(\frac{t}{2} - \frac{\sin 2\omega_0 t}{4\omega_0} \right) \bigg|_{t=0}^T$$

$$= A^2 \left(\frac{T}{2} - \frac{\sin 2\omega_0 T}{4\omega_0} \right) \tag{A.75}$$

For ϕ_1 to have unit energy, A^2 must satisfy

$$A^2 = \left(\frac{T}{2} - \frac{\sin 2\omega_0 T}{4\omega_0} \right)^{-1} \tag{A.76}$$

When $\omega_0 T = n\pi$, then $\sin 2\omega_0 T = 0$. Thus, Eq. (A.76) reduces to

$$A = \sqrt{\frac{2}{T}} \tag{A.77}$$

Substituting (A.77) into (A.75) yields

$$E_1 = 1 - \frac{\sin 2\omega_0 T}{2\omega_0 T} \tag{A.78}$$

When $\omega_0 T \gg 1$, the second term of Eq. (A.78) will be extremely small and reasonably approximated by zero, thus indicating that ϕ_1 and ϕ_2 are approximately orthonormal. In a similar manner, the energy of $\phi_2(t)$ can be found to be

$$E_2 = A^2 \int_0^T \cos^2 \omega_0 t \, dt$$

$$= A^2 \left(\frac{T}{2} + \frac{\sin 2\omega_0 T}{4\omega_0} \right) \tag{A.79}$$

Thus

$$E_2 = 1 \quad \text{if} \quad A = \sqrt{\tfrac{2}{T}} \quad \text{and} \quad \omega_0 T = n\pi$$

$$E_2 \doteq 1 \quad \text{if} \quad A = \sqrt{\tfrac{2}{T}} \quad \text{and} \quad \omega_0 T \gg 1$$

A.4 Derivatives

The following are some derivative forms that often prove useful in theoretical analysis of communication systems.

$$\frac{d}{dx}\sin u = \cos u \frac{du}{dx} \tag{A.80}$$

$$\frac{d}{dx}\cos u = -\sin u \frac{du}{dx} \tag{A.81}$$

$$\frac{d}{dx}\tan u = \sec^2 u \frac{du}{dx} = \frac{1}{\cos^2 u}\frac{du}{dx} \tag{A.82}$$

$$\frac{d}{dx}\cot u = \csc^2 u \frac{du}{dx} = \frac{1}{\sin^2 u}\frac{du}{dx} \tag{A.83}$$

$$\frac{d}{dx}\sec u = \sec u \tan u \frac{du}{dx} = \frac{\sin u}{\cos^2 u}\frac{du}{dx} \tag{A.84}$$

$$\frac{d}{dx}\csc u = -\csc u \cot u \frac{du}{dx} = \frac{-\cos u}{\sin^2 u}\frac{du}{dx} \tag{A.85}$$

$$\frac{d}{dx}e^u = e^u \frac{du}{dx} \tag{A.86}$$

$$\frac{d}{dx}\ln u = \frac{1}{u}\frac{du}{dx} \tag{A.87}$$

$$\frac{d}{dx}\log u = \frac{\log e}{u}\frac{du}{dx} \tag{A.88}$$

$$\frac{d}{dx}\left(\frac{u}{v}\right) = \frac{1}{v^2}\left(v\frac{du}{dx} - u\frac{dv}{dx}\right) \tag{A.89}$$

Derivatives of polynomial ratios

Consider a ratio of polynomials given by

$$C(s) = \frac{A(s)}{B(s)} \quad B(s) \neq 0 \tag{A.90}$$

The derivative of $C(s)$ can be obtained using Eq. (A.89) to obtain

$$\frac{d}{ds}C(s) = [B(s)]^{-1}\frac{d}{ds}A(s) - A(s)[B(s)]^{-2}\frac{d}{ds}B(s) \tag{A.91}$$

Equation (A.91) will be useful in the application of the Heaviside expansion.

A.5 Integration

Large integral tables fill entire volumes and contain thousands of entries. However, a relatively small number of integral forms appear over and over again in the study of signal processing, and these are listed below.

$$\int \frac{1}{x} dx = \ln x \tag{A.92}$$

$$\int e^{ax} dx = \frac{1}{a} e^{ax} \tag{A.93}$$

$$\int x e^{ax} dx = \frac{ax - 1}{a^2} e^{ax} \tag{A.94}$$

$$\int \sin(ax) dx = -\frac{1}{a} \cos(ax) \tag{A.95}$$

$$\int \cos(ax) dx = \frac{1}{a} \sin(ax) \tag{A.96}$$

$$\int \sin(ax + b) dx = -\frac{1}{a} \cos(ax + b) \tag{A.97}$$

$$\int \cos(ax + b) dx = \frac{1}{a} \sin(ax + b) \tag{A.98}$$

$$\int x \sin(ax) dx = -\frac{x}{a} \cos(ax) + \frac{1}{a^2} \sin(ax) \tag{A.99}$$

$$\int x \cos(ax) dx = \frac{x}{a} \sin(ax) + \frac{1}{a^2} \cos(ax) \tag{A.100}$$

$$\int \sin^2 ax\, dx = \frac{x}{2} - \frac{\sin 2ax}{4a} \tag{A.101}$$

$$\int \cos^2 ax\, dx = \frac{x}{2} + \frac{\sin 2ax}{4a} \tag{A.102}$$

$$\int x^2 \sin ax\, dx = \frac{1}{a^3}(2ax \sin ax + 2 \cos ax - a^2 x^2 \cos ax) \tag{A.103}$$

$$\int x^2 \cos ax\, dx = \frac{1}{a^3}(2ax \cos ax - 2 \sin ax + a^2 x^2 \cos ax) \tag{A.104}$$

$$\int \sin^3 x \, dx = -\frac{1}{3} \cos x (\sin^2 x + 2) \tag{A.105}$$

$$\int \cos^3 x \, dx = \frac{1}{3} \sin x (\cos^2 x + 2) \tag{A.106}$$

$$\int \sin x \cos x \, dx = \frac{1}{2} \sin^2 x \tag{A.107}$$

$$\int \sin(mx) \cos(nx) \, dx = \frac{-\cos(m-n)x}{2(m-n)} - \frac{\cos(m+n)x}{2(m+n)} \quad (m^2 \neq n^2) \tag{A.108}$$

$$\int \sin^2 x \cos^2 x \, dx = \frac{1}{8} \left[x - \frac{1}{4} \sin(4x) \right] \tag{A.109}$$

$$\int \sin x \cos^m x \, dx = \frac{-\cos^{m+1} x}{m+1} \tag{A.110}$$

$$\int \sin^m x \cos x \, dx = \frac{\sin^{m+1} x}{m+1} \tag{A.111}$$

$$\int \cos^m x \sin^n x \, dx = \frac{\cos^{m-1} x \sin^{n+1} x}{m+n} + \frac{m-1}{m+n} \int \cos^{m-2} x \sin^n x \, dx \quad (m \neq -n) \tag{A.112}$$

$$\int \cos^m x \sin^n x \, dx = \frac{-\cos^{m+1} x \sin^{n-1} x}{m+n} + \frac{n-1}{m+n} \int \cos^m x \sin^{n-2} x \, dx \quad (m \neq -n) \tag{A.113}$$

$$\int u \, dv = uv - \int v \, du \tag{A.114}$$

A.6 Dirac Delta Function

The *delta function* or *impulse function*, is denoted as $\delta(t)$ and is usually depicted as a vertical arrow at the origin as shown in Fig. A.3. This function is often called the *Dirac delta function* in honor of Paul Dirac (1902–1984), an English physicist who used delta functions extensively in his work on quantum mechanics. A number of nonrigorous approaches for defining the impulse function can be found throughout the literature. A *unit impulse* is often loosely described as having a zero width and an infinite amplitude at the origin such that the total area under the impulse is equal to unity. How is it possible to claim that 0 times infinity equals 1? The trick involves defining a sequence of functions

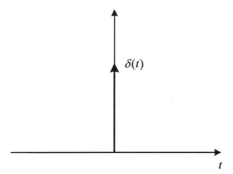

Figure A.3 Graphical representation of the Dirac delta function.

$f_n(t)$ such that

$$\int_{-\infty}^{\infty} f_n(t)\,dt = 1 \qquad (A.115)$$

and

$$\lim_{n\to\infty} f_n(t) = 0 \quad \text{for} \quad t \neq 0 \qquad (A.116)$$

The delta function is then defined as

$$\delta(t) = \lim_{n\to\infty} f_n(t) \qquad (A.117)$$

Example A.3 Let a sequence of pulse functions $f_n(t)$ be defined as

$$f_n(t) = \begin{cases} \frac{n}{2} & |t| \leq \frac{1}{n} \\ 0 & \text{otherwise} \end{cases} \qquad (A.118)$$

Equation (A.115) is satisfied since the area of pulse is equal to $(2n)(n/2) = 1$ for all n. The pulse width decreases and the pulse amplitude increases as n approaches infinity. Therefore, we intuitively sense that this sequence must also satisfy Eq. (A.116). Thus the impulse function can be defined as the limit of Eq. (A.118) as n approaches inifinity. Using similar arguments, it can be shown that the impulse can also be defined as the limit of a sequence of sinc functions (see Sec. A.7) or Gaussian pulse functions.

A second approach entails simply defining $\delta(t)$ to be that function which satisfies

$$\int_{-\infty}^{\infty} \delta(t)\,dt = 1 \quad \text{and} \quad \delta(t) = 0 \quad \text{for} \quad t \neq 0 \qquad (A.119)$$

In a third approach, $\delta(t)$ is defined as that function that exhibits the property

$$\int_{-\infty}^{\infty} \delta(t) f(t)\,dt = f(0) \qquad (A.120)$$

While any of these approaches is adequate to introduce the delta function into an engineer's repertoire of analytical tools, none of the three is sufficiently rigorous to satisfy mathematicians or discerning theoreticians. In particular, notice that none of the approaches presented deals with the thorny issue of just what the value of $\delta(t)$ is for $t = 0$. The rigorous definition of $\delta(t)$ introduced in 1950 by Laurent Schwartz [4] rejects the notion that the impulse is an ordinary function and instead defines it as a *distribution*.

Distributions

Let S be the set of functions $f(x)$ for which the nth derivative $f^{[n]}(x)$ exists for any n and all x. Furthermore, each $f(x)$ decreases sufficiently fast at infinity such that

$$\lim_{x \to \infty} x^n f(x) = 0 \quad \text{for all } n \tag{A.121}$$

A *distribution*, often denoted $\phi(x)$, is defined as a continuous linear mapping from the set S to the set of complex numbers. Notationally, this mapping is represented as an inner product

$$\int_{-\infty}^{\infty} \phi(x) f(x) \, dx = z \tag{A.122}$$

or alternatively

$$\langle \phi(x), f(x) \rangle = z \tag{A.123}$$

Notice that no claim is made that ϕ is a function capable of mapping values of x into corresponding values $\phi(x)$. In some texts such as [5], $\phi(x)$ is referred to as a *functional* or as a *generalized function*. The distribution ϕ is defined only through the impact that it has upon other functions. The impulse function is a distribution defined by the following:

$$\int_{-\infty}^{\infty} \delta(t) f(t) \, dt = f(0) \tag{A.124}$$

Equation (A.124) looks exactly like Eq. (A.120), but defining $\delta(t)$ as a distribution eliminates the need to deal with the issue of assigning a value to $\delta(0)$. Furthermore, the impulse function is elevated to a more substantial foundation from which several useful properties may be rigorously derived. For a more in-depth discussion of distributions other than $\delta(t)$, the interested reader is referred to Chap. 4 of Weaver [6].

Properties of the delta distribution

It has been shown [5–8] that the delta distribution exhibits the following properties:

$$\int_{-\infty}^{\infty} \delta(t)\,dt = 1 \tag{A.125}$$

$$\frac{d}{dt}\delta(t) = \lim_{\tau \to 0} \frac{\delta(t) - \delta(t-\tau)}{\tau} \tag{A.126}$$

$$\int_{-\infty}^{\infty} \delta(t - t_0) f(t)\,dt = f(t_0) \tag{A.127}$$

$$\delta(at) = \frac{1}{|a|}\delta(t) \tag{A.128}$$

$$\delta(t_0) f(t) = f(t_0)\delta(t_0) \tag{A.129}$$

where $f(t)$ is an ordinary function that is continuous at $t = t_0$.

$$\delta_1(t - t_1) * \delta_2(t - t_2) = \delta(t - (t_1 + t_2)) \tag{A.130}$$

where the asterisk denotes convolution.

A.7 Sinc Function

The form $(\sin x)/x$ occurs in communications theory and other fields of engineering so frequently that the function $\text{sinc}(x)$ has been defined to represent this form.

$$\text{sinc}(x) \triangleq \frac{\sin \pi x}{\pi x} \tag{A.131}$$

In signal processing, it is sometimes convenient to redefine $\text{sinc}(x)$ as

$$\text{sinc}(x) \triangleq \frac{\sin x}{x} \tag{A.132}$$

Evaluation of either form is straightforward for values of $x \neq 0$. L'Hospital's rule can be used to evaluate Eq. (A.132) for the case of $x = 0$.

$$\text{sinc}(0) = \left. \frac{\sin x}{x} \right|_{x=0}$$

$$= \left. \frac{\frac{d}{dx}(\sin x)}{\frac{d}{dx}(x)} \right|_{x=0}$$

$$= \left.\frac{\cos x}{1}\right|_{x=0}$$

$$= 1$$

A.8 Bessel Functions

Modified Bessel functions of the first kind

The *modified Bessel function of the first kind* of order n and argument x is usually denoted by $I_n(x)$ and is defined as:

$$I_n(x) = \frac{1}{2\pi} \int_{-\pi}^{\pi} \exp(x \cos\theta) \cos(n\theta) \, d\theta \qquad (A.133)$$

The modified Bessel function of the first kind is used in the analysis of Rice random variables and in the definition of the Kaiser window (see Sec. 10.7).

Identities for modified Bessel functions

Listed below are some identities that involve I_n.

$$I_n(x) = \sum_{m=0}^{\infty} \frac{(x/2)^{2m+n}}{m!(n+m)!} \qquad (A.134)$$

$$I_{-n}(x) = I_n(x) \qquad (A.135)$$

$$I_n(-x) = (-1)^n I_n(x) \qquad (A.136)$$

$$\exp(x \cos\theta) = \sum_{n=-\infty}^{\infty} I_n(x) \exp(jn\theta) \qquad (A.137)$$

$$\exp(x \cos\theta) = I_0(x) + 2 \sum_{n=1}^{\infty} I_n(x) \cos(n\theta) \qquad (A.138)$$

$$\frac{d}{dx}[x^n I_n(x)] = x^n I_{n-1}(x) \qquad (A.139)$$

$$\frac{d}{dx}\left[\frac{I_n(x)}{x^n}\right] = \frac{I_{n+1}(x)}{x^n} \qquad (A.140)$$

$$I_0(x) = \frac{1}{\pi} \int_0^{\pi} \exp(x \cos\theta) \, d\theta \qquad (A.141)$$

$$I_0(x) = \frac{1}{\pi} \int_0^\pi \cosh(x \cos \theta) \, d\theta \qquad (A.142)$$

Evaluation of Bessel functions

For large values of x, $I_0(x)$ can be approximated as:

$$I_0(x) \simeq \frac{\exp(x)}{\sqrt{2\pi x}} \qquad (A.143)$$

For small values of x, $I_0(x)$ is approximately equal to 1.

A.9 Matrix Algebra

Because they involve sequences of sample values and sequences of coefficients that both can be expressed in vector form, virtually all digital signal processing techniques can be presented in the form of matrix equations. However, this approach will not be taken in this book because it would impose on the reader the burden of becoming totally fluent in matrix notation and matrix algebra to gain access to the details of the various signal processing operations. However, statistical signal processing is one area in which matrix formulations prove to be indispensible.

Basic definitions

Matrices are rectangular arrays of numbers and/or variables, usually enclosed by a pair of brackets such as

$$\begin{bmatrix} 1 & 5 & 3 \\ 7 & 9 & 2 \end{bmatrix} \qquad \begin{bmatrix} a & 5 & 17 \\ 0 & b & -1 \\ x+y & 4 & 12 \end{bmatrix}$$

Matrix-valued variables usually are written as bold face Roman letters such as **A**, **B**, **D**, and **I**. When variables are used to represent individual matrix elements, they usually are written as lower-case italic letters with double subscripts indicating the row and column in which the element lies:

$$\mathbf{A} = \begin{bmatrix} a_{11} & a_{12} & a_{13} \\ a_{21} & a_{22} & a_{23} \end{bmatrix}$$

The row is always indicated first; the element a_{21} lies in row 2 column of matrix **A**. When the row and column indices are expressions or multidigit numbers, they are either separated by commas or enclosed in brackets

$$\mathbf{A} = \begin{bmatrix} a_{1,1} & a_{1,2} & \cdots & a_{1,N-1} & a_{1,N} \\ a_{1,2} & \ddots & & & \vdots \\ \vdots & & & \ddots & \vdots \\ a_{M,1} & a_{M,2} & \cdots & a_{M,N-1} & a_{M,N} \end{bmatrix} \qquad (A.144)$$

$$\mathbf{B} = \begin{bmatrix} b[1,1] & b[1,2] & \cdots & b[1,2M] \\ b[2,1] & \ddots & & b[2,2M] \\ \vdots & & \ddots & \vdots \\ b[14,1] & \cdots & \cdots & b[14,2M] \end{bmatrix} \qquad (A.145)$$

When the dimensions of a matrix are stated, the number of rows is always given first. In Eq. (A.144), **A** is an $M \times N$ matrix and in (A.145) **B** is a $14 \times 2M$ matrix. A *square matrix* has the same number of rows and columns.

Def. A.1 Diagonals. The *principle diagonal* or *main diagonal* of an $n \times n$ square matrix consists of those elements for which the row index equals the column index, i.e. $a[i, i]$ for $i = 1$ to n. The *cross diagonal* or *reverse diagonal* of an $n \times n$ square matrix consists of the elements $a[i, n + 1 - i]$.

Def. A.2 Diagonal matrix. A *diagonal matrix* $\mathbf{D} = \{d[i, j]\}$ is a square matrix in which all elements are zero except for the elements along the principal diagonal.

$$d[i, j] = 0, \text{ for } i \neq j$$

A compact way of specifying a diagonal matrix is

$$\mathbf{D} = diag\left(\delta[1], \delta[12], \ldots, \delta[n]\right)$$

where $\delta[i] = d[i, i]$.

Def. A.3 Symmetric matrix. A *symmetric matrix* $\mathbf{S} = \{s[i, j]\}$ is a square matrix for which $s[i, j] = s[j, i]$ or $\mathbf{S}^T = \mathbf{S}$.

A *row vector* is a $1 \times n$ matrix

$$\mathbf{r} = [r[1] \quad r[2] \quad \cdots \quad r[n]]$$

and a *column vector* is an $m \times 1$ matrix

$$\mathbf{c} = \begin{bmatrix} c[1] \\ c[2] \\ \vdots \\ c[m] \end{bmatrix}$$

Vectors are usually denoted by boldface lowercase Roman letters. In many texts, the notation is reserved specifically for column vectors with row vectors being indicated as the transpose of a column vector

$$\mathbf{c}^{\mathrm{T}} = [c[1] \quad c[2] \quad \cdots \quad c[n]]$$

Def. A.4 Vector inner product. The *vector inner product* between two $n \times 1$ column vectors $\mathbf{u}$ and $\mathbf{v}$ is denoted as $\langle \mathbf{u}, \mathbf{v} \rangle$ and is defined as the summation

$$\langle \mathbf{u}, \mathbf{v} \rangle = \sum_{j=1}^{n} u^*[j]\, v[j]$$

$$= \mathbf{u}^{\mathrm{H}} \mathbf{v}$$

Def. A.5 Vector outer product. The *vector outer product* or *dyad* $\mathbf{uv}^{\mathrm{H}}$ of two $n \times 1$ column vectors $\mathbf{u}$ and $\mathbf{v}$ is the $n \times n$ matrix

$$\mathbf{uv}^{\mathrm{H}} = \begin{bmatrix} u[1]v^*[1] & u[1]v^*[2] & u[1]v^*[n] \\ u[2]v^*[1] & & \\ u[n]v^*[1] & & u[n]v^*[n] \end{bmatrix}$$

Def. A.6 Norm. The *norm* of a vector $\mathbf{v}$ is defined as

$$\|\mathbf{v}\| = \left(\sum_{i=1}^{n} |v[i]|^2 \right)^{1/2}$$

$$= \langle \mathbf{v}, \mathbf{v} \rangle^{1/2}$$

Matrix arithmetic

Addition. Two $m \times n$ matrices $\mathbf{A}$ and $\mathbf{B}$ can be added or subtracted to find a new matrix $\mathbf{C}$

$$\mathbf{C} = \mathbf{A} \pm \mathbf{B}$$

by simply adding or subtracting corresponding elements

$$c[i, j] = a[i, j] \pm b[i, j]$$

Matrix addition is both associative and commutative.

Scalar multiplication. Multiplication of a vector $\mathbf{A} = \{a[i, j]\}$ by a scalar s is defined by

$$s\mathbf{A} = \{sa[i, j]\}$$

Conjugation. The conjugate of $\mathbf{A}$ is the matrix $\mathbf{A}^*$ with elements $\mathbf{A}^* = \{(a[i, j])^*\}$.

Matrix multiplication. The matrix product **AB** is defined only for the case in which the number of columns in **A** equals the number of rows in **B**. If $\mathbf{A} = \{a[i, j]\}$ is $m \times n$ and $\mathbf{B} = \{b[i, j]\}$ is $n \times p$, then the matrix $\mathbf{C} = \mathbf{AB}$ is $m \times p$ with elements

$$c[i, k] = \sum_{j=1}^{n} a[i, j]\, b[j, k]$$

Properties:

1. in general $\mathbf{AB} \neq \mathbf{BA}$
2. $\mathbf{AB} = \mathbf{0}$ does not imply that either $\mathbf{A} = \mathbf{0}$ or $\mathbf{B} = \mathbf{0}$
3. $\mathbf{AB} = \mathbf{AC}$ does not imply that $\mathbf{B} = \mathbf{C}$.

Matrix transformations

Transpose. The *transpose* of an $m \times n$ matrix **A** is the $n \times m$ matrix $\mathbf{A}^{\mathrm{T}}$ obtained by interchanging the rows and columns of **A**

$$\mathbf{A}^{\mathrm{T}} = \begin{bmatrix} a[1,1] & a[2,1] & \cdots & a[m,1] \\ a[1,2] & a[2,2] & \cdots & a[m,2] \\ \vdots & \vdots & \ddots & \vdots \\ a[1,n] & a[2,n] & \cdots & a[m,n] \end{bmatrix}$$

The transpose of the sum of two matrices is the sum of their transposes

$$(\mathbf{A} + \mathbf{B})^{\mathrm{T}} = \mathbf{A}^{\mathrm{T}} + \mathbf{B}^{\mathrm{T}}$$

The transpose of the product of two matrices is the product of their transposes in reverse order

$$(\mathbf{AB})^{\mathrm{T}} = \mathbf{B}^{\mathrm{T}} \mathbf{A}^{\mathrm{T}}$$

A symmetric matrix equals its own transpose

$$\mathbf{S}^{\mathrm{T}} = \mathbf{S}$$

Hermitian transpose

The *Hermitian transpose* of an $m \times n$ matrix **A** is the $n \times m$ matrix A^H formed by conjugating and transposing **A** (in either order):

$$\mathbf{A}^{\mathrm{H}} = (\mathbf{A}^{\mathrm{T}})^* = (\mathbf{A}^*)^{\mathrm{T}}$$

The Hermitian transpose operation shows several properties similar to the transpose operation. The Hermitian transpose of the sum of two matrices is the sum of their Hermitian transposes

$$(\mathbf{A}+\mathbf{B})^{\mathbf{H}} = \mathbf{A}^{\mathbf{H}} + \mathbf{B}^{\mathbf{H}}$$

The Hermitian transpose of the product of two matrices is the product of their Hermitian transposes in reverse order

$$(\mathbf{AB})^{\mathbf{H}} = \mathbf{B}^{\mathbf{H}}\mathbf{A}^{\mathbf{H}}$$

Def. A.7 Hermitian matrix. A *Hermitian matrix* $A = \{a[i,j]\}$ is a square matrix for which $a[i,j] = (a[j,i])^*$ for all values of i and j. In other words, a Hermitian matrix is its own Hermitian transpose.

$$A = A^H = (A^*)^T = (A^T)^*$$

A Hermitian matrix is sometimes referred to as a *conjugate symmetric* matrix.

References

1. Boyer, C. B. *A History of Mathematics*, Wiley, New York, 1968.
2. Spiegel, M. R. *Laplace Transforms*, Schaum's Outline Series, McGraw-Hill, New York, 1965.
3. Abramowitz, M. and I. A. Stegun. *Handbook of Mathematical Functions*, National Bureau of Standards, Appl. Math Series 55, 1966.
4. Schwartz, Laurent. *Théorie des distributions*, Herman & Cie, Paris, 1950.
5. Papoulis, A. *The Fourier Integral and Its Applications*, McGraw-Hill, New York, 1962.
6. Weaver, H. J. *Theory of Discrete and Continuous Fourier Analysis*, Wiley, New York, 1989.
7. Brigham, E. O. *The Fast Fourier Transform*, Prentice-Hall, Englewood Cliffs, NJ, 1974.
8. Schwartz, R. J. and B. Friedland. *Linear Systems*, McGraw-Hill, New York, 1965.

Index

A-law companding, 478–483
Adaptive filtering, 3, 403–439
   excess MSE in, 429–430
   learning curve for, 429
   LMS algorithm for, 431–435
   method of steepest descent, 425–428
   misadjustment in, 429–430
   performance measures for, 429–430
   performance surface for, 405–411
   recursive least-squares algorithm for, 436–439
Adaptive linear combiner, 404–405
Aliasing, 116–117
Alternation theorem, 271
Aperture effect, 119
Asymptote, 6, 21
Autocorrelation, 28
   method of linear prediction, 391–395
Autoregressive sequences, 395–396
   mixed, 421–425
   predictable, 417–421
   regular, 411–416
AWGN (*see* Noise, additive white gaussian)

Bandpass noise, 43–44
Bartlett periodogram, 446–449
Bartlett window (*see* Triangular window)
Bernoulli, Daniel, 1
Bernoulli numbers, 498
Bernoulli trials, 356
Bessel filters, 108–112
   frequency response, 109–112
   group delay, 112
   transfer function, 108–109
Bessel function, 200, 507–508
Bilinear transformation, 305–320
   factored form of, 306–309
   frequency warping in, 310–312
   properties of, 309–312
   quantization impacts upon, 318–320
   software implementation, 312–314

Block diagrams, 127–129
Boltzmann's constant, 29, 42
Borel field, 354
Burg method for spectral estimation, 460–462
Butterworth filters, 79–88
   determination of minimum order for, 84
   frequency reponse of, 81–84
   impulse response of, 84–85
   step response of, 85–87
   transfer function of, 79–81

Carrier delay (*see* Phase delay)
Cauer filters (*see* Elliptical filters)
Causality, 50–51
Cepstral analysis, 471–477
Certain event, 354
Chebyshev approximation, 269–271
Chebyshev filters, 88–100
   determining poles of, 92
   frequency response of, 95–96
   impulse response of, 96–100
   renormalizing of, 92–95
   step response of, 100
   transfer function of, 89–90
Chebyshev, Pafnuti, 200
Cholesky decomposition, 397–399
Complex numbers, 491–494
   logarithms of, 494
   operations in rectangular form, 492
   polar form, 492–493
   powers of, 493
   roots of, 494
Companding,
   A-law, 478–483
   μ-law, 483–488
Convolution, 52
   discrete, 127
   integral, 52
Correlation
   of two random variables, 361
Correlogram, 455–456

514 INDEX

Covariance
   method of linear prediction, 396–399
   of two random variables, 361
Cumulative distribution functions, 357

Daniell periodogram, 443–446
Data windows, 184
Decibels, 491
Decimation, 321–322, 325–328
   by integer factors, 322
   by non-integer factors, 325
   of bandpass signals, 325–332
   via integer band sampling, 329–332
Decimators, 333, 337
   multistage, 341–345
   polyphase, 337
Delay response, 61
Derivatives, 501
DFT (*see* Discrete Fourier transform)
Digital filtering, 3
Digital simulation, 1, 2
Digital spectral analysis, 3
Digitization, 113–123
Dirac delta function, 503–506
Dirac impulses, 123
Dirichlet conditions, 14
Dirichlet kernel, 184
Dirichlet, Peter, 14
Discrete Fourier transform (DFT), 153–161
   applying, 157–160
   computational complexity of, 163
   parameter selection, 154–155
   properties of, 155–157
   short-term, 474
Discrete-time Fourier transform (DTFT), 123–125
   convergence conditions, 125
   relationship to Fourier series, 125
   short-term, 474
Discrete-time signals, 120–123
Discrete-time systems, 126–131
   diagramming of, 127–131
   difference equations for, 126
   and discrete convolution, 127
   transform analysis of, 143–152
Distribution
   gaussian, 35–39
   generalized function, 505
   of a random variable, 27
   uniform, 31
Dolph, C. L., 200
Dolph-Chebyshev window, 196–200
DTFT (*see* Discrete-time Fourier transform)
Durbin's method, 385

Elementary event, 354
Elliptical filters, 100–108
   determining required order for, 101–102
   parameter specification for, 100–101
   transfer function for, 103–108
Energy signals, 10–11
Energy spectral density, 24–25
Envelope delay (*see* Group delay)
Equivalent noise temperature, 43
Error function, 35, 37–38
Euler numbers, 498
Euler's identities, 497
Exponentials, 489

Faders, 184
Fast Fourier transform (FFT), 163–179
   butterflies, 169
   decimation-in-time algorithms, 163–171
   decimation-in-frequency algorithms, 171–177
   prime factor algorithm, 177–179
Fast Fourier transform analysis, 3
Fejer kernel, 189
FFT (*see* Fast Fourier transform)
Filters,
   adaptive, 403–439
   analog, 71–112
   bandpass, 76–79
   bandstop, 79
   Bessel, 108–112
   Butterworth, 79–88
   Chebyshev, 88–100
   elliptical, 100–108
   finite impulse response (FIR), 207–223, 225–246, 247–268, 269–285
   highpass, 76
   infinite impulse response (IIR), 287–293, 295–304, 305–320
   lowpass, 71–76
FIR filters, 207–223
   cascade form, 217–218
   constant group delay, 211–215
   design using frequency sampling method, 247–268
   design using Remez exchange, 269–285
   design using window method, 225–246
   direct form, 215–216
   Fourier series method, 225–235
   frequency response, 208–211, 213–215
   half-band, 337–340
   linear phase, 211–215
   linear phase, structures for, 218–220
   multirate implementations, 348–349
   quantization impacts upon, 221–222, 238–246, 266–268

software implementation, 207–208, 222–223
transposed direct form, 216–217
Formant frequencies, 470
estimation of, 475–477
Fourier series, 11–17, 125
exponential form of, 12–13
properties of, 14–16
relationship to discrete-time Fourier transform, 125
Fourier transform, 21–25
discrete-time, 123–125
inverse, 22
notation, 22
properties of, 23
Frequency window, 184
Fundamental frequency, 12

Gauss, Johann K. F., 37
Gaussian distribution, 35–39
Generalized Levinson recursion, 386–387
Golden section search, 258
software for, 262
Gradient estimation, 411
Group delay, 62

Half-band filters, 337–340
Hamming window, 194–196
Hanning window (*see* von Hann window)
Harmonic, 12
Harmonic analysis, 1
Heaviside expansion, 57–58

Ideal sampling, 113, 115–117
IIR Filters, 287–293, 295–304, 305–320
cascade form, 290–291
design using bilinear transformation, 305–320
design using impulse invariance method, 295–300
design using matched z transformation, 302–304
design using step invariance method, 300–302
direct form, 288–290, 292–293
frequency response of, 288
quantization impacts upon, 291–293, 318–320
software implementation of, 293
structures for realizing, 288–293
Impulse response,
of a linear system, 51–52
scaling of, 75
Independent events, 355
Instantaneous sampling, 113, 117–119

Integer band sampling, 329–332
Integration, 502–503
Interpolation, 321, 322–326
by integer factors, 322–324
by non-integer factors, 325
of bandpass signals, 325–328
Interpolators, 333–336
multistage, 345–348
polyphase, 337

Kaiser window, 200–201

Lag windows, 184
Laguerre method, 60–61
Laplace transform, 53–58
inverse, 55, 57
pairs, 54
properties of, 54
relationship to z transform, 147
Lattice filters, 399–401
Levinson recursion, 382–383
generalized, 386–387
in linear prediction, 401
inverse, 401
Linear estimation, 389–390
Linear prediction, 389–401
forward, 390
autocorrelation method of, 391–395
covariance method of, 396–399
using lattice filters, 399–401
Linear systems, 47–69
additive, 48
anticipatory, 50
causal, 50–51
characterization of, 51–53
computer simulation of, 64–69
discrete-time, 126–131
homogeneous, 48
impulse response of, 51–52
notation for, 47–48
step response of, 53
time-invariant, 49–50
Linearity, 48–49
LMS adaptive filtering, 431–435
LMS method for spectral estimation, 464–466
Logarithms, 490

Magnitude response, 61
features of, 71–74
scaling of, 75
Magnitude spectrum, 13
Markov processes, 373–376
Matrix,
autocorrelation, 368
conjugate, 510

Matrix (*cont.*)
  diagonal, 509
  Hermitian, 368
  Hermitian transpose, 511–512
  symmetric, 509
  Toeplitz, 368
  transpose, 511
Matrix algebra, 508–512
Mean, 27
Minimal standard generator, 35
Modified Yule-Walker equations, 386
μ-law companding, 483–488
Multirate signal processing, 3, 321–351
  implementation of lowpass filters, 348–349
  software for, 350–351

Natural sampling, 113, 120
Neper frequency, 53
Network function (*see* Transfer function)
Noise, 27–46
  additive white gaussian, 30
  bandlimited, 30
  bandpass, 43–44
  linear filtering of, 29
  pink, 29
  processes, 27–29
  quantization, 136–142
  temperature, 29
  thermal, 42–43
  white, 29–31
  white gaussian, 30
    simulation of, 39–41
  wideband, 28
Noise equivalent bandwidth, 30–31
Nyquist rate, 116

Parseval's theorem, 17
Passband, 71
Periodicity, 8
Periodograms, 441–455
  Bartlett, 446–449
  Daniell, 443–446
  Welch, 454–455
Phase delay, 61–62
Phase response, 61
  scaling of, 75
Phase spectrum, 13
Planck's constant, 42
Polynomial expansions, 68–69
Polyphase structures, 336–337
Power signals, 10–11
Power spectral density, 25, 28
Probability, 354–355
  conditional, 354–355
  joint, 344–355

Quadrature modulation of bandpass signals, 326–328
Quantization, 113, 114–115, 131–142
  A-law, 478–481
  of digital filter coefficients, 135–136
  impacts upon FIR filters, 221–222, 238–246, 266–268
  impacts upon IIR filters, 291–293, 318–320
  in fixed-point numeric formats, 131–133
  in floating point numeric formats, 133–135
  μ-law, 483–485
  noise, 136–142
  of speech signals, 477–488

Raised-cosine window, 193
Random number generators, 32–35
Random processes, 28, 363–376
  autocorrelation functions of, 365–368
  autocorrelation matrix of, 368
  autocovariance of, 367
  autoregressive model of, 380–383
  autoregressive-moving average model of, 379–380, 385–387
  ergodic, 28, 363, 366
  estimating the correlation matrix of, 372–373
  estimating the moments of, 369–371
  linear filtering of, 369
  Markov, 373–376
  moving average model of, 384–385
  parametric models of, 379–387
  power spectral density of, 368–369
  predictable, 364
  realizations of, 353, 363
  regular, 363
  stationary, 28, 365–366
  uncorrelated, 367
  weakly stationary, 366
  wide-sense stationary, 366
Random sequence, 353
Random variables, 27, 356–363
  central moments of, 359
  correlation, 361
  covariance, 361
  cumulative distribution functions of, 357
  functions of, 362–363
  kurtosis of, 359
  mean of, 358
  moments of, 358–359
  relationships between, 360–361
  skew of, 359
  standard deviation of, 359
  statistically independent, 361
  variance of, 359–360
Rectangular window, 181–186

Recursive least-squares (RLS) adaptive
    filtering, 436–439
Remez exchange, 269–285
Response,
    delay, 61
    magnitude, 61
    phase, 61
RLS method for spectral estimation,
    463–464

Sampling, 113
    critical, 116
    ideal, 113, 115–117
    instantaneous, 113, 117–119
    integer band, 329–332
    natural, 113, 120
    rate selection, 116–117
Schrage's algorithm, 35
Sigma field, 354
Signal flow graphs, 130–131
Signal models, 5–10
    transient, 5–7
    steady-state, 8–10
Signals,
    decaying exponential, 20–21
    discrete-time, 120–123
    mathematical modeling of, 5–11
    power, 10–11
    saturating exponential, 21
    speech, 469–471
    steady-state, 8
    transient, 5, 17–21
    unit doublet, 20
    unit impulse, 20
    unit ramp, 18
    unit step, 17–21
Single sideband modulation, 328–329
Spectral estimation, 441–468
    of noisy AR processes, 466–468
    use of windowing in, 450–453
    using Bartlett periodogram, 446–449
    using Burg method, 460–462
    using correlogram, 455–456
    using Daniell periodogram, 443–446
    using LMS method, 464–466
    using periodogram, 441–443
    using RLS method, 463–464
    using Welch periodogram, 454–455
    using Yule-Walker method, 457–460
Spectral window, 184
Spectrum
    magnitude, 13
    phase, 13
Speech processing, 3, 469–488
    using cepstral analysis, 471–477

Statistical signal processing, 3
Steepest descent adaptive filtering,
    425–428
Step response
    of a linear system, 53
    scaling of, 75
Stochastic process (*see* Random process)
Stopband, 71
Superposition, 49
Superposition integral, 51
Symmetry, 8–10
System function, 147–148

Tapering window, 184
Tapers, 184
Time series analysis, 1–2
Transfer function, 55–57
    biquadratic form of, 67–68
    computer representation of, 62–64
    discrete-time, 147–148
    partial fraction expansion of, 151–152
    poles of, 58–61
    scaling of, 74
    sum-of-powers form of, 65–67
    zeros of, 58–61
Transient signals, 5, 17–21
Transition band, 71
Trial, 354
Triangular window, 186–190
Trigonometric functions, 494–495
    series and product expansions, 497
Trigonometric identities, 495–497

Uniform distribution, 31
Uniform sampling theorem, 116–117

Variance, 27
von Hann window, 192–194

Welch periodogram, 454–455
WGN (*see* Noise, white gaussian)
Wiener-Khintchine theorem, 368–369
Windows, 181–205
    Bartlett, 186
    data, 184
    discrete-time, 183–184
    Dolph-Chebyshev, 196–200
    frequency, 184
    Hamming, 194–196
    Hanning, 192
    in spectral estimation, 450–453
    Kaiser, 200–201
    lag, 184
    quantization of, 201–204
    raised-cosine, 192

Windows (*cont.*)
  rectangular, 181–186
  software for, 183, 184–186, 190–192
  spectral, 184
  tapering, 184
  triangular, 186–190
  von Hann, 192–194

Yule-Walker equations, 380–381
  modified, 386

Yule-Walker method for spectral estimation, 457–460
z transform, 143–152
  bilateral, 143
  inverse, 149–152
  pairs, 148–149, 150
  properties, 148, 151
  region of convergence for, 143–146
  relationship to Laplace transform, 147
  unilateral, 143

ABOUT THE AUTHOR

C. Britton Rorabaugh is a Senior Staff Engineer with a leading aerospace company, and a widely respected expert who has over 20 years of experience in the design and analysis of high-performance signal processing and communications systems. He is the author of the *Digital Filter Designer's Handbook* and *Error Coding Cookbook*.

SOFTWARE AND INFORMATION LICENSE

The software and information on this diskette (collectively referred to as the "Product") are the property of The McGraw-Hill Companies, Inc. ("McGraw-Hill") and are protected by both United States copyright law and international copyright treaty provision. You must treat this Product just like a book, except that you may copy it into a computer to be used and you may make archival copies of the Products for the sole purpose of backing up our software and protecting your investment from loss.

By saying "just like a book," McGraw-Hill means, for example, that the Product may be used by any number of people and may be freely moved from one computer location to another, so long as there is no possibility of the Product (or any part of the Product) being used at one location or on one computer while it is being used at another. Just as a book cannot be read by two different people in two different places at the same time, neither can the Product be used by two different people in two different places at the same time (unless, of course, McGraw-Hill's rights are being violated).

McGraw-Hill reserves the right to alter or modify the contents of the Product at any time.

This agreement is effective until terminated. The Agreement will terminate automatically without notice if you fail to comply with any provisions of this Agreement. In the event of termination by reason of your breach, you will destroy or erase all copies of the Product installed on any computer system or made for backup purposes and shall expunge the Product from your data storage facilities.

LIMITED WARRANTY

McGraw-Hill warrants the physical diskette(s) enclosed herein to be free of defects in materials and workmanship for a period of sixty days from the purchase date. If McGraw-Hill receives written notification within the warranty period of defects in materials or workmanship, and such notification is determined by McGraw-Hill to be correct, McGraw-Hill will replace the defective diskette(s). Send request to:

Customer Service
McGraw-Hill
Gahanna Industrial Park
860 Taylor Station Road
Blacklick, OH 43004-9615

The entire and exclusive liability and remedy for breach of this Limited Warranty shall be limited to replacement of defective diskette(s) and shall not include or extend to any claim for or right to cover any other damages, including but not limited to, loss of profit, data, or use of the software, or special, incidental, or consequential damages or other similar claims, even if McGraw-Hill has been specifically advised as to the possibility of such damages. In no event will McGraw-Hill's liability for any damages to you or any other person ever exceed the lower of suggested list price or actual price paid for the license to use the Product, regardless of any form of the claim.

THE McGRAW-HILL COMPANIES, INC. SPECIFICALLY DISCLAIMS ALL OTHER WARRANTIES, EXPRESS OR IMPLIED, INCLUDING BUT NOT LIMITED TO, ANY IMPLIED WARRANTY OF MERCHANTABILITY OR FITNESS FOR A PARTICULAR PURPOSE. Specifically, McGraw-Hill makes no representation or warranty that the Product is fit for any particular purpose and any implied warranty of merchantability is limited to the sixty day duration of the Limited Warranty covering the physical diskette(s) only (and not the software or information) and is otherwise expressly and specifically disclaimed.

This Limited Warranty gives you specific legal rights; you may have others which may vary from state to state. Some states do not allow the exclusion of incidental or consequential damages, or the limitation on how long an implied warranty lasts, so some of the above may not apply to you.

This Agreement constitutes the entire agreement between the parties relating to use of the Product. The terms of any purchase order shall have no effect on the terms of this Agreement. Failure of McGraw-Hill to insist at any time on strict compliance with this Agreement shall not constitute a waiver of any rights under this Agreement. This Agreement shall be construed and governed in accordance with the laws of New York. If any provision of this Agreement is held to be contrary to law, that provision will be enforced to the maximum extent permissible and the remaining provisions will remain in force and effect.